SONG AMONG THE RUINS

Marco Polo describes a bridge, stone by stone. "But which is the stone that supports the bridge?" Kublai Khan asks. "The bridge is not supported by one stone or another," Marco Polo answers, "but by the line of the arch that they form." Kublai Khan remains silent, reflecting. Then he adds: "Why do you speak to me of the stones? It is only the arch that matters to me." Polo answers: "Without stones there is no arch."

—Italo Calvino, *Invisible Cities,* 1972

SONG AMONG THE RUINS

WILLIAM J. SCHULL

HARVARD UNIVERSITY PRESS
Cambridge, Massachusetts
London, England
1990

Library of Congress Cataloging-in-Publication Data
Schull, William J.
    Song among the ruins / William J. Schull.
        p.   cm.
    Includes bibliographical references.
    ISBN 0-674-82042-8 (alk. paper)
        1. Atomic bomb—Japan—Hiroshima-shi—Research—History.
2. Atomic bomb—Japan—Nagasaki-shi—Research—History. 3. Ionizing
radiation—Toxicology—Research—History. I. Title.
RA1231.R2S384   1990   89-37014
616.9'897'009521954—dc20   CIP

CONTENTS

ILLUSTRATIONS
Following page 118

A Protestant church in Hiroshima after the atomic bombing.
The courtyard of the Hiroshima Red Cross Hospital.
The Hondori, one of Hiroshima's main shopping streets, in 1949.
Frank Poole and Jack Schull at Haneda airport.
A goldfish vender in Nijimura, 1950.
Dr. Tachino and Nurse Minato examining an infant.
The staff of the Genetics Program in 1950.
A shopping street in Hiroshima during the late fall sales.
"No More Hiroshimas."
A torii in Nagasaki partially destroyed by the atomic bombing.
Two children and their mother at a dedication ceremony.
New Year's day in Kure, with families on their way to a temple.
A village scene on the island of Ninoshima in Hiroshima's harbor.
Hermann J. Muller, with senior Japanese geneticists.
The Fujiya Hotel at Miyanoshita in 1951 when Muller visited.
Japanese princesses visiting the ABCC facilities on Hijiyama.
Hiroshima in the winter of 1954, as seen from Hijiyama.
The memorial cenotaph in the winter of 1954.
The living room at 93 Fufugawa in Nagasaki in 1960.
An examination of a little girl in the Child Health Survey.
Dejima as represented pictorially by a Japanese woodblock artist.
Pre-Meiji woodblock print of a Dutch merchant and servant.
An entrance to the Urakami Cathedral as it appeared in 1950.
A farm courtyard on the island of Kuroshima in 1960.
A farm woman carrying nightsoil to the fields.
A Buddhist temple and the Catholic church in Hirado in 1964.

MAPS

PREFACE

The need to set these memories to paper has built steadily over the years. Once begun, thoughts have tumbled out of the closet of my mind like old garments whose warmth had been forgotten. Names of people, places, and events have arranged themselves without effort in a roll call of remembrances. As these took form, so too did the book's purposes, at least the conscious ones. This is not the history of an organization, nor the recounting of experiences related to the particular scientific study that first took me to Japan—an investigation, under the auspices of the Atomic Bomb Casualty Commission, of the genetic effects of radiation exposure. This is, first and foremost, a personal journey.

Travel is often merely a means to confirm our own preconceptions, but it should be more. It can be a way of learning how much we have, and how much we will never have, of finding what we are through the recognition of ourselves, wholly or in part, in others so different. To Japan and its people I owe much, not that they deliberately sought to alter my perspective. They taught through sharing another system of values, another set of human lots and expectations. They whetted my interest in the understated, the astringent, and aroused an appreciation of many subtle resonances of life that might not otherwise have occurred. I grew to recognize the contradictions in our own culture through those so apparent in theirs, and to be wary of the all-too-human temptation to judge others as though this were an inalienable right.

At a time when Japan's economic prominence in the world has occasioned no end of introspection within our own institutions of business, education, and governance and has prompted repeated calls for the emulation of Japanese techniques, it seems worthwhile to attempt to recapture the Japan of the early postwar years from which this miracle emerged. Today's visitors and foreign residents who did not share this time of trial and testing can little appreciate the enormousness of the achievement or the sacrifices that were involved. Physically, the cities and countryside of Japan of the 1990s bear little similarity to the country of the late forties or fifties. Time has also altered styles of life, interpersonal relationships, expectations, and institutions, but how deeply and whether always for the better is arguable.

One of the more enigmatic features of Japan is its capacity to project the appearance of change without changing. There is a timeless quality to the country and its culture that is difficult to fathom and place in the perspective of the alterations seen over the last four decades—alterations that are superficially as dramatic as those which accompanied the Meiji restoration in the second half of the nineteenth century. While this book is not a thorough analysis of this transformation, I do try to describe some of its aspects, with particular emphasis on the years from 1945 through 1965, although more recent events occasionally intrude.

These recollections have passed the scrutiny of many friends with whom some of the experiences were shared, and names, dates, and happenings have been checked wherever possible. But one's memory is a fragile, often perverse, thing. It can err egregiously, and for such errors the blame is mine. To those on whose time and recollections I have imposed so frequently, and to others too numerous to mention who have read this manuscript and commented perceptively, I am indebted beyond my ability to repay. But to Vicki, my wife, who counseled perseverance when my interest waned, and read and reread what I have written countless times for errors, to Susan Wallace, who made of a jumble of words a manuscript, and to Howard Boyer, whose support never faltered, my gratitude is unbounded. Finally, I am also appreciative of the encouragement and suggestions of John Dowling, Judith Dowling, Margaret Dybala, Yuka Fujikura, Dan Graur, Bruce Levin, William Moloney, Lucian Pye, Mary Pye, Emoke Szathmary, and Deborah Theodore.

Throughout these pages I have invariably referred to Japanese individuals by giving their family name first, as is the convention they use, except in the instance of Japanese-Americans, and have italicized Japanese words on their first but not subsequent use. A short glossary at the end of this volume supplements the definitions in the text.

SONG AMONG THE RUINS

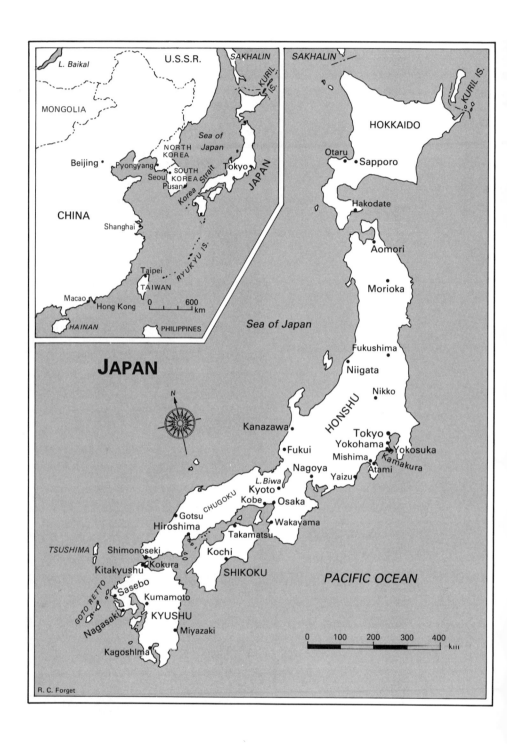

INTRODUCTION

*O*n a sultry summer day in western Japan in 1945, the detonation of a single bomb released moments before from the bay of a B-29, the *Enola Gay*, set our world permanently askew, or so it seems. The process that culminated in this act and thrust us into a new, an atomic, age had begun a number of years earlier. Its apocryphal history commences with Leo Szilard, the Hungarian physicist, who, on a street in London, it is said, suddenly conceived the notion of an atomic chain reaction. Irrefutable evidence for the birth of the atomic age can be found in 1939 and the early 1940s, when a series of seminal experiments—including the production of the first self-sustained controlled neutron chain reaction on December 2, 1942, by Enrico Fermi and his colleagues at the University of Chicago—unlocked the heretofore inaccessible energy contained in the nucleus of an atom. Most people were unaware of these developments and what they foretold. They were unaware, too, of the awesome force bound within a few pounds of plutonium that had been revealed to scientists scarce weeks before in the sands prophetically known as the Jornado del Muerto (Journey of Death), at Alamogordo, New Mexico. The world was preoccupied with a great war, and it was expected, given the destructive power a nuclear weapon could unleash, that atomic energy would be directed toward the ending of that war. Despite this, the news of the atomic bombing of Hiroshima, and then Nagasaki, broke upon a world ill-prepared to understand the enormity of these events. Unfortunately, neither then nor now has our past, our societal structures, or our methods of governance equipped us to cope with the many biologic, economic, and social challenges nuclear energy presents.

Some four decades later, we are more knowledgeable but live no more easily with our atomic handmaiden. Wherein lie our fears? Are we so aware of the frailty of our own humanity? Or is the sheer immensity of the damage that can be done unreal? "Bravo," the largest weapon the United States has tested in the atmosphere, had a yield of fifteen megatons. That is a thousand times the estimated yield of the uranium weapon exploded over Hiroshima. Can we imagine a thousand Hiroshimas or Nagasakis? What do we actually know about the hazards accruing to ourselves as individuals and as members of populations as a consequence of increasing exposure to ionizing radiation? More specifically, what can we or have we learned from the experiences of the only large group of human beings exposed to substantial doses of radiation, Chernobyl notwithstanding—the survivors of the atomic bombing of Hiroshima and Nagasaki? Where does fact end and fancy begin? Dozens of books have been written, but few enlighten. Too frequently they have begun with unrevealed agendas, political or commercial, or seek to place blame, cryptically or overtly, when the societies of which we are members and we all, as individuals, are responsible.

At the time of the bombing, both of these cities were actively and significantly engaged in the Japanese war effort. Hiroshima was a major industrial and rail center and a staging area for many of the troops sent overseas. Nagasaki was not only the site of a large shipyard but also a manufacturer of munitions, notably torpedoes. As the firebombing in the spring and summer of 1945 desolated city after city in Japan, but not Hiroshima, many of its citizens found refuge in the thought that they were being spared because this city was the home of so many Japanese people who had lived in the United States at some time or were relatives of American citizens. It is difficult to limn adequately the subsequent destruction. One hundred thousand human beings perished instantly or within days, and tens of thousands more were injured. The physical devastation that accompanied the bombings beggars description; within moments thousands of buildings were leveled or obliterated, vehicles destroyed, streetcars incinerated, and bridges collapsed or moved sideways on their pilings, rupturing the mains carrying the water essential to the fighting of the hundreds of fires kindled simultaneously. These raged relentlessly across the city, consuming everything that would burn. When the fires were finally curbed, few structures within two kilometers of the hypocenter stood, only the ferroconcrete ones, stark, gutted, and scarred amidst ash and charred timbers that had once been dwelling places resonating to the sounds of life. Cities vibrant seconds earlier were seas of desolation across which the maimed and dying drifted like flotsam, aimlessly seeking a solace not to be found. Above all hung the stench of death and dying. A war-weary people and ravaged economy offered the succor it could; and within days, a few weeks at most, once the more pressing human requirements were addressed, efforts were mobilized to rebuild. Today, both cities are thriving, with few outward signs of their ordeal.

Many of the survivors of the bombings are less fortunate. Scarred physically and mentally, they have more frequently developed leukemia than is to be expected and are prone to other malignancies as well—cancer of the breast, colon, esophagus, lung, ovary, stomach, thyroid, and urinary bladder in particular, as well as multiple myeloma.

On November 26, 1946, in response to the recommendations of the United States' commission assembled to investigate the damage to these cities and their inhabitants and at the urging of his advisors, including James Forrestal, then the Secretary of War, President Harry Truman directed the National Academy of Sciences and National Research Council "to undertake a long range, continuing study of the biological and medical effects of the atomic bomb on man." Forrestal argued that "such a study . . . was beyond the scope of military and naval affairs, involving as it does humanity in general, not only in war but in anticipated problems of peaceful industry and agriculture," and that it "continue for a span of time as yet undeterminable." Specifically identified as matters of concern were "cancer, leukemia, shortened life span, reduced vigor, altered development, sterility, modified genetic pattern, changes in vision, 'shifted epidemiology', abnormal pigmentation and epilation".[1]

Although the National Academy of Sciences had been created by Abraham Lincoln and the National Research Council had been in existence since Woodrow Wilson's time, this charge lay outside their previous experience. Traditionally, the Academy and Council had been independent advisory bodies to the government, not agencies involved in the conduct of field studies. Their task was made more formidable by the fact that Japan was thousands of miles away and governed by a military commander noted for his independent bent of mind. It would be impossible to initiate the studies contemplated without General Douglas MacArthur's support. His endorsement was forthcoming, and in early 1947 there came into existence in Japan the Atomic Bomb Casualty Commission (ABCC), an agency under the auspices of both the Japanese National Institute of Health (JNIH) and the United States National Academy of Sciences—National Research Council (NAS–NRC). ABCC's first acting director was James V. Neel; its beginnings were humble. Eventually, it would grow to over one thousand employees, of whom no less than a hundred were professionals of one sort or another. In the spring of 1947, however, space in Hiroshima was limited, and ABCC began with a dozen or so workers in borrowed rooms in the Red Cross Hospital to the south of the center of the city. Although damaged, this reinforced concrete structure had withstood the atomic blast and subsequent fires.

The initial studies focused largely on damage to the body's ability to form blood cells and on leukemia, but soon many other investigations were under way or planned. These included clinical observations of the survivors themselves, adults and children; examinations of the offspring of the survivors and suitable comparison groups; assessments of sterility and infertility; and steps to take a census of the survivors and to determine the probable doses of ionizing radiation they received. I had been recruited to participate in the genetics studies— to help determine the extent of mutational damage to offspring of people who were exposed to the ionizing radiation released with the detonation of the bombs. We knew, from the studies of Hermann Muller and others, that such damage occurred in experimental animals, and we presumed it would occur also among human beings. We did not know, however, whether the mutability of human genes would be the same as or different from that of animal species that had been studied, nor what form the damage might take. We did know that most newly arising mutations are deleterious, and we assumed that they might be manifested as an increased frequency of birth defects, fetal and infantile mortality, and retarded mental and motor development. In 1949 there were relatively few individuals in the United States trained in human genetics, and I, though only a very recent graduate of Ohio State University, was among that few.

Several years passed before all of these programs were in place and functioning properly. The economic circumstances in postwar Japan were stringent. Housing was scarce; food and clothing were rationed; transportation was limited. As a result, the Commission had to be self-sufficient, and its self-sufficiency had to be achieved in the context of an occupied, impoverished nation. It needed not only its own fleet of trucks, jeeps, buses, and the like but the means to maintain and repair them. To print medical record forms, printing facilities had to be established. Clinical laboratories had to be built and staffed and laundry services organized. Diagnostic tools, such as x-ray, and photographic facilities were unobtainable in a devastated Japan and had to be shipped from the United States. Most importantly, staff, both professional and nonprofessional, had to be identified and recruited. This proved to be a continuous, often frustrating task. Not all war-born animosities had subsided, and many qualified professionals in the United States were unenthusiastic about a position of uncertain tenure in war-ravaged Japan. Moreover, many unnecessary administrative barriers impinged on recruitment. For example, before the Commission's new employees could sail for or fly to Japan, all of us had to undergo a security clearance, although we were to have no access to classified material, even of the lowest order. Perhaps this overreaction merely reflected the paranoia of the McCarthy years, but we felt that it was unnecessary and demeaning. Unfounded suspicions were aroused in the minds of our American friends, neighbors, and employers, for rarely were they told that the clearance investigations were routine. Distrust festered for years, and the implied secrecy at the Commission's inception troubled the eventual acceptance of the scientific findings themselves. This intrusion was made no more palatable when we learned—as we inevitably did, once in Japan—that recruitment there of former U.S. servicemen or Japanese personnel entailed no similar examination of their presumed trustworthiness.

Disgruntlement was inadvertently sown in other ways. Our early relationships with the Medical School of Hiroshima University were strained by distance—Aga, the town where it was then located, is twenty or so miles from Hiroshima—and by the school's uninspired and uninspiring faculty, many of whom were repatriates from lesser medical schools in Taiwan. When we sought help in Tokyo and Kyoto, at Japan's former imperial universities, our actions were interpreted by the faculty at Hiroshima as a slight that caused hard feelings for years. Moreover, our formal association with the Ministry of Health and Welfare (Koseisho) made recruitment from the outstanding national universities difficult, since they were the administrative responsibility of another agency, the Ministry of Education (Monbusho). Bureaucracies seem deeply entrenched wherever they occur but especially so in Japan, and they frustrated efforts to move personnel from one agency to another. We were attractive only to the academically footloose, the masterless, the *rōnin,* who had no alternatives.

Still other issues struck at the very purpose of the organization. Should we, indeed could we, legally treat the survivors? And if so, what should that treatment be? How could it possibly be more than palliative, when we were not even certain what the effects of ionizing radiation were? Moreover, if treatment was undertaken, what would be the longer-term impact not only on the health care of the survivors but on the practice of medicine in these communities? Would it, through the special attention paid to one large group of inhabitants of Hiroshima and Nagasaki, stultify the development of adequate health resources for all? Would it leave the survivors devoid of other means of care once the studies ended, as they must in time? Questions of this nature transcended our own individual persuasions; they involved two governments and two societies more disparate then than now. Some of the decisions are still debated and many seem capricious in hindsight, but they were not lightly nor insensitively reached. The present has no monopoly on concern. Science is just as circumscribed by the political and moral vision of its time as business or any other social institution.

These difficulties notwithstanding, collectively, the information generated through these forty and more years of inquiries provides the single most important basis we have for the estimation of the biological risks associated with exposure to ionizing radiation, whatever its origin, and, ultimately, for the determination of the levels of exposure considered permissible, occupationally or environmentally, nationally and internationally.

PART

I

OCCUPATION

I

HIROSHIMA, 1949

$\mathcal{A}$s the plane droned westward from Minneapolis through the darkened July sky, lulling more experienced travelers to sleep, uneasiness invaded my thoughts and made rest impossible. The events that had brought me to this Northwest Airlines flight en route to Japan, to study the late effects of radiation exposure on the residents of Hiroshima and Nagasaki, spun through my mind like an endless record. Had I done the right thing? I had spent two and a half years with an infantry division in the Pacific contesting Japanese efforts to dominate the Orient. I had lost friends and patched the wounds of women and children who suffered only because they were inadvertently caught in the outburst of senseless violence wrought by the Japanese in Manila and the northern Philippines as defeat grew more imminent. Like cornered animals, the Japanese had vented their fear and frustration indiscriminately, shooting and bayoneting anyone nearby. Even their treatment of their own people bordered on the inhuman. One particularly distressing event was still etched clearly in my mind. When the war ended, our unit, a medical clearing company, was camped outside Alcala along the highway to Aparri in the northern Luzon province of Cagayan in the Philippines. Many sick, wounded, and malnourished Japanese soldiers were brought to us. Animosity quickly disappeared in the presence of their suffering; they were human beings in need, whose only crime had been to follow their own notions of patriotism. One day a tattered, wounded, malaria-ridden Japanese soldier was brought into one of the hastily erected tents where the prisoners were kept. As I began to dress his wounds, his body was racked with a chill and he urinated on himself and on the litter where he rested. A guttural, commanding shout in Japanese arose from the cot behind me. With excruciating effort, the emaciated soldier rose and presented himself to the voice, which had come from a trivially ill, truculent noncommissioned officer with a skin ulcer of the leg. The soldier was rewarded with a further remark, unintelligible to me, and a sharp blow to the head that tumbled him onto his litter. I was sickened. Would my relationship with the Japanese now be inextricably colored by the disgust I had then felt? Can individuals—indeed, can nations of individuals—change, save superficially? Are we all merely mindless captives of our culture and upbringing, with no will or morality of our own? Many Japanese seemed then, and even now, not only ill-informed about the excesses of their military but unwilling to be informed. Pearl Harbor was rationalized by saying that the United States' national policies threatened their country's existence. But how could they account for the savaging of China, Malaysia, the Philippines? None of these nations threatened Japan.

How could I possibly approach this land and these people with some feeling of compassion and willingness to understand? Defeat to a tradition-rich people like the Japanese must be a numbing experience, leaving them bereft of values, of independence, of a vision that motivates daily actions. Was I prepared to deal with the many psychological demands that would confront me in my new position? I had worked my way through several histories, including George Sansom's *Japan: A Short Cultural History* and had studied more diligently Ruth Benedict's *The Chrysanthemum and the Sword,* a perceptive analysis of the cultural forces that culminated in Japan's entrance into World War II. Was this enough? Matthew Perry's monumental five volumes—the official record of his nineteenth-century trip to pre-Meiji Japan—were available, but they were intimidating in their thoroughness, comparable in American literature only with the accounts of Lewis and Clark's expedition of the Rockies. They spoke to another time, another world.

In 1945, Western scholarship about Japanese history and culture was poor. In the early postwar years, much of what we knew about Japan had been written either by nineteenth- and early-twentieth-century visitors such as Basil Chamberlain and Ernest Satow or by missionaries and their children. In the United States, only a very small number of private and state universities—California, Chicago, Columbia, Harvard, Michigan, Princeton, and Yale—had ties of some duration with the island empire and its culture when the war began. Most of these had provided the few cultural authorities and language-training programs that could be mobilized during the conflict. We Americans had been preoccupied with Europe and the mainstream of our social and cultural heritage, to the exclusion of the Orient. Had we been better informed, and less cavalier in our approach to Asian problems, the tragedy that had so recently wracked this region of our earth possibly could have been avoided. No prescience was needed to see the gathering clouds of conflict—the Japanese fear of isolation from the resources so desperately required to feed, clothe, and house their population, the demeaning of the national sense of self-worth, and the often undirected, reflexive American response.

It was early morning when, at last, Japan's green paddy fields could be seen off our starboard wing. In moments, our plane was over Tokyo Bay and on its final approach into Haneda, Tokyo's only airport at that time. Days earlier, Herman Wigodsky of the National Research Council had called to inform me that a representative of the Atomic Bomb Casualty Commission, George Fukui, would meet me in Tokyo. A bespectacled, kinetic Japanese-American with dancing eyes and a perpetually over-burdened look on his face, George carried a placard with my name on it and led me through the entry formalities, conducted under the watchful eye of the Occupation forces and General Douglas MacArthur. One could not enter the country without his approval as the Supreme Commander of the Allied Powers (SCAP), or that of his surrogates. The special passport under which I traveled stated, "This passport, if properly visaed, is valid for travel into any country except Germany, Austria, Trieste, the Main Islands of Japan or Okinawa. Before travel in any of the above named places may be undertaken, an appropriate military permit must be affixed to this passport." It further noted, "Approved for entry into Japan by the Supreme Commander for the Allied Powers as a Representative of the National Research Council for a period indefinite. Entered into Japan on July 11, 1949 at Haneda." We walked briskly from the terminal, an unlikely forerunner of the mammoth buildings that now squat on Narita's former rice paddies and serve as Tokyo's international airport, to a wood-paneled stationwagon, for the drive to Tokyo. Formalities did not cease with the inspection of one's luggage and passport but continued through a number of checkpoints, where we were scrutinized by military police, saluted, and waved on.

The route we followed into Tokyo revealed few overt signs of the destruction visited upon the area a scarce four years earlier. Many buildings were temporary, but there were no open, devastated, blackened expanses to remind the Japanese or me of the numerous incendiary raids and the trauma of their loss. Behind the facade of new construction, however, squalor existed. Ramshackle huts were indistinguishable from the piles of wood fragments, cardboard, and corrugated metal from which they had been constructed. Farmers were better housed, but their lives were not easier. They did eat more regularly than their urban countrymen; rice gruel, often mixed with barley, was their staple, made more palatable by an egg and soy-seasoned pickles. Meat, even fish, was a rarity. Signs of Japan's reawakening from defeat and its attendant demoralization were barely perceptible, and then only to an observer more skilled than myself.

Shortly after the end of the war, the value of Japanese currency had fallen precipitously from 15 yen or so to the dollar to 360, a value that persisted for several decades; cigarettes were a stabler currency. Rationing of rice and other staples continued, bicycles were more common than automobiles, and most commercial traffic moved on *batabatas,* the three-wheeled motorized carryalls that are now objects of curiosity to young Japanese. A bewildering variety of these machines threaded the streets, lurched over streetcar tracks, dodged between pedestrians and cyclists, and made the air purple with partially combusted gasoline from their diminutive engines. The driver, exposed to all of the inclemencies of the weather, sat with legs astride the gasoline tank, feet firmly planted on the supports a few inches above the ground, warping his hand-turned front wheel to left or right.

George spun forth a steady stream of instructions, admonitions, and directions as we drove into Chuo-ku, Tokyo's center, and made our way to the Marunouchi Building, where Brigadier General Crawford Sams, the Occupation's doyen of public health, had his headquarters and where the Atomic Bomb Casualty Commission maintained its Tokyo office. There were orders to be cut so that I could travel to Hiroshima, identification documents to be made, military scrip to be obtained, and a call to be placed to Hiroshima to announce my arrival and to determine whether I should proceed directly there or wait in Tokyo. After the call, George informed me that I was to remain in Tokyo for several days until Jim Neel, who was in Hiroshima but would soon be on his way to Tokyo and the United States, arrived. I was to help him analyze some of the preliminary genetics data for his annual report to the National Academy of Sciences. The prospect of a delay was not distressing, for it provided time to see the city, and I had no notion when next this opportunity would arise.

Once the tasks at the Marunouchi Building were completed, we went to the Billeting Office, since it controlled all accommodations for Occupational personnel in the Tokyo area. Scarcely in the door, I found myself before an overly officious sergeant, who posed one question after another. "Length of stay?" "Purpose?" "Rank?" I was not aware that I had a rank, but George volunteered without hesitation, "Captain," apparently the military equivalent of the civil-service status I would have had, had I been a civil servant. As we left the Billeting Office, George told me that hotel accommodations were assigned on the basis of one's military rank. Mine qualified me for a room in the Dai Ichi Hotel near Shinbashi station, a hotel restricted to captains and majors. More distinguished officers were quartered in the Imperial or Sanno Hotels. I never knew where lesser commissioned ranks—mere lieutenants—were billeted and could not venture a guess about enlisted personnel. Since I was traveling to my duty station, as the military termed it, the cost of the room and the meals I ate were to be reimbursed subsequently. A free billet, however, was not the economic boon the words might suggest—a room in the hotel was only 75 cents a night.

The Dai Ichi Hotel was a cream-colored rectangular box only a stone's throw from one of Tokyo's main rail lines and very near the Far Eastern office of the Stars and Stripes. Close by was the major billet for nurses and officers of the Women's Auxiliary Army Corps (WAACs) who found themselves in Tokyo, and before which each evening there gathered a motley clowder of uniformed cats in an endless season of heat. The Dai Ichi Hotel had been built to house the participants in the 1940 Olympics, which, because of the war in Europe, were never held. The rooms were small and furnished in a Spartan style best characterized as "Occupation Modern." Each contained a single bed or two, a small desk with accompanying chair, a nightstand, and one or two lamps. Brown blankets, familiar to anyone who has served in our armed forces, and sheets starched stiffly enough to stand alone covered the bed. Clothes hangers were few, and each had to serve a variety of needs.

The occupants of the hotel all seemed attired in rumpled khaki with sodden backs and armpits. Only the occasional visiting Australian, in short pants and an open-necked tan shirt with sleeves rolled above the elbows, appeared comfortable in the heat and humidity. A reciprocating electric fan provided what little relief I could find, and no end of daily experimentation went into the task of choosing a spot for it that provided cooling without inadvertently scattering my work all over the floor.

Symbols of the occupied and the occupier clashed at every corner in Tokyo's city center. Jeeps and numerous two-and-a-half-ton trucks, each with its unit affiliation emblazoned on the bumpers, bustled self-assuredly up and down the streets amidst Japanese cyclists making their way about the city. Adult-sized tricycles, whose platform between the two rear wheels was piled toweringly high with boxes, bags, or scrap wood and metal, eased slowly about as the cyclist labored strenuously to move a load equal to several times his weight. Shabbily dressed pedestrians shuffled through intersections, many with a knapsack on their backs or an aging *furoshiki,* a convenient kerchief-like carrier, clutched in their hands. Some were employed, often in clerical or custodial positions, in the various offices of the Occupation, such as the headquarters of the Far East Air Forces (FEAF), which stood on one of the few tree-lined streets of the city center. Near these headquarters were the offices of the Russian presence at MacArthur's court—in an old, dingy red building. In front of it stood surly-looking armed guards, dressed in rumpled baggy uniforms, who appeared to be at peace neither with themselves nor with the throngs passing steadily by.

The walk from the Dai Ichi Hotel to the avenue known as the Ginza was a study in smells. The road under the rail line at the Shinbashi Station reeked of urine and decay, while the small streetside stalls were enveloped in cocoons of rancid oil, hot soy sauce, or the acrid smell of charcoal embers. Since it was summer, here and there stands selling iced watermelons by the slice or whole were surrounded by chunks of rind and seeds spattered about the sidewalk. Ice candy (*aisu kiyande*) was available to slake an obdurate thirst. At every opportunity in those first several days I prowled the Ginza, a less-crowded and more fascinating thoroughfare then than now. Innumerable sidewalk booths warred ceaselessly with pedestrians for the limited space between curb and store. From these little enterprises one could purchase almost anything that could be readily transported, from cigarettes to the treasured possessions of families forced to choose between food and fondness. Although the economic circumstances were ameliorating, they were harsh, and few had more than the necessities of life. Would our lot, had we lost the war, have been as bad? Would we have feared a less enlightened occupation? As I wandered, such thoughts crowded together like frightened people in a storm, each seeking solace or understanding from the other.

I kept to the major streets, not out of fear but out of concern that I would lose my way in the tortured lanes that made up most of the city. Tokyo's streets were then largely unnamed. To aid the occupiers, the Occupation had given to the major thoroughfares the imaginative designations A, B, C, . . . and to the cross streets 1, 2, 3, . . . and so on. Irresistibly, I was drawn to the Central Post exchange, a multistoried, architecturally pedestrian structure on the Ginza not far from the intersection of the road that passed before the Imperial Palace and the Ginza itself. Visitors to Tokyo now know it as the Matsuya Department Store. Before its doors a never-ending series of episodes were played out, some painful—like the eager eyes of the Japanese children who looked longingly at the toys displayed in the windows—and others memorable for different reasons. I was there when Mrs. MacArthur, her son, and his amah appeared to go shopping. The hubbub that attended their arrival could not have been greater if the Queen of England had arrived.

Not far away was the perpetually open Nichigeki, one of the brassy theaters that had imported the striptease to Japan. Its audiences embraced all segments of life, from noisy GIs, whose loneliness often manifested itself in loud whistles and catcalls, to elderly kimono-clad women, who stared in seeming disbelief but fascination at the events on the stage. Without effort or knowledge, the audience provided a more striking performance than the tiny, small-breasted women who unskillfully and dispassionately removed their clothes to the music of a somewhat off-key orchestra. Each group seemed more or less oblivious of the others. When a performance ended and the patrons erupted into the night again, they found themselves on a Ginza far less extravagantly colored than the one in Tokyo today. The forest of neon signs, large and small, that now twinkle and dance their way through the evening dark had not yet come. Sidewalks and intersections were illuminated only by the incandescent bulbs of stores still open, or the infrequent street light.

Among the more conspicuous vehicles on the streets were the military buses that plied their way from housing areas, such as Grant and

Washington Heights, to the part of the city near the Imperial Castle
and the central railway station, where many of the buildings housing
the apparatus of the Occupation were found. Most of the combat
muscle that had poured into Japan in the weeks and months immediately following the surrender was gone, replaced by an army of tech-
nocrats, some uniformed, some not. Many of the skills needed to
maintain the Occupation were provided by civilians employed by the
Department of the Army, so-called DACs; they included teachers to
instruct the children of the Occupation forces and to reorganize the
Japanese educational system, lawyers to help draft new regulations
and to advise Japanese legislators, union leaders to guide the fledgling
Japanese trade unions, social scientists to gauge needs for social reform, industrial appraisers to help preside over the reparations Japan
was obliged to provide, language specialists, and numerous others.
No aspect of Japanese life managed to avoid some scrutiny. Many of
these individuals were drawn from war-inflated branches of our own
government or were on leaves of absence from universities in the
United States. Indeed, on one of my first days in Tokyo, at breakfast
in the Dai Ichi Hotel, I met an absent member of the Department of
Sociology and Anthropology at Ohio State University whom I had
known when I was a student there. He was involved in a series of
social surveys attempting to appraise Japanese reactions to some
pending recommendations of the Supreme Commander to the Diet.

Each hazy morning I would walk to the ABCC's headquarters in the
Marunouchi Building, to see whether Jim Neel had come and to learn
what further might be expected of me before I went to Hiroshima. My
path took me past the old Imperial Hotel and Hibiya Park. A short
distance eastward from the park was the Ernie Pyle Theater, dedicated
to the war correspondent killed on Ieshima in the Ryukyus late in the
conflict. He had been a special friend of the dogface, the infantryman,
sharing his hardships and articulating better than any other his fears
and aspirations. As I picked my way up the wide boulevard that ran
past the pine-sprinkled grounds of the Imperial Palace, once in awhile
I would see a still-uniformed Japanese soldier on the sward, cap in
hand, paltry possessions on the ground before him, bowing to the unseen presence within, whom he had served. Was this gesture in expia-
tion for his failure to win the war, or an apology for surviving? I did
not know. The emperor, the object of the soldier's attention, had re-
nounced his divinity and now traveled frequently about the country in

an aged, two-toned, prewar limousine with the imperial crest mounted on the sides of the rear doors. His presence among his countrymen and his obvious concern for their welfare were clearly appreciated by the Japanese, although some doubtlessly were disturbed by his renunciation of his godliness.

Across the street from the palace, with its murky-watered moat, stood a massive square-colonnaded structure known as the Dai Ichi Building, headquarters of the Supreme commander of the Allied Powers. Tall guards, in carefully pressed trousers, ascots, and glistening boots that bespoke hours of preparation, were frozen in a perpetual salute to the many stars that strode self-importantly into the building. They were the modern fuglemen, to be seen and emulated and to provide the chorus for MacArthur's comings and goings. These were announced with a fanfare suitable to a Caesar whose Rubicon was closer than any of us realized. Confident in his purpose and importance, MacArthur would appear with his jauntily trimmed, gold-encrusted cap and, in the summer, an open-collared chino shirt with its rosette of stars. Colder days might find him in a short leather jacket worn over a studiedly casual shirt. As his vehicle, with its five stars unsheathed, pulled away from before the building, traffic was halted by the military or Japanese police at the various intersections through which he passed on his way to the American Embassy, where he stayed. If MacArthur had personal problems, they were so well sublimated that only a Freud could have imagined them. Although his military skills were grudgingly respected by the troops he commanded, few, including myself as a member of one of his infantry divisions, sincerely admired him.

Japan had few heroes who emerged unscathed in body or reputation from the war. Besides the emperor, there were Mikimoto Kokichi, the founder of the pearl company that bears his name and his ubiquitous bowler, and Yoshida Shigeru, the prime minister, whose years in office saw not only the enactment of the reforms demanded by the Occupation but the transition from reform to reconstruction. A nation devoid of heroes dreams no dreams; its outlets are few. But new heroes arise, and one was General MacArthur. At virtually any hour of the day, a clutch of Japanese waited patiently to see the man they knew as "Macassa," who, whatever his other faults, presided over the most enlightened, informed, and altruistic occupation that has followed a major war. It sought no significant economic advantage nor the possession of lands that had not previously been its own. This is not to imply that it did not make its share of mistakes; it did. Some of these stemmed from ignorance, some perhaps from the inexperience of those who had to govern under circumstances they had never previously confronted. Nevertheless, MacArthur and his staff launched Japan on a new course, the abnegation of war. They introduced land reform, revised the school system, purged those who were the authors of Japan's wartime aspirations, and furthered the cause of unionism.

Even the much-maligned war trials were more humane than some have painted them. Obviously, they began in an atmosphere of mutual distrust; on the one hand, the Japanese indicted for war crimes undoubtedly feared the "winner's justice," but on the other hand, the victorious powers were legitimately concerned that 27 percent of the 132,000 prisoners of war taken by the Japanese failed to survive their captivity, whereas only 4 percent of the 235,000 military prisoners of the Axis powers, Germany and Italy, died. Many more prisoners taken by the Japanese were brutalized, despite repeated calls by the Allied powers to honor the Geneva Convention in respect to prisoners of war. These acts would ultimately become the legal bases for the activities of the International Military Tribunals in the Far East immediately following the war. These tribunals sought an explanation, and punishment where appropriate, for these acts of barbarism, at least as Western nations construed them. But only the most prejudiced can see these undertakings as consciously malevolent. Of some 100 so-called class-A war criminals charged, only 28 were actually indicted, including Tojo Hideki, the general and wartime prime minister. Acquittals did occur, and defense attorneys, whatever their individual views of those they defended, pleaded their cases vigorously.[1]

In its zeal, SCAP promulgated some ill-advised, galling edicts, such as the suppression of the play *Chushingura,* a story of feudal vengeance, presumably because it might foster rebellion. The publication of Japanese studies on the early effects of exposure to the atomic bomb was also proscribed, for reasons not clear. Other equally ill-considered steps included the foisting upon Japanese medicine of an internship system at variance with the nature of medical education and health-care provision in that county. Similarly, the destruction of the then-limited potential for research in high-energy physics was clearly disadvantageous to the education of physicists and the development of nuclear energy as a resource. With the formal cessation of hostilities in the autumn of 1951, these errors could be rectified, and they were. Japan now generates a substantial amount of electricity through nuclear power stations, and obligatory medical internships have not existed since 1963.

Nevertheless, many of the SCAP-initiated changes made possible the "New Japan." It was MacArthur's office that first invited W. Edwards Deming to Japan, the man who, with Walter Shewhart, was the founder of quality control in the manufacturing process, a management means that has permitted Japan in the postwar years to establish itself as a purveyor of outstanding merchandise, despite an international image before the war as the maker of cheap, shoddy goods— the five-cent product in the five-and-dime store. Like many facile impressions, this was at best a half-truth, for products made for the Japanese market have always had an enviable flair, a sense of proportion and quality of manufacture that speaks of craftsmanship and pride.

Upon the arrival of Jim Neel—a spare, fair-haired, bespectacled physician whose intense, driving ambition, vision, and dedication have propelled the genetics studies in Japan for over forty years—I no longer had time for these strolls and musings. I found myself mentally chained to a Frieden calculator, desperately involved in organizing and analyzing the data Jim proposed to incorporate into his annual trip report. It was a good introduction to the duties I was to inherit as head of the Genetics Program in Hiroshima and Nagasaki, and it offered insights into the program's needs more quickly than might otherwise have come. It was a continual briefing. When finally we parted, he for the United States and I for Hiroshima, I was prepared for the next step in this introduction to a new land and people.

*   *   *

The various military airports of Japan were connected by the military air transport system (MATS), but travel to the Hiroshima–Kure area was largely by special trains that the Japanese National Railroads ran for the Occupation forces. Military rail transportation throughout Japan was supervised by the 8010th Transportation Military Railway Service, which maintained some 300 units distributed about the country. The train I boarded at Tokyo Central Station—an intriguing, turreted, hexagonal structure built in 1914 out of red bricks and stones— left in the late afternoon, to arrive in Kure, my immediate destination, some twenty hours later. Our speed, which averaged less than 30 miles an hour, was a far cry from that attained by the trains that race over the high-speed line today at 130 to 140 miles per hour.

In 1949 the line was electrified only to Gifu, near Nagoya. There, our electric engine was replaced with a coal-burning one, and we were again under way. Japan is a mountainous country repeatedly pierced by tunnels. As each tunnel approached, the windows were hastily closed, for once the engine entered a tunnel the cars that followed were instantly enveloped in a cloud of smoke and soot. If the tunnel was sufficiently long, even if all the windows were promptly and tightly closed, the interiors of the cars were perfused with acrid smoke through the canvas cowlings that integrated the train. Air conditioning did not exist, and the small, erratically rotating ceiling fans were a weak defense against summer's heat and humidity. Sweat cloaked one's brow and spread through one's clothing. Sodden shirts served to snare every wayward particle of soot, and starched collars grayed as they wilted. To aggravate matters, at frequent intervals the car boy would sweep the floor; with a short broom he pushed whatever debris had accumulated towards little brass floor wells, where dirt, cigarette butts, empty cigarette packages, and other detritus spilled onto the track flashing beneath. More dust was set into motion, ultimately to settle either to the floor again or onto a sweat-stained shirt.

We rolled our way through a litany of towns too small to warrant a stop, but wherever there was a garrison of troops or an office of the ubiquitous apparatus of Occupation, our engine belched to a halt, enveloped in its own smoke. Generally, I hardly had time to read the name of the station before the train lunged onward. I amused myself by trying to decipher the names written in *hiragana,* the more cursive of the two syllabaries commonly used to render into Japanese those words for which no ideogram (*kanji*) exists or to make accessible names of places and things to the less literate. This form of amusement must have occupied the minds of innumerable other travelers. Stops in the larger cities—Nagoya, Kyoto, Osaka, and Kobe—were longer, and I had an opportunity to step out onto the platform to stretch my legs, or mind, if an English-language newspaper could be found. Mandarin oranges, neatly packaged in clusters of five, and boxed lunches tempted the hungry. Milk (or *gyūnyū* as it was then commonly called) was available for the strong of heart unworried about milk-borne bacteria. Steadily, hour after hour, our train plodded westward past a constantly changing sea of rice paddies. Here and there, farmers tilled their soil with oxen, mud-begrimed to the knees, that trod placidly before a plow. With each step I could imagine the plaintive sucking sound as the mud reluctantly gave up the animal's foot, only to receive it again a few inches further forward. Lotus, with its broad, flat, green leaves and magnificent white, pink, or red flowers, floated on the muddy waters.

We reached Osaka in the early morning. After brief pauses there, in Himeji, with its marvelous White Heron castle in view, and in Okayama, we swept westward on a narrowing coastal plain. To the north could be seen tier after tier of tree-covered mountains, each somewhat higher than its predecessor, bathed in the vapors of morning. Mihara finally appeared, and we were separated from the main line, which turned into the mountains toward Saijo and Hiroshima, as we continued to follow the coastline to Kure, along a path interrupted by tunnels through the foothills that sprawled to the sea.

The five prefectures (Okayama, Hiroshima, Yamaguchi, Shimane, and Tottori) that comprise the region known as Chugoku, as well as the island of Shikoku, were garrisoned by the British Commonwealth Occupation Forces (BCOF), whose headquarters were in Kure, once the clandestine center of the Japanese Navy. Since Hiroshima was within the jurisdiction, ABCC employees were dependent upon the British for housing and other logistical needs. When the British Commonwealth Forces appeared on the occupational scene—some originally scheduled as support units for the invasion and others diverted from their intended earlier roles—their occupying areas had already been designated. The Australians were assigned to Hiroshima and Okayama Prefectures, the New Zealanders to Yamaguchi, and the British Indian Divisions to Shikoku and Shimane Prefecture. Most of these units arrived in the spring and early winter of 1946 to replace the American forces that had moved into the Chugoku area in the autumn of 1945—first the 41st Infantry Division and then the 24th Division and the 308th Station Hospital. When I reached Kure, most Commonwealth troops had returned to England, New Zealand, or India. The garrison that remained was largely Australian—the Third Battalion, the 77th Squadron, and others. Australia's white policy, still intact, was the source of needless friction and discrimination, particularly for the young American-born Japanese physicians and technicians—the Nisei—who had come to Japan to work for the Atomic Bomb Casualty Commission. Change was in the offing, but several years passed before it actually occurred.

Peggy Green, the Public Relations Officer for the Commission, met me at the Kure station and took me to the Sawahara House, where I was to live for the next several months until housing became available and Vicki, my wife, could join me. Residents of Sawahara House were a mixed assortment of Australians and Americans, employees of the Commission as well as members of the military government unit that oversaw affairs in Chugoku. Peggy quickly acquainted me with the places to eat, shop, and add to my growing list of currencies. I now had not only U.S. Military Scrip and yen but also British Armed Forces Scrip (BAFS). Since the BAFS were based on the pound sterling, then worth about $4.86, another rate of exchange had to be carved into my mind. Yen were used on the Japanese market, scrip dollars in the Post Exchange and at the American clubs and commissary, and scrip pounds at all installations under Australian aegis. Guineas, pounds, shillings, and pence surely constituted the most impossible of all currency systems devised by man—twelve of one, twenty of another, and on it went. Since all denominations were in paper, except pennies, one's wallet always bulged more promisingly than one's resources warranted. Peggy's parting instructions to me were how to get to Hiroshima the next day, since I was expected to attend the groundbreaking ceremonies for our new facilities and the evening party arranged by the contractor.

Dawn found me restlessly twisting in bed as my mind groaned on, trying to label and absorb the ceaseless parade of the previous day's instructions and the new sights, smells, and sounds that assaulted every nerve ending. Sleep came in dollops, a few moments at a time. Finally, unable to stifle the turmoil, I arose, dressed, and waited patiently for the jeeps that would take us to the railway station in Kure, to board one of the Commission's buses for the drive to Hiroshima. It was only a few minutes from Sawahara House to the station. Along with Arthur Wright and Geoffrey Day—my new Australian colleagues and fellow employees—I was there well before the convoy of Commission vehicles arrived, standing before the entrance to the rail transit office (RTO). Two of General Motors' "Flxible" buses soon appeared like wheeled matrons and sedately stopped before us. At that time, I was not aware of the rigid order that prevailed among these vehicles. The first bus was driven by a senior chauffeur, Shinji by name, and following it was a second bus with a driver whose credentials were only slightly less impeccable. These two gleaming testimonies to a now-lost American preeminence in the automotive market were followed by several station wagons, which halted here and there to gather employees who lived at other locations in the city or in any of the small towns between Kure and Hiroshima. Our convoy prompted many stares and occasional confusion, as when a cyclist ran into the side of one of the momentarily parked buses, so intent was he on the show we unintentionally provided.

Hiroshima was approached from Kure at that time over one of the few paved highways in western Japan. It connected Kure—the Imperial Navy's major base—with Hiroshima, the headquarters for the armies of southern Japan, and ultimately with Iwakuni, a naval air station. The road twisted and turned, hugging the Inland Sea, unspoiled by the industrial pollution that has since corrupted Japan's coastal waters and driven her fishermen further and further east and south in search of food for their nation. Fishing still occurred close enough to shore that one could hear the chanting of the fishermen as the capstans to which the net cables were attached were turned, and see the beating of the water to dissuade the fish from escaping until the ring formed by the net had tightened.

Here and there the road's concrete surface had been scarred during the war by the armor-piercing shells of a strafing Allied aircraft undoubtedly intent on the destruction of a train on the tracks adjacent to the road. We rolled steadily onward over a spectacular route through little communities with names such as Yoshiura, Tenno, and Saka, where a factory existed that creosoted railway ties. We were stopped briefly as a headkerchiefed, work-clad woman shuffled slowly across the highway from the factory with two or three newly creosoted railroad ties perched precariously on an A-frame on her back. As we waited, we could see still another woman position herself between two men, who laboriously lifted two or three ties, one at a time, from a stack at the factory on the sea side of the road and place them carefully on her A-frame as her knees slowly bent under the gathering weight. These would be carried across to another stack beside the tracks, where eventually they would be loaded on a rail car. The process seemed endless and rarely interrupted, save for the infrequent bus or truck on its way to or from Hiroshima.

Once again under way, we drove to Yano, with its large, private bronze Buddha silhouetted against the sky, dark and forbidding on rainy days but sun-crowned when the weather was fine, and Kaitaichi, where near a small BCOF cantonment the road joined another one,

National Highway 2, which wove its erratic course down the valley
from Saijo. We glided on past Toyo Kogyo—better known in the
United States as the Mazda Corporation, a wartime fabricator of air-
planes but in 1949 one of the larger manufacturers of the ubiquitous
bata-batas—past one of the factories of Japan Steel Corporation and
into the outskirts of Hiroshima. Off to the right could be seen the
Kirin Brewery complex, busily engaged in meeting the needs of the
local population and quenching the heroic thirsts of the Australian
occupiers.

Soon after we entered the limits of the city, our progress slowed
measurably. The route was pocked and pitted. Our drivers, intent on
as smooth a ride as possible, crawled cautiously through the maze,
easing their buses into and out of the larger holes and through the
water and mire beneath the underpass of the Ujina branch of the rail
line. Children, giggling to one another in reinforcement of their brav-
ery, waved enthusiastically to us, and the less timorous shouted
"chewing gum" or "harro." The final few hundred yards of the trip
were along a sea wall, past a series of dilapidated wooden structures
that apparently served as a fish market.

The Commission's central offices were in the Gaisenkan, the Hall of
Triumphal Return, from which so many of Japan's soldiers were dispatched to unmarked graves in south and southeast Asia. It was a
sturdy, boxy, three-storied gray-colonnaded building, impressive for
the day with its drive-through portico and sweeping expanse of stairs
that faced the unusually placid waters of the Inland Sea immediately
across the road. In addition to the administrative office of the Commission, the Gaisenkan contained data-processing facilities, an auditorium, and a dining room. Adjacent to it was a tile-roofed single-
storied building that temporarily housed the clinics and laboratories
until the permanent quarters being built on Mount Hijiyama could be
occupied. Beside this building was a motor pool. Before this complex,
after we had all dismounted, the convoy disbanded to be parked,
awaiting the evening's return.

Prior to August 1945, Hiroshima, although a city of almost
300,000 inhabitants, was known to few non-Japanese. It is located on
a deltoidal plain formed by the Ota River, which arises in the interior
of Japan to the north. To the south is Hiroshima Bay, a part of the
Inland Sea (the Setonaikai), and on the other three sides are mountains. Only three prominences disturb the otherwise flat plain on

which the city is situated; these are Ogonzan, Hijiyama, and Eba, all now the sites of parks. Mori Terumoto, one of western Japan's major *daimyō* (literally, the great names) or military lords, established a castle in the area in 1589. Early maps suggest that at the time he did so, much of the present land did not exist; but centuries of silting and reclamation have considerably enlarged the plain and made movement between one "island" and another simpler. A feudal community slowly grew around the castle, and through the seventeenth and eighteenth centuries, Hiroshima's history mirrors that of most of Japan's castle towns. In 1663 its population consisted of 37,212 individuals, of whom no less than 1,070 were said to be Buddhist priests (*bōzu,* the shaven-headed). Some of the older temples they served, such as Mitaki and Fudoin, still stand, but many were consumed in the fires that followed the atomic bombing.

As time passed, merchants and tradesmen displaced the *samurai* (warriors) as the numerically dominant force, and the city differentiated socially. The road running east and west in front of the castle was Horikawa-cho; and where it intersected with the north-south road, the center of the city developed. Here the wholesale and retail merchants had their shops. Somewhat to the east, near Kyobashi (the location of one of the city's older bridges), tradesmen, especially carpenters, were located. The townspeople (*chōnin*) lived largely to the south and southwest of the castle. To the east and west, in the hamlets of Kawata and Onaga were the *kawata* or *eta,* who by virtue of their occupations as the handlers of the dead, the slaughterers of cattle, the executioners, the tanners of leather—all unclean tasks in a Buddhist society—were the outcasts of the social system; much later, Japan's conscripted Korean laborers were forced to reside here. In the immediate postwar years, these areas could still be discerned, more by the occupations still practiced and the garments of the residents than by differences in the buildings of the inhabitants, but today these distinctions no longer exist. East of the castle, in 1620, the still extant Shukkeien was built—one of the strolling gardens made fashionable by the popularity of the tea cult (*chanoyu*) which the military lord Hideyoshi assiduously cultivated and which became an integral part of life for the *daimyō* in the sixteenth century.

In 1894, during the Sino-Japanese War, the government established the headquarters of the Imperial Armies of the West in Hiroshima, and it began to bustle with land and water transport and other military affairs. The Sanyo Railroad line, which traverses the southern shore of Honshu and connects Kobe with Shimonoseki, had been completed only shortly before and served to make Hiroshima a major rail hub, one which connects the Sanyo Line with the Sanin Line along the Japan Sea coast facing Asia. Prewar Hiroshima was part of a military complex that included Hiro, where one of the naval aeronautical arsenals was situated, Kure, the nation's most powerful naval port and site of its largest naval foundry, Iwakuni, the location of a major naval air station, and the naval academy, which had been established in 1888 on the island of Eta Jima in the bay. Evidence of this military-industrial complex, still visible in the early postwar years, is barely perceptible today.

On my first day in Hiroshima, my thoughts hardly focused on its history but on unquelled anxieties. I was taken to meet the recently appointed Director of the Atomic Bomb Casualty Commission, Lieutenant Colonel Carl Tessmer, a blue-eyed, pipe-smoking man with sharp but kindly features and the closely cropped hair favored by the military. He welcomed me into his office on the second floor, overlooking the harbor, asked a few perfunctory questions, and graciously ushered me on to meet others of the administrative staff. Subsequently, I was guided through the various formalities that would provide me with a monthly salary and a more permanent identification card than the one issued in Tokyo. After these tasks were completed, a jeep took me to the Hiroshima Red Cross Hospital, the home of the Genetics Program.

The Niiseki Byoin, as the hospital is known locally, is a squat, angular, and—to me—stern-appearing structure. It was built prior to World War II at a time when public buildings were commonly surmounted by a squarish tower that extended another story or two above the roofline. This gave the building a ship-like appearance—a mammoth stationary vessel in an ocean of unpainted wooden structures slowly weathering to uniformity. Access to the hospital from the street on which it faced was through a small gate wide enough to admit at most two vehicles at a time. Small stores opened onto the courtyard before the hospital and nestled so close to the hospital gate that the latter was difficult to see from the street until one was virtually opposite it. Through this gate each morning our jeeps, bearing young physicians on their way to the homes of newborn babies, spewed forth like a swarm of marauding bees, to return late in the afternoon.

The walls of the hospital, particularly the stair wells, bore testimony to the bomb's destructiveness. They had been creased and gouged by flying glass. Although the Red Cross Hospital was undoubtedly much better than most of Hiroshima's hospitals of that day, it was depressing nonetheless. Most patients were housed in wards, and their beds were little more than hard wooden platforms on which little relief was found. Care and feeding of the ill were shared responsibilities of staff and family. Rows of charcoal braziers lined the corridors and were pressed into periodic service to heat broths, food, or water for tea. The smell of phenol and partially combusted charcoal lingered constantly in the air—centralized food services would come, but not for several more years. Corridors and rooms were routinely mopped but not cleaned; the infrequently changed mop water left surrealistic swirls on the floors. The terminally ill were generally sent home to die amidst family; all other patients stoically accepted their lot. Many were atomic-bomb survivors here to undergo surgical rehabilitation of their injuries or, as I was to learn, to confront the spectre of leukemia.

As a consequence of the use of polished rice, beriberi—a thiamine-deficiency disease—was commonly diagnosed; tuberculosis was also widespread, presumably a consequence of bodies weakened by undernutrition, if not malnutrition. Many mornings queues of school children would gather before the hospital, one sleeve rolled high, to await an inoculation of the Bacillus Calmette-Guerin (BCG) vaccine, an attenuated strain of the tubercle bacillus, which had been introduced to minimize the ravages of this disease. Apprehensively, under the scrutiny of their teachers, tot after tot extended an arm, grimaced stoically as the needle entered, and scampered away to josh those classmates who had not yet received their shot.

The genetics group had space on two floors of the hospital. On the second, a large room served as the offices of the young physicians who did the home examinations and as a conference room; on the third floor was a smaller main office. In this room I was introduced to Dr. Takeshima Koji, my chief of staff, as it were; Iwamoto Reiko, who was the secretary for the genetics group; and Matsuda Naoko and Morita Midori, two clerks responsible for ensuring the completeness of the data the physicians collected and the preparation of work schedules. Matsuda-san was a more self-assured young divorcée than one commonly encountered in Japan, with a vivacious but horribly spoiled young daughter. Morita-san was an attractively demure young woman of twenty. Koji, a general surgeon, was a sallow, small-statured, shaven-headed man in his mid-thirties. His eyes were his most arresting feature. Warmed by good humor, they danced and twinkled. Neither Matsuda-san nor Morita-san spoke much English, so communication between us proved tedious initially; but Reiko—or Alice, as she preferred—and Koji spoke English well, for both had been born in the United States and had spent most of their early years there. Alice, like many others, had been sent to school in Japan shortly before the war and had been trapped by subsequent events. Koji, on the other hand, had been born in Hawaii, where his father was a very highly regarded Buddhist priest of the Shinshu sect. The family had returned to Japan when Koji was sixteen, to a temple in Saijo, which his father served for the rest of his life.

As I toured our space and was introduced to the young Japanese doctors—largely recent graduates of wartime medical schools—with whom I would be so intimately associated, I extended my hand to each in greeting and friendship. Only subsequently did I learn that the Japanese do not greet one another in this manner, and to put one's hand on someone who is not kin borders on a serious breach of etiquette. With time, it became apparent even to me that the Japanese have an exquisitely developed sense of interpersonal relationships, and a bewilderingly complex vocabulary to express them. But there was no guide to lead one through these complexities, and undoubtedly we Westerners often seemed insensitive or boorish.

Koji took me to the roof of the hospital to point out some landmarks that would give me a sense of direction. As we walked up the steps, I asked him how I should address the clerks. He said the Japanese rarely use first names in addressing one another, and that I should simply say Matsuda-san and Morita-san. When we emerged on the roof, I could see scattered through the city, particularly in the residential areas, numerous fire towers, usually of steel, some roofed, some not, rising slightly above the level of the two-storied buildings that made up most of the city's structures. At the top of each was a bell or a hand-cranked siren that served to sound the alarm, summoning the residents of the area and presumably the professional firefighters. On the roofs of many of the homes was a covered wooden platform, possibly six by eight feet, reached by a narrow wooden outside staircase, upon which clothes and bedding were dried or aired, supported by bamboo poles. Behind the more affluent homes often stood an ostensibly fireproofed warehouse (*kura*), usually constructed of wood swathed with mud or stucco or some other fire-retarding material and covered with a thick tiled or metal-sheathed roof. Fire was an omnipresent danger, and prevention or very early containment were the only effective means to avoid disaster. These precautions notwithstanding, huge areas of feudal Japan's cities were periodically swept by fire. Hiroshima had been no exception. As Koji pointed to the north, I could see the hypocenter of the bombing, located very close to the Commercial and Industrial Exhibition Building. Around it had already sprouted many stores, some with such unlikely names as Bookseller Atom. Most sold distorted glass bottles or burned roof tiles that purportedly were artifacts of the bombing, though the rapidity with which they were replaced made one wonder. Near the hypocenter was a stage on which was held the annual memorial services to those who succumbed in the bombing. Emblazoned across it was the slogan, "No More Hiroshimas." Beyond, one could see the largely denuded grounds that surrounded Hiroshima castle, which had been destroyed save for its moat and huge gray stone walls, infested anew with lichens. To the immediate left, above the jammed streetcars that lumbered steadily past the hospital, stood the Central Fire Station.

Since I had been told not to miss the ground-breaking ceremony (*jichinsai*) for the ABCC's new facility, I returned to the Gaisenkan shortly before noon to eat and join the others for the drive to Hijiyama, where this event was to take place. Hijiyama is a small mountain, more properly a hill, in the east central portion of Hiroshima, about two kilometers from the hypocenter, that derives its name from its paired (*hiji*) peaks. At the time of the bombing, Hijiyama was the site of a small park and a graveyard where the military dead from other wars were interred. Callously, in the opinion of many of Hiroshima's citizens, the graves had been moved to make way for this intrusion. It is difficult even now to know to what extent their grievance rests upon their view of the impropriety of this action (especially given that alternative construction sites were available and even favored by some American authorities) and to what extent it rests upon the fact that the ABCC's buildings on this site are the most conspicuous symbol in the city of Japan's defeat and a relentless challenge to the self-serving rewriting of history that has accompanied Japan's emergence as an international economic power. One senses that it was not impropriety alone, for years later the Japanese experienced no guilt over their own demands to move a cemetery in North Korea occupied by Allied dead to make room for one of their industrial developments.[2] Presumably their own economic well-being was being judged by a different metric.

The ground-breaking ceremony was a Shinto ritual to placate the earth prior to new construction. I am sure much of its drama and beauty eluded me, but I still recall the narrow, winding dirt road, clinging tenaciously to the mountainside, which took us to the summit; the elaborate flowing white robe of the Shinto priest; his black headpiece (*kanmuri*), with its gracefully curved adornment; his somber unsmiling demeanor and the steady cadence of his chants, punctuated by the swishing back and forth of a paper-festooned wand (*gohei*), which simultaneously attracts the gods and exorcises evil. The participants in the ceremony and the observers sat in open tents, protected from the vagaries of summer weather. The focal area of the ceremony was surrounded by walls of cloth of alternating broad white and red stripes, bulging now and then as they trapped a freshet of air. And, of course, there was the ceremonial turning of earth.

Once the ritual was over, we were gathered again into vehicles to go to the Koran, a distinguished restaurant and geisha house in the western part of the city. At the time, the Koran was the loveliest Japanese building I had seen, though the years have taught me that it really was not exceptional. Small courtyard gardens of the type that the Japanese use with great flair surrounded the building, but these were not of major caliber. They were not old and did not have the patina that age brings. To my untutored eye, however, they were enchanting. As our vehicles arrived and we stepped forth, we were ushered into the foyer of the building (the *genkan*), where we removed our shoes and were led to a series of interconnecting rooms, with tables set for the impending party. As I took off my shoes, a brief moment of panic swept across me: Had I put on a pair of socks without a hole? Before the evening was over, I was introduced to *sashimi* (raw fish), *sake* (rice wine), Japanese beer, and the variety of dishes that make up a formal Japanese dinner.

After inexpensive, typically humorous gifts were given to each of us, we were entertained with dancing and singing by the geishas who had hovered over us, serving food and pouring beer. They now unstintingly sought to teach us the rudiments of a few Japanese folk dances, especially that old favorite, the coal miner's dance (*tankō bushi*). Self-consciously and clumsily we Americans aped their gestures and movement. There followed a series of drinking games the geisha used to entertain, most of which were tests of dexterity that lead to the loser's having to drink a cup of sake or a small glass of beer. In a typical one, a sake cup is set on a piece of absorbent rice paper and filled to near its brim with sake. The bottle (*tokkuri*) is passed from participant to participant, each of whom must pour a drop of sake from the bottle into the cup. The one whose drop makes the cup overflow must drink the sake. It amazes the uninitiated how high the meniscus can build before it collapses over the sides of the cup. Another version involves a glass with a piece of rice paper spread across its mouth. In the center of the paper is placed a ten-yen coin, and each participant is expected to burn a visible portion of the paper with a cigarette without dropping the coin into the glass. The more skilled could satisfy the game's requirements as the support of the coin grew progressively more threadlike and tenuous. Obviously, in time, the coin must fall. These games penalize the unsteady of hand and with each loss the odds that one will also lose the next game must increase in some multiplicative manner. Endless permutations on these models exist, but all can lead to the same end, a frightful hangover.

At nine or so the party was over, and we were again ushered into station wagons to return to our quarters in Kure or Hiro. As we drove through the partially sleeping city, we passed little wheeled roadside stands selling noodles, grilled squid, or tea to the few people still on the streets. Before each stand were set several stools, often within a small canopy so that only the legs of both the stools and patrons were eerily visible. Here and there was an open store with the proprietor seated before it clothed in his underwear, wafting the humid summer air across his brow with a fan or animatedly speaking to a neighbor or a tardy customer.

Japanese parties in the immediate postwar years did not normally continue late into the evening. Most, like this one, terminated about nine or so, sufficiently early that the party-goers could catch a late streetcar or train home. Each homeward-bound reveler seemed determined to appear more drunk than any other, irrespective of how much he may have had to drink. Parties were times to tipple heavily and to enjoy the companionship of one's friends and colleagues. Drink also afforded an opportunity to say things one would hesitate to say under other circumstances; it could be an occasion for candor between the employed and the employer, one which invited no subsequent penalty.

As the days passed, I grew to know more of my colleagues. Although the bulk of the ABCC's personnel was Japanese, the professional, managerial, and technical staff were largely Americans—either recruited from the armed services in Japan or from universities in the United States—or Australians, or the occasional Canadian or Englishman recruited in Japan. They included anthropologists, architects, biochemists, biometricians, engineers, illustrators, internists, mechanics, pathologists, pediatricians, photographers, surgeons, and even two geneticists, myself included. The nonprofessional staff who sustained the various support activities was no less colorful or varied. The medical-illustration department consisted of a small enclave of Australians in which sat uneasily one American, a one-time seminarian, whose previous role in life had not prepared him for the ebullience with which he was surrounded. The motor pool, which grew daily, was directed by two recent dischargees from the armed services. They brought to the task a military perspective, with daily inspection of all the jeeps, wagons, trucks, and buses. Brass fittings had to be polished and the engine spotless.

Staff meetings of the Commission were a study in contradictions. The Americans and Australians lounged about comfortably, disdaining formalities, whereas our Japanese colleagues looked like a group of modern samurai—impassive of face, attentive in demeanor, poised as if ready to shout the guttural "ha" of affirmation at a moment's notice, or to audibly suck their teeth, as is the habit among Japanese males displaying their masculinity.

Management of the Commission's business activities was a daunting affair, complicated by our status within the Occupation, the distance intervening between ourselves and the United States, from whence most of our equipment and the like had to come, and rapid growth. The head of business affairs, Milton Evans, was a tall, dark-haired, urbane, and generally unflappable man. His burden was lightened by a small army of able section heads, a business manager, a personnel officer, a chief of the accounting section, a fiscal and property officer, and numerous others. ABCC was a business, and it was run as such.

Progress on the new clinical facilities moved forward under the watchful eyes of the architect in charge, his associate engineer, and

Mick Rappaport, who supervised the arrival and unloading of materials from the United States. A substantial amount of site preparation had to be completed before construction could begin. Scraggly trees, still bearing the scars of the bombing, had to be removed, a number of earthen gun replacements leveled, and several monuments to previous wars relocated. The foundations of a Shinto shrine that had been destroyed by the bombing and a small structure said to have been used as a repository for the remains of the war dead had to be dismembered. On the uppermost level of the hill, an old house, its roof tiles in disarray or scattered broken on the ground about the building, its northwest corner crumpled inward by the blast and now precariously supporting the roof above, had to be razed.

Today, bulldozers, backhoes, and wrecking balls would quickly move the needed earth or level the unwanted structures, but in postwar Japan this was done through the manual efforts of numerous men and women. Armed with picks and shovels, the men worried out basket after basket of earth and stones to be carried away by *mompe*-clad women, some two hundred of them. Mompe are baggy garments, mixtures of skirt and pantaloons, generally made of a blue cotton cloth and normally worn only for farm, construction, or industrial work. Leggings of a sort, split-toed canvas and rubber shoes, and a white cloth or towel carefully tied about the head completed the women's attire. Each woman was also equipped with a strong pole (a *tenbinbō*) from which two shallow reed baskets were suspended with three ropes. When the baskets were filled, she would stoop, rest the pole across her back and shoulders, adjust her position between the baskets to distribute the weight uniformly, and rise. The baskets of earth would dangle a foot or so above the ground, and to prevent their needless rocking and swaying, she would move off with a shuffling gait, a hand clutching the supporting ropes to further steady her burden. Once at the spot where the earth was to be dumped, a quick upward tug on one of the ropes supporting the basket tippled it, spilling out the earth. Back she would return for another load. Hour after hour this continued; and, as the days passed, the top of Hijiyama was slowly remade. As these events were transpiring, the Commission briefly toyed with the idea of building similar structures in Nagasaki, at the site of a former prison in the Urakami valley. These notions were abandoned, but not before the site had been fully surveyed by Mick and some of the staff from the Nagasaki office.

Structures that are to rise more than two stories generally require some sort of scaffolding—commonly pipes bolted together like an oversized erector set—but in Japan in the late 1940s and early 1950s such scaffolding was constructed from long, supple pieces of bamboo. These were laced together with chunks of rope and suitably reinforced to assure the requisite strength. The result was often a work of engineering art, climbing four, five, or more stories. It was a fabulous sight to see the male workers, often bare to their waists, scurrying about the rising scaffold, quickly lacing together the successive pieces of bamboo as they were lifted aloft. Considerable coordination was needed as two men lifted into place a 12- or 15-foot bamboo pole and lashed it securely to the frame already assembled. The network often looked so fragile that its fall, like an overly tall stack of match sticks, seemed imminent. This never happened.

Slowly but steadily the buildings took shape, though undoubtedly not ones the citizens had anticipated. The buildings themselves were designed by Skidmore, Owings and Merrill, a reputable architectural firm, but the shape of the structures they erected have been the butt of endless comments. Superficially, they look like the wartime quonset hut, but they are substantially more—two stories in height, for example, and built to withstand typhoons and earthquakes. It has been said, possibly apocryphally, that the original intention was to dismantle several quonset structures that liberally studded the Pacific islands and bring these to Japan. But before specific buildings could be earmarked, they had either been sent elsewhere or pushed into the sea. Allegedly, the designers, enamored with the functionality of these military buildings, copied their basic style. However, some modifications were necessary. The various buildings had to be connected by corridors, closed generally on the first level but not the second; and to ensure the requisite lighting on the second floor, a bank of dormer-like wood-framed windows extending virtually the length of each building was incorporated into the design. This exposed the supportive metal ribs, compromising the available space and providing obstacles that have barked unnumbered heads. The shape of the buildings is very similar to an inexpensive, popular fishcake in Japan, called *kamaboko*. This has lead to snide remarks about the *kamabokojō* (fishcake castle) or *kamabokotei* (fishcake palace) on the hill.

Initially, bureaucratic machinations resulted in doors, boilers, hardware, and other parts of the structures aging slowly on unloaded ships sitting in Kobe's or Hiroshima's harbor. And because many of the construction techniques were new to the contractor, careful supervision and repeated time-consuming consultation were needed if proper installation and economies were to be achieved. Even so, mistakes in design and construction were made through ignorance and oversight. For example, electrical current in Japan is 100 volts, but the equipment, manufactured in the United States, required 110 volts. As a result, the architects understandably opted to install 110 volt lines, but this entailed the use of transformers. Today, much of the equipment is of Japanese manufacture and is designed to operate on 100 volts, and therefore the buildings are still plagued with the need to transform the current, now from 110 to 100 volts. Similarly, fearful of fire but unaware of the possible carcinogenic risks, the architects specified that all ceilings of the rooms should be sprayed with asbestos, and concern presently exists for the staff that work in those rooms. Moreover, since construction began before the scientific program had been fully developed, the ultimate needs of different aspects of the program were uncertain, and often the space allocated has proven inadequate. This has led to further construction and a continuing need to renovate existing space.

These problems notwithstanding, when the buildings were finally opened eighteen months after the ground-breaking ceremony, they were undoubtedly the grandest scientific facility in Japan, shining with clinical and research equipment. Rapidly the Commission's headquarters became a local showplace, and for two or so decades, tour groups would mingle with the survivors who had come there to be examined.

MACARTHUR'S JAPAN

*M*acArthur's Japan made few, if any, special provisions for civilians, whatever their status elsewhere might have been. Civilians employed by the Department of the Army were governed by virtually the same rules as governed military personnel. Other civilians, including business types from the United States and foreign traders, lived more tenuously. A few hotels and shopping facilities catered to their needs, particularly with foods more exotic than the commissary usually carried. Those of us who were with the Atomic Bomb Casualty Commission fell into neither category. We had some of the perquisites of both, but not the independence of the traders nor the security of Department of the Army civilians.

Housing was the great equalizer among the military and civilians attached to the Occupation. It could not be applied for until one was in Japan; and then availability of quarters locally remained the ultimate determiner. My application had been filed as soon as I arrived in Japan. It was now a matter of waiting. But what was I to do until Vicki could come?

Once our bus deposited me in Kure of an evening, I would generally go to the Australian Officer's Club to eat. A filet mignon with vegetables was about 75 cents; the addition of soup and a dessert would bring the meal to slightly more than a dollar. Good as the food was, and as pleasant the atmosphere and service, after a few weeks it did pall. Occasionally, I was invited to dinner with the families of colleagues at the Commission. Saturdays and Sundays were days of sightseeing and photography, usually in Kure and its immediate vicinity. To vary the agenda, I would take a ferry trip to Eta Jima to visit the training facilities where two or more generations of Japanese naval officers had been housed and weaned. A mixture of Annapolis and England's

Dartmouth, this spacious, well-tended collection of red brick and concrete buildings, dormitories, and offices faced a corner of the Inland Sea, and in their midst was a large, green drill field and several swiveling gun housings that had once served to familiarize plebes with naval gunnery. These facilities had been commandeered by the Occupation and now housed one of the post exchanges of the area. Set midst these buildings was a shrine-like memorial to Admiral Togo, who had defeated the Russian fleet during the Russo-Japanese war and was the ultimate architect of Japan's modern navy. The Occupation closed the shrine, presumably to further peaceful aspirations among the Japanese, and made it off-limits to all of us, occupiers and occupied.

Inclement weather often found me exploring Sawahara House, which—though certainly not the most imposing home in Kure—was nonetheless an intriguing structure, owned by one Sawahara-san. Even cursory inspection made clear that an amateur—or for that matter, a professional—woodworker had much to learn from the Japanese. Splines, double splines, dovetails, mortise and tenon were joints known to Western cabinetmakers, but here they were used with exceptional skill. Some of the exquisite joinery can be ascribed to the lengthy apprenticeships Japanese craftsmen serve and to the tools they use. Saws, which come in a limitless variety of sizes and tooth settings, cut on the pull rather than the push, and planes cut similarly. This action gives their user a finer measure of control; and even simple, inexpensive wooden boxes profit from the skill this affords. Tables, chests, and charcoal braziers were gems of a variety of woods—persimmon, oak, paulownia, and sandalwood. Their smooth, polished surfaces invited the hand to find an imperfection.

Chinese ideograms also began to fascinate me. Presumably each was once a more representative drawing than it may presently be. Imagining this evolution became an amusing pastime. Some origins are simple to see, as in the character for a horse, with its mane, four legs, and a tail; others are more abstract in nature, and it is more difficult to envisage their origin. I devoted my evenings to learning to read and write Japanese. This was a challenging task, for when the Japanese imported their present written forms from China somewhat before and during the sixth century, they not only brought the prevailing reading of the Chinese ideogram but added to it their own expression for the character. As a consequence, most ideograms have two readings, one called the *on,* or Chinese reading, and the other the *kun,* or Japanese reading. A word that requires two characters to represent can have four readings in theory, but generally if the first character is read in the on (or kun) manner, so is the second. As in most languages, there are exceptions, and these can only be committed to memory. Early on, I bought a brush, inkpot, and ink, made either from the soot of burning pine or oil, and began to accumulate old newspapers, which I had been told would provide a suitable practice medium.

My first lesson was to learn how to mix the ink (*sumi*) with water to achieve the correct intensity of color. The second step was to learn to point the brush (*fude*) properly and to practice strokes. My instructor seemed to think I wanted to become a calligrapher, when in fact I simply wanted to be able to write and to read some of the more common characters. Once we had reached an understanding of my limited goals, I was put to writing *hiragana* and *katakana,* the simpler scripts. After these were mastered, I graduated to characters and faced the task of learning the order and relationships between the strokes, top to bottom, left to right. Japanese children learn to write by filling boxes, first large and then progressively smaller ones as skill and a sense of proportion matures, and so did I.

Concurrently, I was taught to use a *kanji* dictionary, which is organized around 217 simple figures or radicals; these may be a single brush stroke or a full character that has meaning in itself but is also the root for a collection of other characters. Under a radical, ideograms are collected on the basis of the number of brush strokes required to fashion them. One must, therefore, be able to count the strokes, but this is less easy than it sounds. The number is dictated by the use of the brush. Since each line does not necessarily represent a stroke, one has to learn how the brush is used. A simple box, which has four sides and happens to be the character for *kuchi,* or mouth, does not consist of four strokes, as might be imagined, but three, for the topmost portion of the box and the right side are written without removing the brush from the paper. If the character is simple, one is unlikely to be off more than one stroke in counting, but as the character becomes progressively more complicated, the extent of the error can grow. As a result, there is an intimacy between the written word and its method of writing. A Ming literatus in China would have known tens of thousands of ideograms, but Japanese newspapers are restricted to the use of 1,800 or so and their combinations. Even college graduates are unlikely to know more than ten thousand unless their education has been in traditional Chinese and Japanese literature.

There is a beauty to the written language that has made calligraphy a complex and expressive art. It can be powerful, a series of explosive, heavily brushed symbols, or exquisitely graceful, a succession of cursively written interconnected symbols. Certain ideograms lend themselves particularly well to these differences, such as *kokoro*, the heart in a spiritual, not an organic, sense. It can dominate or modulate a scroll. Calligraphy also lends itself to aphorism: to the briefly stated principles of conduct that are at the same time inherently attractive. Western printing has no counterpart. Fonts have been designed to enhance the visual appreciation of a book, and certainly the art of printing and book binding is not unknown in the West, but these adaptations are not of themselves art, capable of standing alone save through their rarity. Ideograms are personalized views of life, its meaning, substance, and resonances.

As familiarity grew, I watched the instructor more carefully. After he would point and ink his brush, he paused momentarily above the paper as though collecting his thoughts and brushing the character on an imaginary piece of paper, and then the brush was put to paper and the character emerged in a series of continuous, rapid strokes. Each was completed skillfully, with a terminal flourish, either forcing the brush into the paper or lifting and twisting it lightly to trail a fine ending. The result was a miniature painting so carefully perfect that it appeared effortless. It was this latter quality, the apparent effortlessness, that was so difficult to emulate. Try as I might, my writing seemed heavy-handed, stiff, forced; page after crumpled page accumulated as I worked endlessly on one character. None of my attempts had the grace of the model before me. As frustration built, my instructor counseled patience and always found some means to whet my flagging interest.

As the number of characters I could read increased, so did my interest in the signs of stores about me. Japanese is now customarily read from left to right, but this has not always been so. Prior to the Meiji restoration in 1868, the opposite obtained. The transition in style that began a century or so ago has been a gradual one, impelled by several governmental edicts but passively accepted by almost everyone. Today, one sees few if any instances of the old form; but in the immediate postwar years, many store signs were still read from right to left. It was always a challenge, especially with those signs that bore only the owner's name, to know how to read the characters. Immediately before the Red Cross Hospital was a store that catered to the needs of patients and their families. I spent many a moment gazing out of the window of my third-floor office and wondering whether the proprietor's name as written on the sign was Murata or Tamura, alternate readings of the two kanji in the sign. Both are legitimate surnames and thus neither could be excluded. The store is gone, and I still do not know what the owner's name was.

*   *   *

We stood there somewhat self-consciously, Frank Poole and I, between the terminal and the fence that controlled access to the tarmac, as the Northwest Airlines DC-4 wheeled its way to a stop and the stairs and baggage-handling gear moved forward cautiously while the engines were feathered. It was September 19, 1949; the day was sunny, but a faint autumnal bite filled Tokyo's air. Frank, an affable, blue-eyed North Carolinian with a soft drawl that perplexed his Japanese colleagues, was wearing a cord suit with one of the garish floral ties that typified the time. My tan suit with solid red tie was more conservative, but its wide lapels, like those on Frank's suit, gave my attire a somewhat comical appearance, at least by today's standards. We clutched welcoming corsages in hand and were ecstatic about the approaching end to our bachelor existence.

As members of the Occupation, we were allowed to greet our wives before they cleared Haneda's formalities, but we opted to wait behind the barrier with everyone else as the passengers disembarked. When Vicki and Harriet appeared at the head of the stairs, we waved furiously. In a sea of undulating hands, they did not see us immediately. Finally, the worried looks that creased their faces when first we saw them disappeared—we were seen. Customs, Immigration, and Quarantine were swiftly completed, and Vicki left the arrival area clutching a special passport that now bore the signature of Corporal James L. Lightfoot, the Allied Powers Supreme Commander's surrogate.

George Fukui had already arranged accommodations at the Dai Ichi Hotel and return passage on the train. Fortunately, we were to have a day in Tokyo before the wearisome trip to Nijimura, the community where our newly assigned house was located. Vicki was especially tired—the product of over thirty hours of flying and apprehension. After a suitable rest, we set off for the PX on the Ginza. We walked the short distance from the hotel, sharing the sounds, smells, and activity of the city. We strolled briefly along some of the intriguing little streets that ran parallel to the Ginza, an area of many bars and small eateries, most of which sported, to Vicki's surprise, the ubiquitous "Off Limits" sign to which I had already become accustomed. George had seen to a temporary identification card for her so that we might pass the gimlet-eyed guards at the door of the Exchange, who carefully identified anyone not in uniform. It was hardly a wonderland for someone so recently removed from the Mid-west, but to me it already seemed like Gimbel's and Macy's rolled into one. Not all suits were extra large, and socks in sizes smaller than sixteen could be found. The floors devoted to cameras—including the German ones, Contaxes and Leicas, which were still more highly prized than their Japanese imitators—were always crowded. Binoculars were cheap and obviously a critical need for everyone, their purpose more dimly perceived than the presumed need. Vicki stopped briefly in the women's department but saw little of interest except the yard goods, where silks were plentiful and cheap. Indeed, the choice of women's apparel was very limited, and most women had their dresses, blouses, skirts, and suits custom-made. Seamstresses were numerous, inexpensive, conscientious, and generally skillful, at least in the mechanics of sewing. When the seamstresses were left wholly to their own devices, however, nothing seemed to fit especially well, and patience and a knowledge of precisely what one wanted were required. As yet Japan had evolved no Mori Hanaes, nor the interest in haute couture that presently exists.

On the Ginza again, Vicki and I walked northward to a department store called Takashimaya. Immediately across the Nihonbashi—a bridge that is the figurative and literal center of the country, since distances are measured from its midpoint—was Mitsukoshi, the oldest of Japan's extant department stores, distinguished by its famous twin bronze lions. Both of these stores have again established themselves as worthy of international comparison, but in 1949 their shelves and counters held far less of interest. One corner of Takashimaya's basement sold antiques on consignment, something that would now be beneath the store's dignity. Mitsukoshi was only slightly less threadbare—no designer boutiques to browse, just simple, utilitarian garments stacked on its counters.

Most fascinating to us was the basement, where Japanese department stores customarily have their food departments. Strange smells from an exotic array of edibles unknown to us drifted through the air and drew us from corner to corner, as we ogled counters of prepared fish, dried squid, and mushrooms. Many Japanese too poor to purchase these items stared quietly, as if feasting vicariously. Others bought modest quantities, one or two of this or that, carefully counting the change into the clerk's hand as though even one ten-yen piece was precious.

During the Occupation years, most Japanese appeared marginally nourished but probably were not. The impression was heightened by their dress, particularly the men's. The belt loops on their pants were commonly placed an inch or so below the top, and the prevailing practice was to wear pants and a belt several sizes too large. Some of this may have been to accommodate the bulk of the *haramaki,* a waistband worn to keep the stomach warm, but the excess cloth gathered in a ruffle about the waist and a belt whose end was in the middle of the back certainly contributed to the appearance of undernourishment. Few people could afford separate summer attire, save possibly for the inexpensive cotton kimono (*yukata*) or the *jimbei,* a kind of lightweight undergarment worn by men. Warmer weather found many of the men in open-throated white shirts and pants and, in 1949, a boater—the flat-topped, modestly brimmed straw hat that was once the mark of summer in the United States. If it got sufficiently hot, men whose work kept them in the sun often wore nothing more than a loin cloth (*fundoshi*). Women, especially the younger ones, were more circumspect, but it was not uncommon to see elderly women bared to the waist, particularly in the countryside.

*   *   *

Nijimura, where Vicki and I would live for two years, is no more. Its two-hundred-odd houses, some single-family dwellings but mostly two-storied duplexes, have disappeared, razed to make way for an industrial park. This tile-roofed, rainbow-hued village on the outskirts of Hiro had been designed to be self-contained; it had its own store, fire station, post office, barber shop and hairdresser, administrative offices, tennis courts, and movie house, where Vicki and I saw many of the earlier classics of the Pine Ealing Studios, including the especially memorable *Whiskey Galore.* This movie, based upon a novel by Compton MacKenzie, was shown in the United States under the name *The Tight Little Island;* words like whiskey could not appear in American movie titles in those days.

When we moved to Nijimura, life was untroubled—the war had been won and we lived secure in our atomic defense. Nights were for parties, the movies, or study. Our house, a pale blue duplex, had three rooms downstairs—a living room, a dining room, kitchen and bath—and, upstairs, a second bath and three bedrooms, one of which was used by the live-in maid. Like the other two-storied dwellings in the community, it had a fire ladder attached at right angles to the building reaching to the second floor. Since the ladder was made of wood, it gave a sense of security without the fact. We did not worry about the appearance of our furniture, for we all had the same—oak chairs, beds, tables, bureaus, and bookcases. Our ashtrays were all made from turned aluminum with a wide, inwardly bent lip, impossible to empty conveniently or completely. On the bottom of each was stamped "Allied Forces JPNZ 4656," the cryptic meaning of which has remained a mystery. I have presumed that the NZ implied New Zealand Forces but am not confident of this. Each house was fully furnished with dishes and flatware, linens, blankets, and drapery; the only objects we were told to bring was a good bottle and can opener and a paring knife. Pictures and the inevitable, quickly collected Japanese objects personalized our homes. The grass around the houses was tenderly if somewhat unsuccessfully nursed, the sandy soil making its cultivation an indifferent thing, but numerous azaleas brightened the village in the early spring.

Domestic help was inexpensive and plentiful; every home had at least one maid and most had a cook as well. We too had a brief fling at being a two-servant family. Shortly after Vicki arrived, as we were still settling into our home, she thought it would be good if we hired a cook. So we set about finding one. Prospective employees were obtained through the employment office in the village; one merely called and stated whatever skills or other attributes were wanted. In due course, a person with the presumed qualifications appeared; if they proved unsatisfactory, they were returned and another would come.

Our cook, a good-humored, late middle-aged man, arrived one morning on a squeaking bicycle, complete with his own uniforms. Little did we know that his sole experience had been as a short-order cook in an enlisted man's mess. He introduced himself and immediately took over the kitchen. A flurry of phone calls ensued; these resulted in our electric stove being adjusted and several other changes made in what he considered his domain. The first several days went rapidly. Vicki suggested three or four uncomplicated menus, and these were reasonably well prepared. She had been startled, however, to find him staring quizzically at a recipe through two separate pairs of glasses, one perched on the end of his nose in front of the other. This did not seem promising, but he did so well in other respects that we thought nothing of it at the time. He had managed to sharpen all of the knives to a razor's edge and had unquestionably put the kitchen into fine shape.

It was only slowly that we discovered he could not cook much except eggs, bacon, and hamburger without very explicit instructions. And it was clear he could not put together a menu on his own. Reluctantly, we decided he would not do, and we returned him to whence he had come. He was not particularly surprised, nor did he seem displeased by this turn of events; I suspect it had happened before. Indeed, as a short-order cook accustomed to an efficient kitchen, this may have been the function that suited him best—to get the kitchens of others in order. Throughout the remainder of our days in Nijimura, every time we picked a paring or butcher knife from a drawer and felt its fine edge, we thought of him.

Our maid, Hasegawa Ayako, was an effervescent young girl, and we were her first employers. She learned quickly and was so out-going that soon, throughout Nijimura, we were simply known as Ayako's family, for more villagers knew her name than ours. She was delighted when we entertained and would carefully starch and press her blue uniform in preparation for the arrival of our guests. As each one came, she would greet them, smiling broadly and winsomely, and if it was winter, relieve them of their coat, carefully placing it in a closet before returning to the kitchen. Once the last guest had arrived and we were all seated at the dinner table, Vicki would ring a small bell to summon her, and through the partially closed kitchen door we could hear a giggle, barely suppressed, as she readied the dishes for the table. Parties seemed to be the spice that gave purpose to her employment. It was customary to give one's maid one day off each week, but she often remained with us, as though ours was a happier environment than the one to which she would return. So much a part of us did she become that on the unexpected death of her father, she sought solace with us in addition to her family. Unfortunately, we could do little more than share her tears. Ayako was with us until we returned to the United States in 1951, and we continued to correspond for some years thereafter. When next we saw her, she was a young matron, a bit more subdued than we recollected, with a small but growing family. Her eyes remained no less lively and her smile brightened her face as glowingly and as often as we remembered.

Shortly after her arrival, Vicki quickly enrolled in flower-arranging classes, took lessons in painting and Japanese, and began teaching English to a small group of Japanese professors at the University of Hiroshima's Medical School in Aga. She quickly discovered that Japanese standards of flower arranging make those of the West seem untidy, an artless clumping of blossoms. Astringency and orderliness dominate the former. Ours place all the burden of attractiveness on the flowers themselves and do little to harmonize their beauty with the structure of the world; color substitutes for a perception of nature's integrity. Most Japanese schools of flower arranging follow rigid rules in the placement of the individual elements that comprise the arrangement; unfailingly, there is a high, middle, and low perspective and perfection is achieved through the deliberate use of the imperfect—a bent stem is preferred to a straight one, an awakening bud with its promise of beauty to a blossom in its prime. The vase or bowl is not merely a vessel for water, a means of support; it is chosen to achieve an overall effect. Bark, twigs, and, in the Sogetsu school, even plastic objects are used to stunning effect. Each arrangement celebrates the mysteries of the season or the one to come rather than the glories that have been or are soon to go, and some arrangements last for weeks, slowly evolving, changing constantly.

Every six weeks or so Vicki's routine was interrupted by the arrival of Takenaka-san, a dapper jeweler of fifty or so. His home and shop were in Nagasaki, but he visited most of the military housing areas between Kyoto, where his jewelry was made, and Nagasaki. He always came early in the evening, after notifying us of his coming; and smilingly, expectantly, he exhibited his strands of cultured pearls, gold rings, and broaches, or took orders for specially designed pieces his Kyoto artisans would fashion. He would not have known how to conduct a hard sale; his jewelry spoke for itself. He was confident of its quality and fairness of price, and so were we.

Virtually every morning an elderly fishwife, garbed in peasant dress over which was drawn a white smock (a *kappōgi*), a straw hat carefully tied to her head, would set her pails of shrimp and boxes of fish about before our door for Vicki to select our needs. Most of our other foodstuffs arrived on a special commissary train. Weekly, on Wednesday, punctually at 6:03 A.M., the train arrived in the Hiro railyards and remained there until 12:59. It offered full food facilities and postal services, including mailing packages and selling stamps and money orders. Wives had to go to the train to select their groceries, meat, and sundries. While some of the green produce originated in the United States, much more came from hydroponic farms the Occupation managed, several in the mountains not far from Hiroshima. Purchases were packed into footlockers labeled with each family's name and house number. These were loaded on a truck, and sometime in the afternoon the footlockers and their contents appeared at one's house. Since the train came only once a week, and one could never be certain that a specific item—a cake mix, for example—would be there, most families slowly but purposefully accumulated a pantry full of cans, boxes, and whatnots. It was not hoarding, for no two families accumulated the same things; it was providential shopping. One did learn to improvise and to supplement one's foodstuffs from the Japanese market and the Australian commissary. Strictly speaking, we were not allowed to shop in the latter, but through our commissary train we had access to things the Australians did not have and vice versa, so a great deal of bartering occurred. Especially favored were the Bols liqueurs with which the Australian store was liberally stocked, or the bocks-beuteled Rheingold, a fine German-like Australian wine. Scotch, bourbon, brandy, gin, and other liquors could be bought at our Officer's Club and were so inexpensive that our liquor cabinets looked like commercial bars.

Fruits, berries, and grapes could be purchased on the Japanese market when they were in season but not always without some anxiety. Concord and muscat grapes, like the pear-like *nashi,* were especially good. The Japanese variety of strawberries were large, tasty, and faultless and an irresistible temptation. Strawberries are hard to wash effectively without severe bruising, however, and since they were grown in earth fertilized by night soil (human excrement), they were a likely source not only of bacteria but of protozoan parasites. Each season we faced the same dilemma: Should we wash them as well as we could and take a chance, or be cautious and merely envy others? Some of our more industrious neighbors tried growing strawberries in their own gardens, with indifferent success. Hydroponically raised ones were available through the commissary train, but they seemed tasteless. Finally, the Commission's parasitologist, Frank Connell, who was no less tempted than the rest of us, suggested a simpler solution. He first washed the Japanese berries thoroughly with water, dipped them briefly into mercury bichloride, and washed them again slowly in running water. The process seemed reasonable, and to my knowledge no one who used it became either poisoned or infested. But we still had the feeling that we were playing roulette with our intestinal systems.

Most Japanese homes, certainly those in areas remote from the centralmost portions of the cities, were not connected to sewers during the Occupation. Each house had a cistern that collected its human wastes, and these had to be emptied periodically. The excrement itself was usually collected by an *owaiya*, a night-soil man, or by farmers, who spread it on their fields or stored it in well-like structures on their farm grounds for subsequent use. At one time it had been customary for the householder to sell the excrement to the owaiya, who would in turn sell it to farmers, but as cities grew and it became more troublesome to get the fertilizer to the farmers, it was necessary to pay to have the night soil transported away. Little automation of this process occurred prior to the 1950s. The excrement was literally dipped from the home cisterns and poured into wooden pails, and these were placed generally on a horse-drawn wagon to be transported out of the city. The process was necessarily slow, and the horse and wagon would block the narrow street until the collection was completed.

As we in the Genetics Program searched out the homes of newborn infants in Hiroshima, we often encountered these horse-and-wagon barriers. The loss of time was troublesome enough, but we had to wait in an atmosphere blue with the odors of putrefaction. Our driver would cautiously seek to back away from the noxious portion of the cloud mushrooming up from the cistern, but often this was not possible, for the streets were sometimes so narrow and tortuous as to be negotiable in only one direction. One tried to turn off one's olfactory processes, but this was rarely successful. Soon everything about one smelled, or so it seemed, and were it not for the commonness of the event, one would have been more embarrassed upon encountering others. Sympathy was not openly expressed, but the absence of comments bespoke understanding.

Of greater moment was the role night soil played in the perpetuation of the parasitic infestations that prevailed in Japan. Virtually everyone had one or more of the roundworms—ascaris, hookworm, or strongyloides, or protozoan parasites such as *Entamoeba histolytica,* or more serious still, one of the liver flukes, notably *Schistosoma japonica.* All could be severely debilitating and some fatal. Japanese public-health authorities and physicians were aware of the dangers these infestations posed, but they also knew of the nation's food needs and the dependence of the latter upon a cheap, nitrogenous fertilizer. To ban the use of night soil would solve one problem but exacerbate the other. Substantial effort was invested, therefore, in other solutions to halt the cycle of parasitism and to find an alternative fertilizer. Among the former was a redesigning of the cisterns in which the farmers stored night soil. A multichambered variety was developed that separated the eggs of the various worms through a flotation process, so that uncontaminated excrement could be drawn off for use in the fields. This proved effective, but the cost of the redesigned cistern was too much for any of the farmers to bear. Fortunately, more progress came from another quarter.

Throughout much of the last half of the nineteenth century and the first half of the present one, nitrogenous fertilizers came mostly from the exploitation of huge fields of guano or mineable nitrate deposits. Both of these sources had contributed importantly to the economies of Chile and Peru; indeed, the nitrate deposits were a significant factor in the war Bolivia, Chile, and Peru waged from 1889 to 1893. However, during World War I an electrolytic process was developed in Germany for the artificial production of urea, the simplest nitrogen-containing organic compound and an important constituent of night soil. This lessened agricultural dependence upon the nitrate deposits in South America and elsewhere, but the cost of fertilizers was prohibitive for the small Japanese farmer. World War II saw further development of this process and a sharp reduction in cost. Once aware of this progress, a number of Japanese chemical companies sought the right to use the new method. Soon inexpensive fertilizers were available, and the growth of farm cooperatives (*nōgyo*) provided the means to buy in quantities that reduced costs still further. This solved the farmers' needs and simultaneously eliminated the threat of parasites.

However, the night soil remained, and its disposal was no less essential. Progress occurred here as well. No longer is it necessary to empty the household cisterns manually; today, the excrement is pumped into vehicles with closed containers that look like diminutive gasoline trucks. Collection is much quicker but unfortunately not entirely odor-free. More cities have developed closed sewer systems, particularly in their densely settled centers, and wastes are filtered and managed more conventionally.

In the postwar years, Japan's farmers struggled ceaselessly to feed a hungry nation grown larger by repatriation and a surging birth rate. Rice was the staple, but no dramatically high-yielding strains existed, and rice cultivation was labor-intensive. Planting of the rice seedlings occurred as soon as sufficient rain came to flood the previously plowed paddies. Mompe-clad women did most of the planting, generally in a cooperative manner. A group of six or eight would set out the new plants in rows so uniform as to please the eye and amaze those of us unfamiliar with rice cultivation. As seedling after seedling was planted, new clusters of plants were brought to the bare-legged women at work in four or five inches of water. A hole was made in the mucky bottom of the paddy with the index and middle finger, and a plant was edged into the hole by the thumb worked across the palm of the hand. Each planter labored in cadence with the others, and the new rows grew as though sprouted. Field after field appeared where only days prior little was to be seen; but to use the spring and early summer rains effectively, the planting season was necessarily short. It was difficult work, and age brought only an arthritic back, the penalty of years of stooped labor.

Tillage, fertilization, and other chores continued until the first heads of grain appeared in late summer and early autumn. Slowly these turned golden, the awns grew conspicuous, and the plant struggled to remain erect. As the grain ripened, scarecrows populated the fields—some straw-filled images of farmers at toil, others simple disks strung on a rope or wire bouncing saucily above the fields, impelled by each breeze. The weeks immediately preceding harvest were vulnerable ones. A driving rain accompanying autumn's typhoons could leave a field of rice prostrate and severely damaged, often beyond salvage. These events encouraged a certain stoicism that seemingly characterizes farmers everywhere.

At mid or late autumn, once the paddies were dry, harvesting began; the fields were again aswarm with men and women armed with small sickles to sever the rice stalks an inch or so above the ground. The cut stalks were carefully bundled and transported to the spot where the grain was separated from the stalk. This threshing was often done with the aid of a very primitive but effective little machine, not much more than a rotary drum from which projected studs that literally tore the heads of grain from the stalks. Power to rotate the drum was provided either by a foot-operated treadle or through a wide belt attached to a small two-cycle gasoline engine. Commonly, this engine was a multipurpose one, mounted on two modest-sized wheels that could be directly engaged and connected to a small wagon or even a small tiller through a round shaft that allowed the rigid axle of the unsprung wheels to tilt independently of the object being towed. Surprisingly uneven terrain could be negotiated, albeit slowly. Guidance was achieved through a pair of handles on which were levers to regulate the speed and to act as a brake, much like a motorcycle. The final winnowing of the rice was done by hand with the aid of the wind. Grain and chaff were placed in shallow woven cane baskets and rhythmically tossed into the air. Occasionally, to aid in winnowing, a large hand-operated rotary fan, spun steadily by one of the harvesters, would be seen. As others tossed the grain into the air to fall onto a carpet of mats, chaff would be haphazardly distributed just beyond the edges of the accumulating pile of grain. Periodically the grain was transferred to woven straw bags for transport. It was a tiresome but essential process, however managed. Most of this work has now been ingeniously mechanized. Little rice combines that cut, separate, and bag the grain roam the autumn fields, leaving a spoor of bundled stalks in their wake that are hung to dry on temporary racks and used as fodder and packing material.

Grain surpluses have become a recurrent political and economic problem in Japan, but in the late 1940s and throughout the 1950s one crop was hardly harvested before another was planted. In much of Kyushu, the southernmost of Japan's four main islands, this second crop was rice, but in the Hiroshima area barley or winter wheat was planted. The demands of these plants are sufficiently different from those of rice that one intensive period of labor was soon followed by another. The paddies were divided lengthwise into rows, perhaps a meter or so in width. Alternate rows were dug down 6–10 inches and the dirt heaped on the adjacent row. This effectively lowered the water table for the rows that received the added soil. Across these rows was planted the second cycle of cereals to be harvested prior to the planting of the spring rice. This method of soil preparation is no longer seen, and in fact few areas in the Hiroshima region even attempt to produce two cereal crops a year. The almost religious notions that governed the raising of the rice seedlings, their transplantation, and the ultimate harvest of the grain have largely disappeared, save as provocation for periodic festivals (*matsuri*).

* * *

Each day seemed to introduce Vicki and me to another facet of Japanese life. Very soon after her arrival we encountered the tea ceremony, a quintessential part of the culture.[1] However, as we learned, there was not just one tea ceremony but many. At least two major divisions exist—omaicha and osencha—and within omaicha there are many slightly different schools. Of these two divisions, it is said that omaicha is for the individual who loves ceremony, whereas osencha is for the lover of tea. There is no similarity between the two, save that both ultimately culminate in a cup of tea.

Osencha is made with dried, intact tea leaves infused with water and is brown and mild. Normally, the tea service consists of a small tea pot, a container for used water, and five small cups, only slightly larger than those in which sake is served. The service usually sits on an elliptical tray, ordinarily of cherry wood. The pot is rinsed with hot water and into it is placed enough tea to make a single cup for each person; boiling water is then added. The suspension is allowed to steep, and the brewed tea is carefully poured equally into as many cups as are needed. Each portion is little more than a single swallow; it is a tea to be tasted with the tip of one's tongue rather than gulped. The tea, like the service in which it is brewed, is chosen with finicky care. Tea from Uji, a type known as *gyokuro*, made from the tenderest leaves of old tea plants, is most highly regarded. If additional tea is needed, it is brewed afresh. The last few drops are customarily poured onto the tray or into the accompanying wooden saucers and thoroughly rubbed into the wood which, with time, takes on an admirable patina. As for the tea service itself, enthusiasts particularly prize Hagi pottery, for with continued use its gray-white glaze, with just the faintest hint of color, will turn a marvelously rich robin's-egg blue, presumably through some hydration process brought about by minute amounts of tea oozing through the fine checks in the glaze. When first introduced to this process, I was so intrigued that I immediately purchased a service. Faithfully each day I made, but did not always drink, the tea. After several weeks, when there was no perceptible change in the color of the cups and it seemed clear that years would be needed, my interest faded.

Omaicha uses a finely powdered green tea, mixed and whipped into a suspension with hot water; to my palate it is much too grassy-tasting to be enjoyable, but one can become enthralled by the ceremony attending its preparation. Externals are most important: the studied grace of the preparer of the tea, the exquisite, often extremely costly bowls in which it is served, the separate house in which the ceremony customarily occurs. Even the conversation is wreathed in protocol. Women and men are schooled in the ceremony for years. Participants are expected to divest their minds of mundane matters, concentrating on the beauty of the ceremony and the surroundings in which it takes place. To aid in the rightness of mind, a tea house is entered through a small, special door too low to enter standing; one literally crawls in— presumably a humbling experience. Inside, one's eyes are irresistibly drawn to the austere, astringent beauty of the interior, with its charming alcove, stunning flower arrangement, and hanging scroll. Embedded in the floor is a charcoal brazier on which water is heated for the preparation of the tea. Edward Morse, whose book on Japanese houses was probably the first careful study of these structures in a Western language, wrote ecstatically and aptly of the studied beauty of the tea house and its marvelous cabinet work.

Although any modest number of individuals can participate, tea is served to just one guest at a time in the order of their social station or importance as perceived by the host. Before the tea is actually made, a sweet is set before each guest on a thin tissue of paper, along with a small, carefully cut twig with which to lift the sweet to the mouth. Again, the actual brewing begins with a rinsing of the cup in hot water drawn from a cast-iron kettle; the hot water is then discarded. The kettles, particularly those from Kyoto or Japan's north, Morioka, are highly prized works of art, with elaborate stippled or raised patterns cast into their surfaces. A small amount of powdered tea is ladled into the cup; the ladle itself, housed in aged, hollow bamboo is also of carefully curved bamboo, and the container of the tea, the *chaire*, which may be of lacquered wood or clay or porcelain, is only slightly less costly than the tea cup itself. As each utensil is used, it is carefully returned to its place before the preparer. Hot water is added and a bamboo whisk used to froth the suspension. The motions are guarded by convention, including even the manner in which the whisk is removed from the cup to avoid dripping. Once prepared, the tea is set before the guest, who, if there is more than one, will turn to the person on the right and express his or her gratitude at being served, and then to the one on the left and excuse himself or herself for drinking before they are served.

Ritual does not cease here. The cup must be lifted to the lips in a particular manner and carefully rotated a quarter turn three times. One drinks a single swallow, sucking loudly to extract the last foamy drop. One then bows forward slightly and inverts the cup, holding it only a few inches above the *tatami* (the thick, rectangular straw-mat flooring), to study the name of its maker, its shape, and glaze. Appreciative remarks are expected and are easily volunteered, for the cups can be truly exquisite. A knowledgeable eye can identify the cup's maker just by its appearance. Among traditional potters, cups for the tea ceremony are viewed as the highest calling of their craft, equalled possibly only by flower vases. An exceptionally attractive tea bowl made by a famous potter can cost thousands of dollars.

Tea-drinking in Japan is, however, more than merely the occasion for ceremony and ritual; it has its commonplace uses, too. Simple green tea is the glue that binds visiting neighbors, accompanies business transactions, and, within an office, serves the functions of our "coffee-break." The utensils used under these circumstances lack the elegance or value of those associated with osencha or omaicha, but the lidded cups often are attractive statements of the potter's art.

Some eight miles to the south of Matsuyama, on the fourth largest of Japan's major islands, Shikoku, is the small town of Tobe. Although the town has now turned to folk-art pottery, it has had a long ceramic history, and in 1949 was producing lovely porcelain. Once when Vicki and I were touring the area, we made arrangements to see the potters at work. Just outside Matsuyama, the road, bordered with endless paddies, was largely dirt or gravel and well rutted. Our taxi finally jostled its way into a small rural town, the streets of which were a succession of small stores and business establishments selling or manufacturing porcelain. We went to one of the larger of these, and after a few moments in the office were taken to the building in which the pots were actually thrown. A row of potters sat on a raised platform, possibly a foot and a half or two feet above the dirt floor of the building; they were dusted with a patina of clay and immersed in the clammy smell of wet earth. Each addressed a small square opening, perhaps three feet or so in dimension, in which was mounted a foot-kicked wheel. Most of the potters sat hunched on a thin, worn cushion (*zabuton*), with one leg folded beneath them and the other dangling in the opening before them. On the floor were their wooden shoes (*geta*) and beside them the simple tools of their trade: a length of string and variously shaped pieces of wood or bamboo used to fashion the interior of a pot. Periodically, and almost soundlessly, a bare foot stroked the wheel's shaft to maintain its spinning momentum. The steady cadence of their work was broken only by a call for more clay or the plop of a new mass on the wheel.

We stopped before an elderly potter, dressed in the remnants of a worn khaki army uniform, who was throwing tea cups. Out of the mound of clay before him arose a cup; its base was quickly pinched, and the cup was separated from the clay mass from which it sprung with a piece of twine. Now and then, he rinsed his hands in a bowl of water that sat beside him and worked the excess moisture into the clay. Each successive cup appeared to have been cloned from the previous one; if there was variability in size or shape, it eluded the eye. The potter interrupted the rhythm of his activities just long enough to demonstrate to us more slowly how his strong, gnarled hands pulled from an undifferentiated lump of clay a cup. As he did this, he grinned and chattered on to us in the local dialect. Save for his service with the Japanese army, he had been a potter since his apprenticeship was finished almost four decades earlier.

Slowly we followed our guide to the area where the slip and painted decorations were added to the cup prior to its firing. Again, the skill and dexterity of the workers was awe-inspiring. A few quick flicks of a small, pointed brush culminated in a classical scene—pines and a Fuji-like mountain. We were encouraged to try our hands at the decoration of a cup. While Vicki wrote "Schull" on one of the cups, I could not resist the opportunity to write our name in katakana. We were told that our efforts would be fired and subsequently sent to us. However, we were so impressed by the skills of the potter and the artists who decorated the cups that we ordered two dozen for our use while in Japan. When some weeks later a shipment of white porcelain cups reached us in Nijimura, we were surprised to find that on precisely half of them the name Schull appeared, in blue, emulating exactly

Vicki's cup, and on the other half it had been rendered in katakana. Clearly, somewhere, something was lost in translation.

*   *   *

When the weather was fair, Vicki and I would spend a portion of our free time at one of the six or so clay tennis courts in our village. Harry Hopwood, Australia's great tennis coach, was alive and had imbued the Australians with a love of the sport that many played well. Save on the hottest of days, it was easy to find a partner for volleying or a serious game. An alternative to tennis was a visit to the beach at Karuga, a sandy cove just on the outskirts of Yoshiura, a small seaside town immediately adjacent to Kure. The beach was sheltered and the water pleasant except when an invasion of Portuguese men-of-war occurred. A beachside resthouse was there, and a few hundred yards away was the Karugasō, a charming inn and restaurant the Occupation had commandeered.

The more adventurous among us would often go to Yae in a small valley in the mountains to the north of Hiroshima, to the horse races, if such they could be called. These were organized by a group of Australians, inveterate gamblers all, with the aid of local farmers who were willing to see their nags run down a dirt straight-away.

Inevitably, Vicki and I came to know many of the Australians who lived in the village. Gus Homeng, a Chinese-Australian, was a loquacious, multilingual warrant officer who served as one of the supervisors of the local labor office. A witty person with an entrepreneurial bent, Gus was usually engaged in some enterprise of marginal legality. He was also an excellent cook, who prided himself on his Chinese and southeast Asian dishes and could regale his guests for hours with one humorous story after another. John Harnetty was the source of local news for the BCON (British Commonwealth Occupation News), the service newspaper for the Commonwealth forces. He was a garrulous florid-faced former war correspondent, a fixture at every cocktail party, cigarette and drink in hand. Charles Astley and John Buckley, captains both, one in Ordnance and one in Signals, were our tickets to a surfeit of parties at their respective cantonments. Like most of their fellow officers, they had seen numerous years of military service in the Middle East and Asia and had a certain disdain for the formalities of garrison life; they were merely waiting for their units to be demobilized.

The tedium of peacetime garrison life had taken its toll among the Australians. News of home came only through barely audible radios or the BCON. Wives accustomed to doing their housework now only managed their household staffs, and husbands spent too much of their time at the club. Yet the social life at Nijimura seemed precious when we compared it with the isolation of North Camp, which was a mile or so to the north of Nijimura geographically but much further culturally. In this collection of two dozen or so modified Japanese houses lived all of the families of the Nisei assigned to military government or employed by the Commission who had been excluded from Nijimura because of Australia's "whites only" policy, as well as those medical officers with the Armed Forces who were assigned to ABCC. Present also were quarters for bachelor officers, a small post exchange, and clubs for officers and enlisted men, with their phalanxes of slot machines. The clunk of the descending arms that activated them and the tinkle of cascading chips hung in the air at most hours of the day and night.

Vicki and I generally spent our weekends in the Kure–Hiro area, with compulsive trips to the post exchange at North Camp, Eta Jima, or Empire House in Kure, the Australian equivalent of our post exchange, to see what new had arrived. Despite the economic stringencies, shopping districts in Kure, Hiroshima, Nagasaki, and similar cities bubbled with activity during the Occupation. Merchants displayed their wares in bins, on tables, or on hangers before their stores, and small private enterprises filled every possible nook and cranny along the busier streets. On the Hondori—Hiroshima's then roofless main shopping street—shoes could be shined or repaired, a letter written, odd used garments purchased, or a name stamp made. For a fee, one could also challenge the shogi player, who welcomed all comers. As the seasons changed, so did the themes in the store windows. Autumn saw the sales of Ebisu, the god of wealth in the heptad of mythical gods, and soon thereafter the year-end brought the shōchikubai, the artful arrangements of pine boughs, bamboo stalks, and plum-tree sprigs set beside the entry to a store or home seeking longevity, prosperity, and constancy for the occupants.[2] Most of the few cars on the road displayed boughs of pines, twisted straw rope, and an orange, representing continuity of good health, on their hoods or affixed to the radiator.[3] Displays everywhere sought to separate the unwary from his or her year-end bonus. Sales came and went, but most shoppers were cautious and did as much looking as buying. Impulse purchases were a luxury limited to a few, mostly the members of the Occupation. However, there were other ways to be parted from one's money.

Here and there along a busy street, stationed to catch the curious, a trainer with his *bunchō*, a drably colored but intelligent little sparrow-like bird, would be encountered. For ten yen, the bird would reveal your future. You handed the coin to the trainer, who, in turn, gave it to his bird. Coin grasped firmly in its beak, the bird hopped unhesitatingly to a small wooden, generally unpainted, replica of a Shinto shrine. To the left of the steps to the shrine was a wooden receptacle into which the bird dropped the coin. On then to the steps, and a small sash attached to a tiny bell suspended beneath the eaves of the diminutive shrine. Grasping the sash firmly in its beak, the bird would ring the bell three times. After this performance, he would open the door to the shrine and select a small piece of folded paper from among many, close the door with his beak, and hop back to the trainer, who would take the paper—your fortune—and give it to you. Each of these steps was thoroughly familiar to the Japanese, whose own behavior at a shrine would be similar, and as they watched, a half-uttered "kawaii ne" (cute) escaped many lips. These endearing beggars are things of the past; at least, it has been years since I have seen them on the streets of the cities of western Japan.

Occasionally one would come upon a strolling flute player (*ko-musō*, a buddhist priest of the Fuke sect of Zen Buddhism), with his basket cap and flute. Most were dressed in black with white short socks (*tabi*) and wore geta; they supported themselves from modest donations. Or more frequently, a clutch of pilgrims was encountered, all in white, staffs in hand, wearing the pilgrim's hat, a broad shallowly curved headpiece of rush extending virtually to the shoulders. Many would pause briefly before stores or shops to recite prayers and seek funds for their continued pilgrimage. Winter saw their white garments covered by a long, knee-length black robe for added warmth.

Often, too, as we walked Hiroshima's streets, we would meet a *kamishibai*. These once-familiar storytellers are relics now, too simple for TV-surfeited children. In postwar Japan, however, they were pied pipers. Most practiced their skills with a bicycle on which was mounted a drum, generally beside the rear wheel, and, on the luggage carrier, a small box-like stage. A casual wheeling through the streets with a steady tattoo on the drum was sufficient to assemble a swarm of eager faces and eager minds. At a spot free of traffic and danger to his small charges, often under the eaves of a roof, the curtain to the stage was lifted. Then, with the aid of gestures, numerous voices, and, most important, the painted removable cards the diminutive stage housed, his story unfolded. It might be a tale of feudal heroism, or an evil ronin, or merely a fantasy. Spellbound, every young eye on the stage, the tots watched as good and evil battled. Passing adults were no less enthralled. The storyteller's reward was a few coins, for the picces of candy hc sold.

*   *   *

Thus, after Vicki's arrival, my life in Japan, even occupied Japan, assumed a welcome routine. Through most days of the year, the geographic limits of movement my job afforded were from Nijimura to Hiroshima. Workdays began when I and those of my neighbors who worked for the Commission boarded the Hiroshima-bound bus at the fire station in Nijimura for the hour-and-a-half bus ride. Homeward-bound trips, especially on Friday evenings, were occasionally sparked by an unexpected cocktail party, thrown by some rider who had thoughtfully prepared a thermos of martinis. At about 5:30 in the afternoon the bus would deposit us at the fire station again. Shinji, our driver, would spring quickly and lightly out of the bus, and with his hat in his immaculate white gloves, bow slightly and bid each of us good night as we descended.

If, for whatever reason, one of us missed the morning convoy from Nijimura and Kure to Hiroshima, there were a limited number of other ways to get to work. A ride might be hailed with someone who drove his own car to and from Hiroshima, but there were not many people who did. Arthur Wright, who had been a fellow-occupant of Sawahara house and a former member of the Australian Special Intelligence Branch, drove periodically, but a ride with him was not relaxing. Rather than use the coastal road out of Kure, he would drive through the kilometer-long tunnel that connected Kure to Yoshiura. This dreary, dark hole did not inspire confidence. Water dripped incessantly from the unsheathed roof, and loosened stones plummeted down now and then to shatter on the pavement below or on top of a passing vehicle. Moreover, Arthur had launched a one-man crusade to increase motor awareness among the Japanese and would stop to lecture a child or adult who wandered into the street without due regard to traffic. Motorists who passed on a curve, including bus drivers, would be forced to the side of the road, where he would harangue them in Japanese for an eternity, it seemed. Judging from the state of the road discipline in Japan today, I would say that Arthur's efforts were largely unrewarded.

Passage to Hiroshima could, of course, be purchased on one of the rickety charcoal-burning, methane-powered buses, but the twenty miles could have been walked almost as rapidly, for at unscheduled moments the boiler mounted on the rear had to be stoked with fuel. The most practical alternative to the convoy was one of the frequent local trains, with their gleaming brass fixtures and ancient engines that burned cheap bituminous coal gouged from hundreds of mines in northern Kyushu. It was always a wonderment to see how in this world of soot and grime the cloth gloves of the station master, who welcomed and dispatched each train through his railed fiefdom, could remain so white. Railroading in Japan is a serious business, whose protocol must be punctiliously served, and I have come to suspect that the station master changed his gloves frequently. Noise was no less a part of this environment: the sudden release of steam as an engine slowly began; the clanging of bells announcing the imminent appearance or departure of a train, generally overburdened with people; the public address system over which alternately crackled announcements of the arrivals and departures and the strains of "You Are My Sunshine" or "Buttons and Bows" or some other equally incongruous song.

Once in Hiroshima, one emerged from the train into a pushing, jostling crowd of commuters and was swept out into the street. There stood a large first-aid station warning the unwary Australian, and presumably American, soldier of the dangers of venereal disease and the need for prophylaxis. It was hardly a gracious introduction to the city. There remained, however, a ride on the crowded streetcar to Ujina and the hall of the Triumphal Return—the Commission's temporary headquarters. At each stop, crowds squeezed on and off the tram. Over their heads could be seen the tall, round, red metallic mailboxes, reminiscent of those of England, that served the nation's postal system.

We had to step down when the stop at the Kaigandori was reached, for the Hall of Triumphal Return was to the east rather than the west, the route the streetcar followed. A five- or ten-minute walk followed to reach the Commission's structures. Both the ride and the walk were pleasant in the autumn, spring, and summer, provided there was no rain; but in the winter one confronted an endless host of white-masked faces, behind each of which lurked a cough. Upper respiratory infections were so widely prevalent throughout the cold weather that we grew to view them as merely a routine part of life in Japan.

Snotty-nosed, drably dressed children were ubiquitous in Hiroshima. A bright scarf or hair ribbon was the sole relief from frequently tattered and patched garments of dark browns, grays, and blues and the anonymity these afforded. Bare feet, purple in the winter with the cold, were shoed with patched canvas sneakers or geta, so worn that little remained of the uprights that lifted the wooden platforms above the ground. Every few yards one encountered a small tot wrapped in a padded robe with an even smaller member of humankind strapped to her back. Winter saw each sniffle echoed by another from the depths of the robe. In summer, clothed often in little more than their runny noses, the children found amusement sucking on ice candy, playing with a pail in the water and mud amidst which their homes sat, or, if older, amusing themselves with *janken,* the game of paper-scissors-stone that required nothing but their hands. The seeming irrepressibility of these children made one wonder how deeply they were scarred by the circumstances. Although Japan's birth rate had already begun to decline by 1950, families were still large, and each older child, particularly the girls, aided in the rearing of her younger siblings. Despite chores that the present generation would consider burdensome, neither play nor cheerfulness were compromised by these tasks. Many an infant could be seen bobbing on the back of an older sister who was jumping rope with her friends. Often as we passed, they would pause briefly, smiling hesitantly and shyly in our direction as though they wanted to say something, but either they knew not what or else courage failed them.

Inevitably, work was reached and these sights and sounds had to be placed in abeyance again.

3

SURVEYING THE
CHILDREN

A more-or-less continuous surveillance of the children born in
Hiroshima and Nagasaki had been under way since 1948. Initially,
Japan's postwar rationing system was the source of information about
new births. In the course of the war, Japanese authorities had found it
necessary to promulgate a law that made it possible for pregnant
women, on the registration of their pregnancies with the appropriate
local office, to obtain rationed food and clothing to sustain themselves
and their as yet unborn offspring through gestation. The Atomic
Bomb Casualty Commission, aware of this fact, in 1948 initiated a
program wherein these mothers-to-be enrolled their pregnancies with
our offices at the same time they registered with the municipal authorities.

Most pregnancies terminated at home in the presence of a midwife,
and newly parturient mothers were reluctant to take their infants out
of the house until they were thirty days or so old, at which time the
child was taken to the shrine or temple to be introduced to the gods.
As a consequence, the surveillance program depended heavily upon
midwife participation and home visits. Throughout the years to follow, our activities were intimately joined with those of the three hundred or so midwives and their associations in Hiroshima and Nagasaki, and more specifically with Yamamoto Setsuko and Murakami
Tei, the presidents of the two midwives' associations.

Most midwives then in practice were licensed under a system that
required a candidate for admission to a school of midwifery to have
completed a *kotōshogakkō* education (about eight years, corresponding to a grammar school). Although it was not required, some candi-

dates had also completed *chūgakkō* (middle school, about four years, corresponding to high school). The course in midwifery itself covered two years. One either spent both of these in a school, or one year in school and the other obtaining experience in the obstetrics-gynecology department of a hospital or in association with a senior midwife. A prefectural examination had to be passed for licensure. Not uncommonly, a licensed midwife married soon after completing school and did not return to the practice of midwifery for some years.

Each midwife was remunerated for the births she reported, but beyond weighing the infants with scales we provided, the midwives, whatever their qualifications for the practice of obstetrics, could not be expected to carefully describe congenital malformations. It was essential that the Genetics Program develop a system whereby each newborn infant was seen by a physician as soon as possible after birth. For this we turned to the numerous young unemployed doctors in Japan. The training of physicians had been markedly accelerated during the war to meet the country's perceived military needs. To many of the twenty or so medical schools that then existed another had been attached to make possible an increase in enrollment, and the curriculum was shortened from the seven years that had prevailed to five. With defeat and the economic stringencies it brought, many of these hastily trained young physicians were unemployed and were too impoverished to launch a private practice.

Once a pregnancy terminated, if the infant was stillborn, or grossly abnormal, or died shortly after birth, or was handicapped in some other manner, the attending midwife informed us immediately by phone or by visiting our offices. If the newborn was apparently normal, as was generally the case, we were notified on a more leisurely schedule, usually once or twice a week. In either event, a physician accompanied by a public-health nurse, was sent to the home to examine the child. All of these aspects of the genetic surveillance program were in place when I arrived. My task was to see that they were properly implemented and, once the necessary clinical support was available, to institute a second examination at or near the end of the first year of life.

The initial examination made within the home was the typical physical assessment that most American parents have come to expect; it included auscultation, neurologic evaluation, and the like. Not all major malformations—those that are incompatible with life, that are life-threatening, or that seriously compromise the infant's ability to grow and flourish in society—can be recognized shortly after birth in a purely clinical examination. Many heart defects or failures of motor or mental development are not recognizable until later—thus the need for a second examination, sometime between the eighth and tenth months of life, when the infants could be brought to the clinical facilities maintained by the Commission. On this occasion, American pediatricians were available to supervise and consult with our young Japanese colleagues.

The home visits, although essential, were time-consuming and tiresome. Most of each day was spent in search of the children to be examined. Though the designation of streets and home address, particularly in the larger cities, has since changed, in those years streets in Japan were generally not named, and consequently house addresses described an area rather than a specific location on a street. Within an area, houses were not numbered consecutively along a road but in the order in which they were built. An address was merely a clue to search in a particular region of the city. Our strategy was either to locate a grocer who might know the family of the infant or, better still, the local police box. Although the Occupation had tried to alter the nature of Japan's police, the effort to decentralize their control was only partially effective, and few happenings of life escaped police scrutiny. Within each police box (hatsu shu sho) was a map that identified every house and its occupants within the jurisdiction of the policemen who staffed the box. We often stopped to consult them and their map in our search for a particular home. Usually, if a perplexed look flushed across our faces as we oriented ourselves, one of the policemen would volunteer to show us the way. We rarely availed ourselves of this offer, for fear that if we arrived in the tow of a policeman, it would prove embarrassing to the occupants and entail no end of explanation to their neighbors. Armed with the information to be gained from their map, we were usually able to find the appropriate home, but a substantial walk might still confront us, for in many areas of Hiroshima and Nagasaki the roads were little more than narrow paths down which even our jeeps, goatlike though they were, could not go.

We could generally verify that we were at the correct home before we knocked, for normally beside the entrance to a Japanese dwelling is a small wooden plaque, a door plate (hyōsatsu), on which the occupant's name is inscribed. We would rap on the door, or more commonly push aside the sliding door to the entrance foyer, which was rarely locked, step inside, and self-consciously call out, "Gomen kudasai" (Please pardon me). Usually, this would be answered from somewhere in the rear of the house with a "Hai, dozo" (Yes, please). Shortly thereafter the housewife would appear, and we would explain our purposes. She would bring her baby and lay it on the tatami, and the examination would begin. This rarely required more than ten or fifteen minutes in the summer, when the infant could be quickly and safely disrobed. In the winter, in an unheated home, it took a few minutes longer. We were loath to expose the baby to any more of the chill in the air than absolutely necessary, and it was generally swaddled in layer after layer of clothing, which took time for the mother to remove. Diapers, at least the disposable kind, were unknown, and most infants were wrapped in several layers of absorbent cloth covered with a sheet of rubber to prevent the soiling of the outer garments. A tiny, colorful kimono or quilted robe warded off the cold.

Curiosity would warm the tiny eyes desperately trying to focus on this intrusion into a world of limited faces, and some grasped errantly at the object that disturbed their routine. Japanese infants were, indeed still are, exposed to so much loving attention within the family that they are rarely apprehensive around strangers at this age. However, if the baby was frightened or if our visit coincided with the first pangs of hunger, a breast gorged with milk would placate. Normally, the nurse helped undress and dress the infant, carried on a continuing, calming conversation with the mother, and handed the physician his stethoscope, otoscope, and other instruments as he needed them. The mothers were encouraged to ask questions about their babies, their care, and whatever else concerned them. Usually the physician responded to these as he recorded his observations on the child or as the examination itself unfolded.

When a pregnancy ended with a stillbirth, or when a child died shortly after birth or was grossly abnormal, a more detailed record— we called it our long form—was completed, and blood was drawn from the mother to test for venereal disease, which was common and can be the cause of certain types of defects. This form identified the nature of any abnormality, often with drawings, and recorded more information on the household and the course of the pregnancy.

Shortly after my arrival it had fallen my lot to revise this form and to have new copies printed. As yet the Commission did not have its own printing facilities, and jobs such as this had to be contracted on the local market. To printers who read no English this was a redoubtable undertaking. When the first proofs of the new form were returned to me, there were numerous typographical mistakes—misspelled words, letters upside down, or erratic margins. I carefully noted all of the corrections and added comments to the printer in the margin. When the second proofs were returned, the changes I had indicated had been dutifully made, but all of my marginal comments had also been set and appeared on the form where I had made them. With each succeeding galley, I managed to expunge one or more of these misunderstood comments until finally a clean proof existed. It came none too soon, for we were rapidly depleting our limited stock, since the examining physician completed a similar form on every infant, normal or abnormal, whose registration number ended in zero, to provide comparative observations.

Each of these long forms would be reviewed by Koji or myself or both of us. As I sat hunched over my desk perusing each, it was difficult to suppress a giggle at the quaint spellings that occurred—no end of children were found to have nuchar (nuchal) hemangiomas or mongorian (mongolian) spots. I dared not laugh, for my Japanese was undoubtedly no less hilarious. Still others would have explanatory comments in German—the clinical language in Japan—which had to be studied before the malformation could be coded. On some forms it was necessary to convert into grams an old Japanese measure of weight, the *mommé*, then still used by some of the midwives.

Typically, the physician and nurse were not aware of the exposure status of the parents and did not specifically inquire into this, for the information had been obtained at the time the mother registered her pregnancy. This "blinding" was done to forestall possible biases in response; it is common for mothers of abnormal children to search their recollections for events that may have precipitated the abnormality. This search tends to be self-fulfilling and has been the basis of many errors in the past. Epidemiologists try to avoid such possibilities through prospective studies; that is, through careful inquiries which precede the events of interest so that the persons interviewed do not unwittingly attempt to "explain" the events. Thus, the data on radiation exposure were obtained before the pregnancy in question terminated and before the prospective parents knew whether their child was to be normal or abnormal. Sometimes it was impractical to keep the physician in ignorance of the mother's or father's exposure, for occasionally it was necessary to verify an observation recorded at the time of the registration or to resolve an inconsistency in a mother's or father's statement on the births of two different registered children.

Upon the completion of the examination, as the physician and nurse left, they gave a bar of face soap, usually Ivory, to the mother to use on the baby. Gentle soaps were difficult to obtain in Japan in the early postwar years, and skin rashes from harsh ones were common. This little gift was in the tradition of reciprocity that governs gift-giving in Japan. The rules determining the practice are complex, the occasions many. There are *ochūgen,* mid-summer gifts; *oseibo,* year-end remembrances; *oyuwai,* celebrations of special occasions like the birth of a child or a wedding; *omimai,* gifts to someone who is ill or recovering; *okaeshi,* the return gift in appreciation of a gift received; *omiyage,* souvenirs given to commemorate a trip; and *otsukaimono,* which are presented to someone when requesting a favor of them. Not all gifts are of equal importance or value, and their choice is a source of frustration to one not familiar with the cultural imperatives. The occasion, the status of the recipient in relation to one's self, and the length of the relationship are all important considerations. Gifts sent in thanks by newlyweds, or bereaved families, or in return for a good-will present are selected with a view toward price; custom demands that they should cost about half the value of the gift received. Other gifts need not be as expensive, and are commonly fifteen to twenty percent of the price of the original one. Presentation is also important, and department stores are particularly adept at wrapping a gift properly and seeing that it is delivered promptly if it is not personally presented.

In these years, some six to seven thousand infants were born annually in Hiroshima and Nagasaki. To examine this number under the circumstances described, we had to send six examination teams into the field seven days a week, on fair days and foul. Our blue jeeps were known everywhere; indeed, on numerous occasions, while we were visiting one home, a neighboring housewife would ask why we had not come to see her new son or daughter. She was quickly assured that we would be there shortly, that the notification of the termination had not reached us but surely would in a day or two. Most infants, save those who were stillborn or died prematurely, were seen six to seven days after their birth. Only on the first few days of the New Year were our jeeps absent from the streets, and then an emergency service was in place if it was needed.

Completeness of this system of registration and the accuracy of the diagnostic information were constant concerns. To ensure the former, we periodically compared our data with the birth records of the city and encouraged the midwives to report all terminations, including those of unregistered mothers. A program of autopsies of stillborn infants or those dying shortly after birth provided some insight into the diagnostic data, and we hoped that the clinical experiences of our young colleagues would sharpen their medical acumen.

Still other troublesome issues kept arising. One particularly vexing matter was the changing attitude toward limiting the growth of the Japanese population. The repatriation of so many Japanese from overseas and the booming postwar birth rate were taxing the nation's already strained economic resources. Families were encouraged to have fewer children, and to this end in 1948 the government had liberalized the legal basis for the artificial termination of pregnancy. This act allowed a pregnancy to be interrupted if its continuance was likely to lead to a severely handicapped member of society (the so-called eugenics clause), posed a serious threat to the mother's health, or imposed an intolerable economic burden on the family and the country. Although use of the provisions of the law was voluntary, the circumstances under which each of these alternatives could be invoked were defined. To abort a pregnancy for economic reasons, the family had to be currently receiving governmental assistance. To utilize the eugenics clause, a family had to have the likelihood of an untoward pregnancy evaluated by a committee of physicians. However, threat to the mother's health was a clinical judgment of her personal physician. Since any pregnancy entails some risk, most of the abortions performed were authorized under this last provision, the easiest to satisfy.

Pregnancies interrupted under the law were to be reported to the local health offices. It was very difficult, however, to obtain a reliable estimate of how many pregnancies were being interrupted. The women involved were generally reluctant to state whether they had or had not had a pregnancy terminated, and the physicians did not want to discuss the matter, since many did not report to the tax office the payment they received for the abortion. Finally, after strenuous efforts to assure a sample of physicians who we knew were performing abortions that the information would be known only to us, a few con-

sented to provide us with the numbers of pregnancies they had interrupted over a period of several months in 1950. To our amazement,
their numbers, when extrapolated to the city, suggested that almost as
many pregnancies were being interrupted in Hiroshima each month
as came to term! This had awesome implications for the study: a fall
in the birth rate could be precipitous enough to compromise our ability to collect sufficient data to evaluate the radiation hazard or, worse
still, could introduce biases that would be difficult to manage, particularly if pregnancies were more likely to be interrupted if one or both
parents had been exposed. It was obvious that this development had
to be monitored carefully and continuously to appraise its possible
effect on the study and the conclusions to be drawn from the data. It
turned out that economic considerations and not the fact of exposure
seemed to be the major determiner of a family's decision to abort a
pregnancy. Largely through the interruption of pregnancies, Japan's
crude birth rate fell from thirty or so per thousand population in 1950
to about sixteen in a period of roughly five years.

At the peak of our activities, some fifteen to twenty young Japanese
physicians were employed in these examinations in each of the two
cities. Although most served on a full-time basis, each physician spent
half of his time in the field and the remainder in continued training
and rotation through clinical services, either the Commission's or one
of the local hospitals. Koji and I spent many days in the field, too; we
sought to ensure uniformity in the examinations and to share with our
colleagues their trials and tribulations.

One slowly grew accustomed to the drab sameness of the exteriors
of the houses we visited, but not to the pleasantness of the small interior gardens so commonly encountered. Never large, each using limited space adroitly, they had their own individuality, alternately
brooding in the rain or fog or glistening in the morning sun. Amidst
carefully disposed, moss-embossed stones would be found an azalea
of special beauty, a carefully protected spider chrysanthemum, or possibly an agapanthus, depending upon the season. After the showers of
summer, dragonflies rose again to resume their flight, or posed delicately on a leaf or blade of grass to shed their shimmering wings of
droplets of water, as if aware of their beauty.

Many a day my backside felt bruised and blue from the pounding
it took over unpaved and irregular roads. As anyone who has ridden
in the military version of the jeep knows, it is not noted for the

smoothness nor comfort of its ride, particularly in the rear seat, where protocol demanded I sit. Despite these hardships, everyone worked unstintingly, persuaded of the importance of what we were attempting to do. We could reassure; we could advise; we could help. We could see through the lives of innumerable individuals what war had wrought. By the time this phase of the study ended in the early spring of 1954, over 77,000 infants had been examined; even now when I walk the streets of Hiroshima and Nagasaki and see young adults forty or so years of age, I wonder how many of them we saw in the first few days of their lives. Would that I could communicate to them the aspirations and concerns written on their parents' faces on those occasions.

Before the second examination of these children could occur, an interminable series of meetings took place with the administrative and public-health authorities of the city and the midwives' associations to secure their approval. These introduced me to the subtleties of Japanese negotiations. Usually, Koji and I would call upon the appropriate individual after an appointment had been made. We would be ushered to a corner of the office, where there would be an overstuffed settee and a low table surrounded by several chairs, all shrouded in white slipcovers or antimacassars. One chair would be conspicuously placed at the head of the table, and here would sit the official. His cohort of assistants were usually disposed along one side of the table, we along the other. We would hardly have exchanged introductions and calling cards and been seated before cups of green tea appeared; all of the cups were alike, save the official's. It was not necessarily larger than the others but was always differently marked, a measure of his position, like the seat he took. Initially, my mind was too busily engaged in trying to sort out all of the titles of his assistants—the *buchō* (department head), the *fukubuchō* (assistant department head), the *shitsuchō* (literally the room head), and the *kachō* (section head)—to notice this mark of status, but with time it became clear that Japan is replete with these symbols of position. It is as though one could not tell the players without a program or the symbols of office. After the obligatory casual remarks about the weather or the season, we presented our case, to the accompaniment of a series of noddings of his head and an occasional "saaa." A few simple questions were generally asked, his approval given, and we were on to the next appointment to seek further consensus. At first this ritual seemed like an incredible waste of time, but as I grew more familiar with Japanese circumstances, its strengths emerged. Everyone was informed, each knew his or her role, and the implied contract, once negotiated, was meticulously honored.

*   *   *

To process the information these various aspects of the study generated was a formidable task. Each record had to be checked for completeness, and if it was the second registration of a pregnancy within the same family, the response to the questions on different occasions had to be collated to identify discrepancies and resolve them. Most of these chores were done by a small army of clerks who worked under the direction of the head of the data-processing group, Richard Brewer, a heavy-set, self-assured vital statistician with a gimpy leg. This group was reasonably well equipped with IBM machines for creating, verifying, and manipulating punched cards, once codes had been developed to convert the data to numerical form and the information had been coded. To expedite this process, our questionnaires and clinical forms were specially designed and constructed. Considerable thought went into each step, for the aim was to be able to recover as much of the basic data as possible through the use of the punched-card equipment, without having to refer constantly to the original record. Punched-card machines, however, were far less flexible than contemporary computers, and very much slower, particularly in their management. Each card was limited to eighty entries or columns of information. To record the information on one questionnaire, often more than one card was required, and each had to be identified as to whether it was the first, the second, or a subsequent one. To minimize errors in data entry, we had to verify a card once it was punched, a process very similar to a second punching of the original information. If the key strokes of the verifier did not match those on the original card, the card was rejected, pending manual verification of the entry.

Information was retrieved by stacking the cards in hoppers that held a thousand or so, and each was mechanically passed through the machine, counted, and sorted into separate bins through the electrical sensing of the locations of the holes that had been punched. A hundred or so cards could be fed through the sensor each minute, but at this speed cards frequently jammed if they were slightly bent in handling or storage. The machine then had to be turned off and the card, or shreds of it, laboriously removed. If the jammed card was damaged, as was commonly the case, a new one had to be made and the process repeated. Humid weather warped the cards and made sorting and counting a frustrating, time-consuming affair.

Since it was impossible to hire workers experienced in the use of these machines in Japan, lengthy periods of apprenticeship and learning were inevitable. A number of the Japanese employees became exceptionally skillful and have become bulwarks in the management of the data. Some, like Omae Kokichi, began with the most unlikely background; he had originally been hired as a houseboy but had higher aspirations. Through dint of ability, perceptiveness, and hard work he rose to be head of the processing unit itself. Others found their niches in administration and supervision of the myriad small details that had to be continuously verified.

As our data base grew, literally hundreds of thousands of cards had to be sorted and counted. To obtain a specific piece of information, an operation requiring two or more steps was often needed. First, the cards were passed to isolate all of those with an entry in a specific column, and then these were passed through the machine again to count the various alternative punches in that column. Most of the statistical processing of the data was done on calculating machines, the old mechanical varieties, for the punched-card equipment could not readily perform such operations as multiplication and division. They could sort, count, add, and subtract. Through an ingenious use of a process known as progressive digiting—based on the principle that multiplication is merely successive additions, and division is successive subtractions—multiplication and division could be laboriously done, but only on a very limited number of variables at one time. Statistical uses of the data had to be simple of necessity and would be considered inelegant at best by contemporary standards. Complex analyses, which only require milliseconds on present computers, such as those which entail the simultaneous scrutiny of six or seven variables, required hours when done manually.

A problem all of us came to know well was how to retain our scientific skills under such tedious circumstances and routines. We had little time for, or access to, current scientific publications; although our institutional library became relatively serviceable with time, it was a source of continual frustration, not because the librarian and her assistants were not helpful and acutely aware of the library's limitations but because the journals were always late, issues failed to arrive, volumes published prior to the Commission's founding were unavailable, and a timely purchase of new books was virtually impossible. We shared with one another the journals to which we subscribed individually and the books we had brought, or subsequently purchased, from the United States. However, this was a poor substitute for the currency of thought that comes from the frequent perusal of new journals. A system of seminars, journal clubs, and programmatic discussions was instituted; but save for outside visitors who came to speak on a topic about which they were exceptionally knowledgeable, it was difficult for us to prepare adequately. Staff seminars were, as a consequence, extremely uneven in content, most interesting when they dealt with some aspect of the study. Even outsiders were often a disappointment—either they had prepared themselves poorly or they approached us patronizingly. It was a time of ambivalence; one was flooded with different and rewarding cultural stimuli that prompted healthy self-inspection, but there was a gnawing sense of loss of skills that were central to the lives we had chosen for ourselves.

Soon after I assumed direction of the Genetics Program, Koji suggested that I call upon some of the more prominent physicians in the city who were contributing to our work. Possibly, the most unusual physician I met on these visits was Dr. Shima Kaoru, a wiry, hyperkinetic little man with a black Hitlerian mustache that only partially served its purpose, to hide his repaired harelip. A widely traveled man who spoke good German and passable English but who could have communicated with a Martian, he was the director of a surgical clinic that he pridefully identified as the Mayo Clinic of Hiroshima. This was not arrogance but aspiration; he sought to emulate professionally and in sense of purpose the standards set by the Mayo brothers.

A visit to Dr. Shima's clinic was a professional rebirth; his enthusiasm was boundless and his intellectual curiosity even greater, if possible. He taught himself to be a creditable painter by copying photographs he had taken. More importantly, he was obviously a trusted and skillful surgeon. Virtually every visitor was drawn into his museum of the pathology his knife had treated. Stomach cancer was and remains the predominant cancer among the Japanese; it accounts for almost forty percent of all fatal malignancies. Partially and totally resected stomachs floated in preservative, and he delighted in describing each case—his diagnosis and surgical procedure. His empathy was no less great for those other Japanese born with harelip, with or without cleft palate; he shared with them a communion. Today, restorative surgery is much better than what he could practice, but it is questionable whether personal involvement could possibly be as great.

Dr. Shima's hospital, a wooden and brick structure, was less than a hundred meters from the hypocenter of the atomic bombing, and only fortuity led to his survival. He had been called away the previous day to visit several patients of Dr. Ueda Tokuji in Mikawa near Miyoshi, some miles from Hiroshima. Like most other Japanese of the time, his usual transportation was a bicycle, and he delighted in describing his tribulations with his recalcitrant two-wheeler. However, on this occasion, because of the distance, he had obtained train tickets for himself and one of his nurses, Matsuda Tsuyako, on the black market through a friend. He had long anticipated that Hiroshima would be bombed but, like others, had assumed it would be a fire bombing and had prepared his patients to evacuate their central location if need arose. When he learned of the atomic bombing of Hiroshima, he and his nurse returned to the city and the rendezvous point as quickly as possible, first by train to Yaga near the outskirts of Hiroshima and then on foot the rest of the way. Their path was repeatedly barred by the exodus of survivors, some injured, some dying; but seeking to discharge their obligation to their patients, he and his nurse steadily strove to reach the spot, along the banks of the Ota River, where the patients had been told to gather. They reached it, but no one was there, and so they decided to go on to his hospital. Enroute, the enormity of the destruction became more apparent, and their way was repeatedly barred by fires. Emotionally and physically spent, they lay down among the dead and dying and wearily fell asleep. On the following day, they reached the hospital. Nothing remained but the two concrete pillars that had flanked the entrance, and these had been partially driven into the ground by the blast. No sign of the patients nor staff could be found, only the charred body of his former head nurse. Reluctant to abandon hope, they continued to search, and even posted a notice on a board propped in the ruins, but to no avail. Some eighty persons had died in the hospital.

*   *   *

Although Nagasaki was a part of my administrative responsibility, it was several months after my arrival in Hiroshima before I visited the city for the first time. Orders were issued, and I was soon on one of

the Occupation forces' trains out of Kure. The train's route followed
the coast to Hiroshima and then joined the main line to Iwakuni, To-
kuyama, Ogori, and Shimonoseki. At Shimonoseki, the steam engine
was replaced by an electric one for the run through the Kammon tunnel under the straits to Moji, where a steam engine again assumed the
task of pulling the train on to the west and south. Kokura passed and
the train paused briefly at Yahata, the location of a huge iron and steel
works that had produced more than half of the iron and steel that
drove Japan's war effort. The complex's importance had made it one
of the potential targets for the bomb that fell on Nagasaki. Foundry,
mills and all, had been designated for reparations; and as we paused,
a small army of Occupation personnel were inventorying its contents
and estimating their worth. Before this monumental task was completed, SCAP intervened and ended the dismemberment of Japan's
limited remaining industrial potential. When the Korean war erupted,
the mills were soon engaged in supplying materials for the United Nations' forces. Finally, we reached Hakata, the southern terminus of the
main line and a part of the city of Fukuoka, where I had to transfer to
another military train destined for Sasebo, a former Japanese naval
base.

Tosu, Saga, and Hizen-Yamaguchi were intermediate stops; at
Hizen, the cars for Nagasaki were separated and hooked to a regularly
scheduled Japanese train bound for Nagasaki. While the military train
proceeded on to Sasebo, the Nagasaki-bound one followed a single
track along the serpentine coast of the Ariake Umi, the sea that thrusts
itself into Kyushu from the west, at an ever-slowing pace, with numerous stops on sidings to allow northbound trains to Hakata to pass. At
Isahaya, we paused briefly to discharge a few vacationers on their way
to the hot springs at Unzen, once a much-favored summer escape for
members of the foreign communities in Nagasaki and Shanghai. The
train was soon moving again. The route it followed to Nagasaki took
a more northerly direction through the Urakami valley than the one in
current use. Once through the last tunnel, the train gathered speed
and, bathed in its own smoke, raced its way toward the bay.

It was fall, and mandarin oranges dotted the slopes of the valley like
ornaments on a festival of trees. Our train thundered through Michi-
noo and Urakami, small communities to the north of the city, and into
a rebuilt Nagasaki station, the end of the line. The original station,
with its slate-tiled mansard roof and fachwerk-like exterior punc-

tuated by sets of tall windows, was consumed by the fires that swept out of the Urakami valley following the bombing. But the terminal yards, with their complex, radiating, fan-like pattern of rails designed to assemble cars and reposition the engines, survived. There were no Occupation installations of consequence in the city, only a small military government team. As a result, the traffic of Occupation-related personnel into and out of Nagasaki warranted no more than a single car, sometimes only a part of a car, appended to regularly scheduled trains.

On arrival I was met by James Yamazaki, a pediatrician and head of the local ABCC staff, and taken to the Bachelor Officers' Quarters, familiarly known as the BOQ, maintained by the military government team on the hillside above the recently occupied facilities of the Commission. These were housed in a drab building, the Kyoiku Kaikan, which had previously belonged to an educational association. Relatively little open space surrounded the building, and our vehicles, largely jeeps, huddled close together in a small parking area to the rear of the building on a narrow road adjacent to the Nakagawa River. When the ABCC studies were initiated in Nagasaki in 1948, suitable space was so limited that work had to begin in the Shinkozen Primary School. But by the autumn of 1949 when I visited, activities in the city had been consolidated at the Kaikan. Jim Yamazaki, the son of a highly regarded minister in Los Angeles, was a stockily built, affable person with a lively mind and a warm, open manner. He had been a German prisoner of war after the unit he served as a physician had been overrun in the Battle of the Bulge. Under his leadership, with the aid of Stanley and Phyllis Wright, two pediatricians from the University of Rochester, the program in Nagasaki took on new dimensions and broadened rapidly.

The living quarters to which I was taken, the commandeered home of the Matsuda family, were entered through a massive, unpainted wooden gate built in the classical Japanese style which opened onto a slab-stone-covered courtyard girded by handsomely pruned pine trees. The courtyard culminated in a gray tile-roofed portico before the front door. On entering the home and passing through the foyer, one found oneself in a huge Western room with wood balustraded staircases along the walls leading to the second floor. On either side of this central hall were large rooms used as a dining area, reading room, sitting room, and the like. The papered walls of the hall bore two huge framed paintings, possibly six by eight feet in dimension, that reached to the ribbed plastered ceiling above. One, I recall, was a synthesis of the famous landmarks of the city: Oura Church, Sofukuji, Dejima, the harbor with sailing ships, and, in the far distance, the sun, either rising or setting adjacent to a snow-capped mountain, presumably Fuji.

At the head of two sets of stairs that fused through a landing midway to the second floor, and down a corridor to the right, was an attractively paneled room that served as a sitting area, and still further was a small, brightly painted room handsomely done in the Chinese manner, which one entered through a circular door. It may have originally been the location of the household altar, or merely a room for entertaining the Matsuda's Chinese guests, or simply an affectation. The individual guest rooms were a mixture of Western and Japanese styles, grouped about a garden with a small waterfall and exquisite stone lanterns in the rough-hewn *yamadōrō*, or mountain, mode.

Across the street from the clinical facilities, adjacent to a small *shōyu* (soy sauce) factory, was a two-storied clapboard building that somewhat later would house the U.S. Information Services Library. A few blocks further northward, in an area subsequently occupied by the Nagasaki Atomic Bomb Hospital, was the military government compound, an unprepossessing set of structures housing offices, a barbershop, a very small canteen, and a motor pool.

Rentals were paid on the use of commandeered housing such as Matsuda Hall; these were collected by the Occupation forces and given to the Japan Procurement Agency, a branch of the Ministry of Finance, and through that agency to the individual property owner. It was commonly believed that the Japanese government had done very well in this arrangement, through tardy adjustments to the owners as the yen devalued in the first year or so following the cessation of the war. Agreements with the owners had been denominated in yen rather than dollars, whereas the agreement between the Occupation forces and the Japanese government promptly reflected the changing exchange rates. Housing constructed specifically for the Occupation, as in communities such as Nijimura, was paid for either from war reparation funds or by the occupying powers. In either event, the payments encouraged construction and boosted employment during a time when substantial numbers of Japanese were unemployed. These facilities, too, were rented to individuals, and the funds were given to the procurement agency.

Evidence of the destructiveness of the atomic bomb was still prominent in Nagasaki in 1949, particularly in the Urakami valley. Houses were relatively sparse, their newness stamped on them. Here could be seen half a torii, still miraculously balanced; the canted chimneys of the university power station; and the gutted, burned-out buildings of the medical school, concrete walls akimbo, windowless, marked by fire, and set in a sea of pilings for wooden buildings that were no more. Nearby, the stark remnants of the red brick walls of Urakami Cathedral reached imploringly heavenward, and the statuary that once framed its entrance was strewn about. In 1949, 240,000 individuals made their homes in the city; today, Nagasaki is a community of 450,000 or so.

The precise numbers of individuals who succumbed and survived the atomic bombing of Hiroshima and Nagasaki are unknown. It has never been established how many persons were in these cities on those fateful days. Many, such as school children or troops that would normally have been there, were elsewhere, either to avoid the anticipated fire bombing or on maneuvers, and there were other individuals in Hiroshima or Nagasaki who would not usually have been there. However, it would be a rare city or town that could establish with certainty the number of individuals within its limits at a particular time. One might have imagined that some effort would be subsequently made to determine the numbers of dead and injured, but so extensive was the damage that this proved impractical. The dead, often burned beyond recognition, needed to be hastily buried for public health reasons, and those who survived long enough to go somewhere else to die were difficult to count.

Between August 1945 and the first postwar census in Japan in 1950, numerous attempts were made within these cities and their immediate surroundings to identify survivors. These efforts commonly took the form of local censuses, including the 1946 Hiroshima A-Bomb Casualty Census, the 1948 Hiroshima City Rice Ration Census, the 1949 ABCC Radiation Census, and their Nagasaki equivalents. All were restricted geographically and necessarily incomplete. At the time of the 1950 national census, at the urging of the Commission, a supplementary schedule was introduced that provided the nearest approach to an enumeration of the entire population of survivors. This enumeration had, however, certain important limitations. Since it was conducted more than five years after the event, it provided no basis for determining those survivors who succumbed between 1945 and 1950. Nor did it include those Koreans or American Japanese who had been repatriated in the interim, nor the Australian, Dutch, and Indonesian prisoners of war exposed in Nagasaki who were returned to their homelands with the cessation of hostilities. Moreover, the subject was not asked exactly where he or she was at the time of the bombing but rather, "Was any member of this household in Hiroshima or Nagasaki at the time of the A-bomb?" The only information actually collected consists of the person's name, residence on the day of the census, sex, and date of birth. These supplementary schedules identified nation-wide some 284,000 survivors, of whom about 159,000 were in Hiroshima and about 125,000 in Nagasaki at the time of the bombing.

Quite unexpectedly, among this number was no less than a dozen, possibly 18 or more, individuals who survived both atomic bombings. Could this possibly be? What circumstances could have led an individual who survived Hiroshima's destruction to choose to go to Nagasaki, of all places? Should they be construed as the world's luckiest persons, for they had survived the atomic devastation of two cities? Or the most unlucky, for of all possible places they might have been in Japan on August 6 and 9, they were in these specific places?

Nishioka Takejiro, in August 1945 the publisher of Nagasaki's principal daily periodical, the *Minyu* (The People), was one of these double survivors. Aware of the threat that bombing posed not only to the cities of Japan but to their newspapers, he had gone to Tokyo to seek the approval of the Ministry of Justice to use convict labor in setting up an underground printing facility that might withstand the bombing. An articulate, persuasive, and influential man, he secured the permissions of the Home Ministry and the Ministry of Justice. As he was returning to Nagasaki it grew increasingly apparent to him that if a shelter was to be built in a month, as anticipated, more help was needed and the most likely source was the Army's labor battalions. Among his friends was Field Marshall Hata Shunroku, whose headquarters were in Hiroshima, and he decided to call upon him to see if Hata would use his good offices to persuade the Army's commanding general in Kyushu, Yokoyama Isamu, to provide the needed help. The train on which Nishioka was traveling was delayed by an air raid and had just reached Kaitaichi, some eight or ten kilometers to the east of the hypocenter, when the bombing occurred. Only the engine was allowed to proceed into Hiroshima, but again his standing gained him the opportunity to continue. However, even the engine was soon stopped by the swarm of survivors leaving the city. Resolutely, he continued on foot to the field marshal's headquarters, which, because of its location, still stood; but once there, he could not find the object of his visit nor even learn whether he was alive. Defeated in his purpose, Nishioka returned to the railway station to gain passage to Nagasaki as quickly as possible.

Although he managed to board the first train out of the city, his experience and the sights he saw on his walk back to the station were devastating. He passed tens of charred and maimed bodies, and those who were alive were aimlessly driven, seeking solace and security. On his return to Nagasaki, on August 7, he immediately sought out the governor to apprise him of what he had seen. He urged the governor to prepare the city, for the existing censorship regulations prohibited him from describing Hiroshima's devastation in his newspaper. On the following morning, August 8, he and others in the publishing community were inspecting a possible site in Oura for the proposed underground printing facility when he became ill, an event he associated with his exposure in Hiroshima. Early on the following morning, after a night spent in Okusa, he boarded a local train for Isahaya, where he hoped to find transportation to Unzen. This proved more difficult than he imagined, but finally a taxi arrived—ironically, one from Nagasaki. Initially the driver was reluctant to take him to Unzen but, upon learning the identity of his proposed passenger, agreed. They had stopped briefly in Obama, some twenty miles to the east of Nagasaki, when Nishioka noticed the tell-tale cloud above the city. Immediately, he and his driver sought protection in an air-raid shelter, and shortly thereafter a sharp, high wind passed over, rattling adjacent houses to their foundation. Once this had ceased, they began the return to Nagasaki, but by the time Isahaya was reached they encountered a steady stream of injured and dying individuals streaming out of Nagasaki. Nishioka continued toward the city. Each successive mile made more transparent the destruction.

Eventually, some ten years after the bombing, Nishioka became the governor of Nagasaki Prefecture, but throughout the years that remained to him, his life was inseparable from that of the tens of thousands of other survivors.

A further effort was made by the Japanese government to identify survivors at the next decennial census in 1960. New, in the sense of previously unacknowledged, individuals identified themselves as such. This was troubling. On the one hand, it could be imagined that a census, such as that in 1950, conducted under Occupation auspices, might fail to record those individuals so embittered by the event that they would do nothing to further initiatives they construed as supportive of the Occupation. On the other hand, in the interim between the two censuses a law had been promulgated which granted substantial medical benefits to the survivors. This law, as it was initially phrased, limited these benefits to individuals within three kilometers of the hypocenter, and we knew that no small number of individuals who, prior to the availability of these benefits, had stated that they were beyond the three-kilometer limit now asserted they were closer. What were we to believe? Ultimately, there is essentially no way of proving that a given individual was or was not present at the time of the bombing. Corroboration comes only from one's sense of the truthfulness of the individual, or the words of others, but even these could be self-aggrandizing, a cabal of opportunists. Save where very explicit information to the contrary exists, we accepted individual assertions as to exposure. Some were undoubtedly wrong, but we presumed that most responses were credible, indeed accurate.

To further the prosecution of the war, Japan's military had pressed into service every able-bodied person it could. Koreans were forcibly mobilized, transported to Japan, and obliged to serve in a variety of capacities—generally, as unskilled or semiskilled labor in war industries. Many of these involuntary immigrants had been brought to Japan under the National Mobilization Plan in 1939, or through further conscription that occurred in 1942, or finally through the National Compulsory Work Order of 1944. Prisoners of war were drafted for these same purposes. It was inevitable that some would share the fates of Japanese residing in Hiroshima and Nagasaki. Their numbers, particularly in the case of the Koreans, are imperfectly known and vary with the investigator who has sought to reconstruct their circumstances. It is generally presumed that at the time of the bombing approximately 50,000 Koreans were in Hiroshima and, of these, 5,000 to 20,000 died. They had been employed at a variety of locations in the city, including the Mitsubishi Munitions Plant in Kannon and the military transport base in Ujina, but most lived within one to four kilometers of the hypocenter. In Nagasaki, the number of Korean A-Bomb survivors has been given as 11,500–12,000, 13,000–14,000, or more than 30,000, and it has been estimated that possibly 1,500–2,000, or 3,000–4,000, or even 10,000–20,000 died. The highest numbers, those advanced by the Nagasaki Prefectural Korean A-Bomb Survivors Council, seem improbable. However imperfect the information may be, it makes clear that a substantial number of Koreans were exposed to radiation and that ten percent or so of these individuals failed to survive their exposure. Most were located at various construction camps throughout the city, particularly around the Mitsubishi Shipyard and the Mitsubishi Torpedo Factory. Those in the vicinity of the former would have received exposures of less than one rad, whereas those near the torpedo factory could have been exposed to lethal or near lethal doses.

The Koreans' lot in Japan has not been a pleasant one. The Japanese Ministry of Justice's administrative edict of April 19, 1952, deprived most of citizenship, including those who were born in the country and know no other land, as it did the peoples of all former Japanese colonies, including Taiwan. Discrimination is rampant in Japan, and all sorts of disparaging names are used to characterize Koreans as a group. The first nationwide postwar census of A-bomb survivors did not count Koreans. Had their number been included, it would likely have been underestimated, for many Koreans, free to return to their homeland, had done so prior to 1950. Nor have they been eligible since 1950 for the benefits that accrue to Japanese survivors under the Atomic Bomb Sufferers Medical Treatment Law. The Korean episode is a sorry page in Japan's history, one which compromises the country's leadership role in world peace and nuclear disarmament.

At least 170 Allied prisoners of war were also exposed; among this number, according to the Nagasaki A-Bomb Disaster Documents, were 24 Australians, 16 British, and 130 Dutch, including Indonesian members of the Dutch Expeditionary Force. Eight of these individuals, all ostensibly Dutch but probably Indonesian, died. Most of these prisoners were within the precincts of their prisoner-of-war camp (Fukuoka Camp 14) when exposed. This camp was near the Mitsubishi Electric Works, at a distance of about 1,400 meters from the hypocenter. An unshielded individual at this distance would have received a dose of a half gray or so.[1] Little is known about what has subsequently happened to most of these individuals; however in 1974–75, Chee and Ilberry attempted to study the Australian survivors cytologically. Of the 27 subjects they discovered (note the discrepancy with the Japanese records), 5 had died prior to the initiation of their study, and 1 was unavailable. Among the remaining, 18 were at Camp 14, but one was in a concrete air-raid shelter at the time of the bombing. Of the 17 who received significant exposure, 3 exhibited evidence of residual radiation damage to the chromosomes of their white cells. Three of those not surviving died of cancer; two had fatal malignancies of the stomach, and one had a fatal cancer of the bladder. Whether these were radiation-related is not known. We do know that Japanese survivors have an increased risk of death from a malignant tumor and that cancer of the stomach is one of those sites of malignancy that is clearly elevated. The evidence for cancer of the bladder is strongly suggestive, if possibly not quite as compelling.

There is no evidence that any American prisoners of war were exposed to significant radiation in Nagasaki. Some prisoners were in the vicinity of the city but at distances far too great to share in the hazards of exposure. This was not true in Hiroshima, but the number who may have been exposed is not clear. Unlike the situation in Nagasaki, there was no prisoner-of-war camp within Hiroshima, although the military command there was in charge of several such installations elsewhere and some prisoners (possibly 20–25) were actually in temporary custody in the city, presumably waiting to be transferred to permanent camps. Some years ago, as an outgrowth of his efforts to determine the number of American prisoners of war killed in Hiroshima, Barton Bernstein, a professor of history at Stanford University, published an article in *Inquiry* (August 1979) entitled "Hiroshima's Hidden Victims."[2] He details a sorry story of ineptness and insensitivity, of dissembling and quite possibly outright malfeasance on the part of Pentagon officials in the concealment of information on American POWs. Especially poignant are the words in a letter of Lieutenant Thomas C. Cartwright, pilot of *Lonesome Lady,* a B-24, and a member of the 494th Bombardment Group of the 866th Bombardment Squadron. On October 26, 1945, seeking information on his crew, he wrote to the War Department, "I am writing you concerning information of my crew who went down with me on Honshu July 28, 1945. I saw five of them alive and well in a city jail in Japan on July 29. I have not seen or heard from them since because I was moved to Tokyo. None of the five men have been reported found as yet. Considering all circumstances and possibilities I have come to believe the city I last saw the men in was Hiroshima, Honshu. Of course I don't know if the men were left there or even for sure that the city was Hiroshima but I think the number of facts pointing towards the possibility warrants an investigation."[3] Cartwright enumerated the *Lonesome Lady*'s crew but never received a reply to his queries. His letter was received, for it remained a part of his military records for years. Why was it not answered? Was it callousness, lack of interest, bureaucratic bumbling, or fearfulness?

*   *   *

At the conclusion of World War II, some 6 million or so Japanese civilians and soldiers were scattered throughout the Orient, from Bangkok to China and Karafuto. Few were welcomed where they were, including thousands of individuals who had been born overseas and had never known Japan. Many of these spoke and read Japanese poorly, had little sense of Japan's culture, and doubtlessly would have been pleased to remain in Malaysia or wherever they were raised. This was not to be. Reason succumbed to vengeance, and a mammoth effort was mounted to return them all—civilians and soldiers—to their homeland as rapidly as feasible. Everything that would float was pressed into service, and between October 1945 and May 1952 over 6 million individuals, roughly equally distributed between civilians and soldiers, were returned to Japan. Most, some 4.6 million, were repatriated in the first year after the cessation of hostilities, but a few trickled in as late as 1951 and 1952. Russia expelled over a million people of Japanese ancestry from Karafuto, the Kuriles, Sakhalin, and Siberia, and similar numbers were returned from Manchuria and China.

Dozens of relocation centers were established throughout Japan to accommodate the orderly reception of these repatriates. As they reached the camps, they were registered and deloused, their health status was checked, and they were returned to their place of family registration, their *honseki,* with a minimum of clothing and supplies, to make their way in a shattered economy. All had returned to Japan with only the luggage they could carry. Whatever assets they had accumulated overseas vanished with Japan's defeat. No lands nor jobs awaited them, and houses were virtually impossible to buy or rent. The national, prefectural, and municipal governments did what they could, but their resources were limited. Those who were whole of mind and body accommodated to the circumstances. But major problems of readjustment were experienced, and some people remained in temporary barracks, subsisting on relief allowances and part-time labor for years.

Among these repatriates were thousands of maimed veterans—the armless, legless, and sightless—who were cast upon a society disillusioned with itself. Little was done for these embarrassingly conspicuous reminders of defeat. Many swallowed what semblance of pride-in-self and uniform that remained and begged. At numberless street corners in the business areas of the cities, one would see a soldier with the familiar peaked tan cap. His uniform was usually covered with a clean, white, smock-like garment, and the claw that served for an arm clutched a metal cup. Equally common, stumps, where legs had once been, were exposed, or a poorly fitting prosthesis could be seen. If the veteran was able, he might be singing to the crowd or playing a musical instrument; often there was a carefully brushed statement of his need on a poster.

Each encounter with one of these unfortunates was a wrenching experience for me, and doubtlessly for him, for quite different reasons. His embarrassment and need cut me to the quick. I did not know whether to place something in the cup and rob him of whatever self-esteem he might still have, or to turn and go the other way, as though he did not exist. Some solved my dilemma by withdrawing the cup they proffered to passers-by as I neared. Thoughts of my own wartime experiences rampaged across my mind. An especially disquieting one returned repeatedly. In the early summer of 1945, when finally the infantry division our medical battalion served, the 37th, broke through the Balete Pass in northern Luzon into the upper reaches of the Cagayan valley, it pressed forward rapidly toward another pass, the Oriung, to forestall another protracted defense. The small city of Bayombong lies between these two passes, and here the Japanese Army had established a Line of Communications Hospital to support the troops defending the Balete. Quickly the town and the hospital were overrun. So complete was the rout that the withdrawing Japanese had been unable to evacuate their bed-ridden wounded, many of whom were amputees. We found them. Each one had been shot to death lest they fall to us. As I stood in one of the wards and looked at the dead stretched before me on their hard wooden pallets, nothing seemed more barbarous. What manner of mind could so callously take these lives which scarce hours before had been so carefully nursed? I was no less repelled at the sight of our own troops rummaging through the pitiful personal belongings of the dead in search of battle flags, photographs, or other souvenirs. Wasn't this too a desecration of human values? Now as I encountered these hapless victims of war, I wondered whether that earlier act had been so heartless.

One of the relocation centers through which this flood cascaded was located on Ninoshima, or Kofuji (little Fuji), as it is sometimes called, an island in Hiroshima Bay, administratively a part of the city. Two thousand or so individuals, distributed over several small hamlets located in different parts of the island, lived there—some farmers, some fishermen, and some employees of the relocation center. Since Ninoshima was a part of the city, we examined their newborn infants, too. Several dozen births occurred each month, but transportation between the island and Ujina, Hiroshima's port area, precluded daily or even weekly visits by our examiners. Births accumulated until a sufficient number were known to warrant a full day's work, and then arrangements were made through the Japanese Maritime Bureau to visit the island. The Bureau would provide a boat to take us to Ninoshima and at a predetermined time would pick us up again at the wharf at the relocation center on the opposite side of the island from where our visit began. Koji and I, along with several other physicians and nurses, would make the trip together. Once on the island, we scattered in different directions so that all of the infants could be seen as expeditiously as possible. Urchins followed us happily from the moment we arrived until we departed hours later. Most people on the island were pitifully poor. Many of the victims of the atomic bombing had been buried here in a large mass grave immediately following the holocaust. Their remains rested in a cave designed to store munitions.

Repeatedly we saw preventable, but unprevented, blindness. Trachoma, a viral disease of the conjunctiva and cornea of the eye, was widespread and the cause of the sightlessness we encountered. Now it is rarely seen, even in undeveloped countries, for it can be readily prevented by medication. Even then the sulfonamides were very effective but not widely available in Japan.

We would reassemble at noon near the peak of the saddle over which one walked to pass from the western side of the island to the eastern side; there we would eat our box lunches and discuss the cases we had seen. It was an attractive, restful spot. Miyajima and numerous other islands could be seen, as well as the traffic of boats into and out of Hiroshima's harbor. Moreover, one's nostrils were not constantly assailed by the smell of night soil enveloping the island. It took a sturdy stomach not to be roiled by the odor that clothed the farmers and their wives as they passed us on route to their fields with swinging pails of sloshing excrement. Although we could climb high enough to avoid these odors, it was difficult to eat under the watchful, hungry eyes of our small followers, and we shared as liberally as our skimpy lunches allowed.

Only a trickle of repatriates now passed through the relocation center, and it was hard to envisage the hubbub that prevailed in that first awesome year, 1946, during one of the more rigorous winters of the last half century in Japan. We had seen the occasional apprehensive face staring toward the mainland, undoubtedly searching for the future. An encyclopedia could not catalogue the hardship and heartbreak that these rude structures had seen. It was with some relief that we would see the small Maritime Bureau cutter round the end of the island and approach the wharf.

*   *   *

More quickly than seemed possible Christmas was upon us, and even the most inured expatriate experiences moments of nostalgia for home and one's culture at this season when most families gather. To offset these feelings of homesickness and alienation, we imported a bit of our own traditions. Scrawny Christmas trees sprouted in corners of our living rooms, and we shopped the Post Exchange assiduously looking for suitable gifts for one another. Wrapped as gaily as possible and sitting beneath our tree, they looked somewhat forlorn until other packages arrived from home through the Army post office. The tempo of parties quickened, and eggnogs became the order of the day. There was no snow, only a chill wind to mark winter's presence.

It seemed appropriate to Wayne Borges, a pediatrician with the Genetics Program, and me to have a tree in the Genetics Department at the Red Cross Hospital and to host an office party, to which the hospital's director was invited. Suitable trees were hard to find, and ours looked like something one would see on a trash heap a few days after the end of the holidays rather than at their inception. Cotton batting from the hospital stores was laced along the branches to simulate snow, popcorn was strung, lights and a few hand-made ornaments were added for a bit of color, and on the top was placed a large paper star. Over the door to the conference room we hung a small branch of mistletoe, which occasioned no end of giggling when the nurses learned of its purpose and history. Amid much laughter and embarrassment, the physicians, male and female, and the nurses hugged and kissed as we had led them to believe was proper. Impoverished clothes, aged but clean, did not dampen the enthusiasm with which carols were sung, nor the exchange of small, inexpensive gifts, nor the consumption of food and drink. Now that Christmas has become almost as large a commercial event in Japan as in the United States, virtually devoid of religious significance, this early, simple party lingers in my mind not for its elegance—there was certainly little of this—but for the moments of warmth and camaraderie we shared when these were more precious than tangible gifts.

The New Year, even at this time of impoverishment, saw a resurgence of things Japanese. Most businesses were closed, and for days prior to the year's end the thump of wooden mallets in a pestle hung in the air as families prepared the seasonal sticky rice cakes (*mochi*). Temples were crowded with people in their best, albeit marginal, finery. It was cold, or at least cool, in Chugoku in January, and the temple-goers were dressed for the chill. Beneath a threadbare overgarment, the conservative colors of a man's *haori* (cape) and *hakama* (a pleated skirt), or the bright edges of his wife's kimono, appeared. The children were often garbed more flamboyantly than their parents, who could ill-afford an equal display of finery. Feathered boas, usually white and timeworn, floated lightly on the shoulders of the women, resurrected from some dwindling store of clothing. Ancient cameras, sparingly used because film was expensive, dangled from the necks of many of the men and were clicked mostly to record the children and their dress. Shrines and temples surged with people, as though a better future was to be found in the propitiation of the gods. Incense hung heavily before the places of worship, and the ritual claps (*hashiwade*) drummed through the air like distant thunder. It was a time to be thankful that matters were not worse, not that they were better.

4

AN HONORABLE
SEND-OFF

*I*t was an awkward time to leave the office and a tiresome drive to Koi, but Mrs. Yamamoto, President of Hiroshima's Midwives Association, rarely called except when something important had occurred that needed my attention. Koji, who was with me, said that one of our young physicians had intimated to the parents of an infant he examined that the attending midwife had mismanaged the delivery. Word of this reached the midwife, who immediately phoned Mrs. Yamamoto. To make matters worse, the offended woman was herself on the association's governing board. It was hard to tell whether we were headed toward a storm or simply a need to serve protocol.

Yamamoto Setsuko was a tiny, wizened woman who usually wore the small-patterned, conservatively colored kimonos seemly for her age; she was in her late sixties. Her hair was drawn sharply back from her face, whose tiny features were distinguished only by their commonness. Her skin, etched by years of use, was lined and swarthy, yet she appeared so fragile one was fearful for her with each movement. Soft-spoken, her speech studded with honorifics that kept my dictionary pages aflutter, she seemed like everyone's grandmother. But the twinkle lurking in her eye, and the vigor with which she managed the midwives and their affairs, revealed no frailty. Although we met formally more or less regularly, now and then I would encounter her on the street with her little black bag that carried the tools of her profession. She had delivered more infants than most obstetricians will deliver in their professional lifetimes, and now, as befitted her experience, she was sought by other midwives as a consultant.

Mrs. Yamamoto lived in the westernmost section of the city, one not quickly reached from the Red Cross Hospital. When finally Koji and I reached her home, we were cordially welcomed, as we always were, and invited up to the pleasant second-floor room she used whenever we visited. As soon as we were seated, she served us tea, and after the obligatory pleasantries and inquiries into health had been exchanged, began to explain what had happened as it was reported to her by the offended midwife. It was evident that whether the physician had or had not been correct, he had committed a tactical error that we could not ignore, for the cooperation of the midwives was central to our whole program. I told Mrs. Yamamoto that the physician was a young man, recently employed, whom I knew to be considerate and not given to insensitive statements. While I did not know all of the facts, I was sure that he meant no slight. However, if she thought it necessary, we would see that he apologized to the midwife.

With her case stated, Mrs. Yamamoto could be generous. She said that she appreciated what we proposed to do but that she did not believe so drastic a step was necessary. She would call the midwife and tell her of our visit and concern and assure her that she, Mrs. Yamamoto, was confident that this would not happen again. Koji and I vowed to ourselves that at our next weekly staff meeting the matter would be raised; and while no culprit would be identified, our physicians would be cautioned to be careful of what they said to the families they saw, particularly in their remarks about the attending midwife.

Once the purpose of our coming had been satisfied, Mrs. Yamamoto became the solicitous host. She insisted that we have more tea while she prepared something sweet for us. She rose, opened a window, and from beneath the eaves of her house snatched two peeled persimmons that had been drying there. As she brushed off several stubbornly tenacious flies, my thoughts must have betrayed me, for she carefully noted that we need not worry about the fruit, it had been sterilized by the sun. I was less than confident of this and managed to leave as much uneaten as good taste would allow. Nothing untoward subsequently happened; perhaps she was right. But this was not the first occasion on which she had uncannily anticipated an uneasiness on my part. On another, she had offered Koji and me *yokan,* the rice-encrusted sweet with a soybean-paste center. Although these can be extremely pleasing to the eye, pressed in various shapes and dyed in different colors, I have never been able to cultivate a taste for them. As she set these particular ones before us, we were informed that when the emperor had visited Hiroshima, he had specifically asked for this variety. Painfully, I ate mine; there was no other option, for to have refused would have been to question his taste.

Despite the importance of their work, our young physicians were engaged in a tedious task. Day after day they bounced over Hiroshima's pitted, unpaved streets in search of the homes of newborn infants, most of whom were cherubically healthy and medically uninteresting. This tedium was interrupted only by the release-time given for the purpose of study and the limited experiences they might win by offering themselves as unpaid aides in Hiroshima's hospitals.

As I grew to know them better, I began to wonder whether they felt ambivalent about being employed by an institution construed by many of their fellow Japanese to be a part of the Occupation. Were they ever chided by their friends or the families they visited because of their employment? Occasionally, albeit rarely, a family would refuse to have their new infant examined; and, as the contacting physician, our young doctors were the ones to whom the family stated their opposition to the Commission. Were the physicians uncomfortable when this happened? What did they say in response? Unfortunately, neither their command of English, nor mine of Japanese, made it possible to explore this matter directly, and it did not seem to be one properly broached through an interpreter. It is doubtful, too, whether they would or could have been forthcoming—I would probably have been lost in ellipticisms that sought not to offend nor to lose face. This chain of ruminations invariably led me to wonder about my colleagues, the young Nisei, at the Commission. How did they feel? Many had had their careers interrupted by forcible relocation or internment in America; their families had lost their possessions or been obliged to accept inequitable settlements. Still others had been drafted into service with our armed forces as physicians, even at a time when their families were still in relocation centers. Were they in Japan searching for more hospitable roots, or had they sublimated this bleak chapter in their lives and were now as enthralled as we were by a different land and a different culture? Did they feel that they found an acceptance in Japan not extended to them in their own country? This seemed improbable to me. Many spoke little or no Japanese, a fact that was a constant source of embarrassment to some of them, for the Japanese presumed they could, indeed, should. Each undoubtedly had his or her own reasons, and no single explanation could account for everyone's attitudes or presence with the Commission.

I did talk about these matters with Koji, but his case was a special one—he was a Nisei by virtue of his birth in Hawaii but was no longer an American citizen. His service with the Japanese army had automatically led to the forfeiture of this status. Since we shared a common language, he knew my graces and foibles and was candid when I sought explanations of things that puzzled me. It was hard, however, to know to what extent his views could be extrapolated to the other physicians.

As Koji had recognized long before I arrived, a regular schedule of meetings, social as well as research, among colleagues is the cement that binds. To serve one of these ends—the social—we at the Commission formed a club to which everyone contributed a small amount each month. These funds financed regular group gatherings or outings to places of beauty or consequence in the neighborhood. One of these was a trip to Sandankyo, an especially scenic area on the upper reaches of the Ota River in the interior of Hiroshima Prefecture, some 45 miles from the city of Hiroshima. The scenery is particularly fine in the autumn, when the gorge is like a gigantic palette, splattered with the reds and golds of maples and katsura.

One Saturday we chartered a bus of our own to go there; the regularly scheduled ones, because of frequent stops, took almost three hours to reach Togochi, the entrance to the gorge itself. Everyone was to be responsible for his or her own beverages and food. However, Wayne Borges and I were aware of the limited funds available to most of our staff and so decided to bring hot dogs, marshmallows, and potato chips for all. Meat was a scarce item in the Japanese diet at that time, something eaten possibly once a week and in minuscule quantities. Because most of our young colleagues were smokers, we brought ample packages of cigarettes. It would have been impolite to smoke without passing the package to everyone. Each person who helped himself to a cigarette usually raised it toward the forehead briefly in recognition of the gift it constituted; now and then a second would be requested. Japanese cigarettes—Ikoi, Peace, and Shinsei, all products of the Japanese Tobacco Monopoly, were the more popular brands at that time—were customarily sold in packages of ten, but single cigarettes could be bought at tobacco stands at three or four yen apiece. Many smokers were reduced to buying two or three at a time; a package, at thirty yen, was a luxury few could afford. A small number of cigarettes also made one more aware of the need to ration oneself. Still others bought the most expensive cigarettes they could, on the thesis that by so doing they would smoke fewer but enjoy the better tobacco more.

We met in Hiroshima at the Red Cross Hospital, where our bus was waiting, and departed in high spirits. Out of Hiroshima, we drove into the countryside toward Kabe, climbing slowly but perceptibly into the mountainous interior of the prefecture. Most Japanese enjoy choral singing, and we were soon lustily warbling, undoubtedly off-key, such universal favorites as "Old Black Joe," "My Old Kentucky Home," "Swanee River," and the like. Our Japanese colleagues had learned the English words in school, although their accents often made it seem as if the songs were sung in some other language. This was of no consequence as we crooned our way up the Ota River valley. Shortly before noon, we reached Togochi, to learn that the gorge was eight miles long and could be visited only on foot. We had come this far and were not to be stopped by inconvenience. Food and beverage were distributed so that each of us carried similar loads as we set out to enjoy the beauty of the gorge and to pick a suitable picnic site.

A dirt path led us alongside the gorge, never far from the sound and sight of water cascading over boulders. Occasionally, a particularly massive outcropping of granite barred this simple walkway. To continue, we either summoned a boatman, stationed for this purpose, who poled us through the narrows, or picked our way along a concrete path anchored to the stone wall, suspended over the rushing waters. Maples, glorying in their autumnal colors, dotted the precipitous slopes and poked their crowns through bowers of pines, whose green glowed as the light streamed through the rustling maple leaves above. Ferns drenched in the spray of water carpeted the soil, and here and there a stunted maple clung precariously to life in the cleavage of two giant stones. Its fragile leaves contrasted with the forbidding inanimateness of these remnants of earth's birth. It was breathtaking, literally and figuratively. The grandeur was palpable.

We walked and stared in awe like uncertain visitors in a cathedral, until finally pangs of hunger forced a halt. We were not far from the river bed, and we thought that a rocky stretch there might provide not only a safe place to start a fire but rocks for sitting. Quickly twigs and pieces of wood were urged into a fire, and the smell and crackle of burning pine saturated the air. Many of our clerks, nurses, and physicians drew forth their box lunches, which contained a few rice balls, several crisp slices of pickled radish, and possibly a small piece of fish, and began to eat. Finally, the fire grew large enough to accommodate the marshmallows and hot dogs. Most of our colleagues had never roasted either and had to be shown how to find a suitable green stick, sharpen its end, and impale and roast the food. We had no inkling of how popular the marshmallows and especially the hot dogs would be, though, given the cost of meat, we should have known.

As darkness and satiety robbed the gorge of its magic, we wearily retraced our steps to the bus. On the trip home, as we hurried down a graveled road in the general direction of Hiroshima, suddenly a large plank the driver had either failed to see or thought he could safely straddle was driven completely through the floor of the bus. A horrendous rumbling noise ensued as the plank was dragged across the pebble-strewn road. Ashen-faced, we poured forth from the bus to see what had happened. Fortunately, the plank had penetrated the bus at a spot just in front of the stairwell and had injured no one. We surmised that the right front wheel of the bus had passed over the plank and tilted it upwards, causing it to be caught on some protrusion from the underside of the floor and be driven upward by the forward motion of the bus. We had come to anticipate flat tires—they were the normal perils of the roads—but this was something more threatening.

It took considerable effort to extract the plank from the bus; and for some time after the ride resumed, our singing was dampened. Finally, one of our more irrepressible physicians launched into "My Old Kentucky Home" once again, and by the time Hiroshima was reached virtually everyone who was not already dozing had added his or her voice to the songs.

Travel in Japan was often an exercise in survival. Auto travel was especially unnerving. It was not just the custom of driving on the left rather than the right, nor the peculiar sense of right-of-way, where, however slightly ahead an adjacent car might be, if ahead at all, one yielded as the driver moved into your lane, nor the fact that a beeping horn was interpreted as an acknowledgment that one was seen rather than a signal to move out of the way. Most disconcerting was the night-time practice of turning off the car's headlights at a distance of 50 to 100 yards prior to actually passing an oncoming vehicle. Needless to say, both cars were expected to follow this convention, and an eerie apprehension prevailed as the final yards of separation were approached. Equally disquieting were the types and placement of turn indicators. Rarely did taillights blink an intended turn; more commonly, a small, illuminated mechanical arm suddenly appeared from some place mid-car and remained there jauntily until the turn was completed. To one unaccustomed to this system, it led to a certain wariness. One studied silhouettes of autos much as we had once studied the silhouettes of aircraft to discern friend from foe.

Along the streets, bicyclists, often riding with only the left hand on the handlebars, their bike canted a bit to the left to offset trays piled high with bowls of noodles or other wares precariously balanced in their right hand, were a constant threat to the equanimity, if not well-being, of pedestrians. Yet most walkers seemed to ignore their presence, and accidents rarely occurred as bicyclists threaded their way through crowded streets to make deliveries. Still other cyclists, generally elderly men, pedaled so slowly that it was a continual source of amazement to me that the bicycle remained upright; it seemed to defy the law of gravity. Occasionally, a motorcycle was seen, usually an aging variety. One day, near the Fukuya Department Store in Hiroshima, a biker spun out of a side street directly in front of the jeep in which I was riding; we struck him, knocking his machine and him down and breaking his leg. Once our driver had called an ambulance, as I stood by self-consciously, he thoroughly castigated the rider, blaming him for his own predicament and showing little commiseration for his injury and pain.

Rail travel had its peculiarities, too. Travel beyond 150 miles or so in Japan generally meant a night or a part of a night on the train. Few of the regularly scheduled Japanese trains had berths; even reclining seats were a luxury that did not exist until the early 1950s. The rigid backs and hard seats made prolonged travel uncomfortable. Upon boarding, Japanese travelers shuffled out of their shoes and sat on the train seats on their heels, much as they might sit on a tatami. Overnight travel proved an object lesson in the capacity of the Japanese to accommodate to almost any set of circumstances. As twilight approached, men would stand and nonchalantly begin to undress and fold their clothes, oblivious to others about. Suit coats were carefully folded to discourage creases, then shirts and ties were hung on the omnipresent wallside hook, and finally pants were removed and folded. In the winter, this left most men in their long underwear; in the summer, in a cotton jimbei. Off came their shoes, and then they settled into the seat, much like a pet nestling into a favorite sleeping spot—a turn here, a wiggle there, a thrusting of the hips into the cushion, and finally sleep.

Daylight brought new dilemmas. As a group, the Japanese are not notably hirsute, and so shaving was less of a problem than it would have been for an equal number of Americans or Europeans. But the washing facilities were limited, and trains, locals in particular, had essentially no such provisions. In many stations there were clusters of basins along the platform that offered a place to wash, brush one's teeth, or shave, provided the stop was long enough. At junction points where trains were assembled and dismembered, there was usually time for passengers continuing onward to perform their morning ablutions hastily.

On long trips the clickety clack, clickety clack of the train's wheels would drone on over intersections of rail, and a soporific lull would eventually settle. Even the flashing fields of rice failed to penetrate this thoughtless stupor after a while. A gust of air and sound occasionally edged its way into an idling mind as the car door opened to admit a passenger or, more commonly, one of the uniformed girls, employees of the National Railroad, who sold food and beverages as the train moved. "Uisuki" (whiskey), "obiru" (beer), "Osake ikaga dessho ka?" (Do you want to buy sake?) This refrain proceeded them down the lurching aisles and echoed back as they opened the door to the next car and bowed to its occupants as they had bowed to those in the car from whence they had just departed. Most seemed ill-at-ease in their uniforms of solid blue or tan edged in white, with their headbands askew, one stocking higher than the other. Their pigeon-toed shuffle betrayed the recency with which they were recruited from the countryside. Nevertheless, their appearance and earnestness eased the monotony. Soon after the train's departure from each station, the conductor would appear, his arrival preceded by the melodic click of the punch with which he marked each newly seated passenger's ticket. Although his uniform often showed the results of long use, it was invariably clean and tidy. Now and then his hand would rise perfunctorily to straighten the red armband precariously held in place with a large safety pin that specified his office, both in Japanese and English.

He would pause briefly, bow slightly, and request one's ticket. A quick inspection, a punch, a slight bow of the head, an expression of thanks, and on he continued, wrapped in his sense of office.

*   *   *

Shock gradually descended. It was June 25, 1950. The North Koreans had invaded the South; most of us in Japan, certainly in Hiroshima, had no sense of the debacle to ensue. Our forces were steadily pushed back toward Taegu and Pusan; a catastrophe was looming if additional troops were not quickly found to oppose the North Koreans. Japan was scoured; able-bodied men who had for years done little more than implement SCAP's directives found themselves given guns and hurried to Korea. For a moment, I thought that I was to be one of this hapless bunch. At one time early in World War II, I had been a member of the Enlisted Reserve Corps of the army, and when the military government learned of this I was summoned to explain my present status. Fortunately, when I returned from the Pacific in 1945, my discharge from the armed forces was unconditional; there was no obligation for further service. I could not be summarily armed and sent to face Kim Il Sung's invaders. It was a tense, unwelcomed moment until matters were clarified, and it certainly provided the impetus to watch more closely the unfolding events on the Asiatic mainland. We had already seen the routing of Chiang Kai Shek's forces in China and the rise to power of Mao Tse Tung and his Red brigades. Was this to be Korea's fate too?

When this undeclared war began, there arose an almost palpable sense of renewed purpose in our community of Nijimura, as wives rolled bandages and visited the wounded at the hospital to lend cheer, to write letters, and simply to stifle the pain of homesickness and injury. Their spouses were charged with developing the base support for the growing number of Commonwealth troops dispatched to Korea. Some of the units, such as Australia's 77th Squadron, were more actively engaged in flying sorties, particularly in the early, disastrous days after the invasion of South Korea had occurred.

We continued our genetic work in Hiroshima and Nagasaki, and it was difficult to believe that a few hundred miles away people were again being maimed and killed. Slowly the tempo rose. Troops began to arrive from Hong Kong, Canada, and England, as well as the United States, to reinforce the garrisons in Japan and Korea. Since the region in which we lived and worked was under British occupation, we saw more of the comings and goings of units like the Northumberlands, the Camerons, the Coldstreams, the Irish Hussars, and the Canadian Black Watch than our own divisions. Some of these regiments were sent directly to Korea and, as a unit, were never in Japan, but their wounded filled the hospital in Kure, and their glazed faces walked the streets in search of "rest and recuperation." We grew to recognize their uniforms and insignia. The Northumberlands, for example, wear their regimental insignia on the front and back of their caps in recognition of Waterloo, where in one desperate moment they formed up back-to-back to meet Napoleon's forces. Then there were the Seventh Hussars who, a century earlier, had made the fateful charge at Balaclava from which so few returned. Their steeds, Centurions, no longer neighed but roared, a thousand horses strong.

With the invasion at Inchon on September 15th, matters seemed to improve. Our troops sprinted northward, overrunning Pyongyang, but then came the debacle at the Yalu River and a steady retreat before the hordes China had committed to the defense of her politically aligned neighbor. An unsteady stalemate ensued in the vicinity of the old line that had divided the peninsula into a communist north and Syngman Rhee's ostensibly democratic south. Neither side seemed capable of dislodging the other.

Most of us with homes increased our entertaining to provide some respite from the horrors of Korea. On one of these occasions we met a former pipe-major of the Hussars. As a would-be piper, I considered this to be an unexpected boon. Months earlier, Wayne Borges and I had purchased chanters and books of bagpipe music from Australia and were trying to teach ourselves how to play. Retrospectively, we should have bought pipes at the outset, rather than just the chanters, for our resolve would have been greater and the learning easier. Bagpipe music is written without rests, for the piper can gather breath while merely pressing on the bag. There is no such reservoir with a chanter, and as a result, one slowly turns purple as successive measures march past. It is impossible to play anything but the very shortest of pieces.

Each evening Vicki's patience would be tested by my efforts to master the demiquaver and the semi-demiquaver and to grace notes with skill and flair. Crochets, minims, and semibreves became part of my vocabulary, and I tried to read everything Compton MacKenzie had ever written. Possibly this was my ultimate undoing. There in bold-face, in the book *Faeries of Scotland,* Mackenzie wrote that to be a piper required seven years and seven generations. Patience might give me the seven years, but only through reincarnation could I summon up seven generations. I still have my chanter; and, like the ex-smoker who feels the need to fondle a cigarette, I finger it now and then, but today even the reed is mildewed. On the evening we met the pipe-master, he tried to bolster my sagging spirits and promised to identify a teacher from among the Commonwealth troops with pipe bands. He even tried to whet our taste further by dancing the Scotch sword dance. Nothing came of this, for I decided it was easier to cultivate a taste for that other flower of Scotch culture, the malt whiskeys. Even the emphysema of age cannot dim this enjoyment.

*   *   *

Masuo Kodani and I were not asked, we were summoned, to Tokyo in the spring of 1951. The geneticist, Hermann Muller, the 1946 Nobel laureate in Physiology and Medicine, was to spend a week or so in Japan on his return to the United States from India and the First Congress for Cultural Freedom. Muller, one of the truly outstanding biologic minds of this century, had in the 1930s been an outspoken advocate of the Marxist experiment in Russia and spent several years working with fellow geneticists there. As the purges that destroyed Russian genetics and claimed the lives of such distinguished investigators as Vavilov and Levit began, Muller grew progressively more outspoken about the paranoia that seemed to have prompted Stalin's directions. The rise of Lysenkoism—a rediscovery of the disproven notions of the nineteenth-century French biologist, Lamarck, who contended that environmentally induced structural changes in organisms are inherited—and the ostracism of Timofeef-Ressovsky, Muller's old colleague, made the rupture complete. Muller's alienation was devastating, for he spoke as one of the disenchanted, one of that small body of former adherents whose acquaintance with Marxism was firsthand. Undoubtedly, the Supreme Commander, MacArthur, saw in Muller's visit a means to blunt further Marxist thought in college circles, particularly in Tokyo and Kyoto.

At the outset of the Occupation, MacArthur had sought to honor freedom of dissent, and as part of this process had allowed the Communists, many of whom the Occupation had released from the jails into which they had been hurled by Japan's wartime thought control police (*kempei tai*), to enter the political process. From the beginning the Communists seemed bent on mischief. They made every effort to penetrate the newly established unions and to exert political pressure wherever possible. The economic circumstances of the time—unemployment, food shortages, and rationing—favored their activities and they were vigorously asserting themselves. Matters culminated first in Communist-inspired violence on May Day 1946, which MacArthur deplored, and then in a threatened general strike in February 1947. This was too much for MacArthur; he issued a peremptory message to the party which "directed them to desist from furtherance of such action" and made it clear he would brook no further suborning of his goals. As a result, there now existed an uneasy truce; the party was clearly bent on enlisting more converts but, save for the excesses of the more radical students, was attempting to do so without incurring MacArthur's wrath.

Kodani and I were certainly not the only geneticists with the Occupation, but our position with the Commission made us more conspicuous, perhaps. Orders to travel were cut, and we arrived in Tokyo in time to meet Muller's flight. His prominence prompted others to be there, too, members of the Occupation forces as well as the Japanese scientific community. The man who marched down the ramp was short-statured and projected a fidgety nervousness that, I was to learn, could obscure his intellectual gifts. In a public address that he gave in Tokyo, his mind seemed to outstrip his lips' ability to articulate his thoughts. He would speak rapidly and then lapse into a series of stuttered ah's, like a computer refilling its buffer, only to disgorge again all that it had accumulated. However, as Muller spoke on his view of intellectual freedom and its abridgment in Russia, his sincerity and depth of feeling were not merely illuminating, they prevailed. This genuineness emerged repeatedly in the days ahead.

Emperor Hirohito was a serious student of biology; his interests focused on the bryozoa, a group of small, moss-like colonial animals. Once he learned of Muller's impending visit, he extended an invitation to visit the palace and the imperial laboratories. When the time came for the visit, I thought surely I might get to go too, since I had been Muller's alter ego for several days. This was not the case, and my contacts with the imperial family have been limited to an infrequent word with those members who, at one time or another, visited the Commission in Hiroshima when I happened to be there.

Muller's next stop was Mishima and the recently established Japa-

nese National Institute of Genetics. One of the olive-drab military sedans, usually reserved for senior officers, and an Army driver were
placed at his disposal. We started from Tokyo so late in the day, however, that an overnight stop in Miyanoshita was necessary. The road
from Tokyo to the west then in use followed the old Tokaido, the route
over which for centuries traffic from the west had entered the city.
Graceful, rustling cedar trees bordered the road to the left and right,
and as we rode my mind toyed with images of racing footmen contesting for the center of the road, announcing the approach of a daimyō
and his entourage with imperious shouts of "Shita ni iro! Shita ni iro!"
(Be down! Be down!), lest some peasant have the temerity to gaze on
the great lord and not grovel on the ground as he or she was expected
to do. Banners aflutter, limned by the neighing of the beautifully caparisoned horses and the sounds of the harness rubbing gently one
piece against another, the entourage passed across my mind's eye enroute to the daimyō's fiefdom or on its return to Tokyo to comply with
the orders of the shogun. We climbed slowly into the foothills surrounding Mount Fuji to Miyanoshita. This small spa is the site of the
Fujiya Hotel, a fascinating relic of the late nineteenth century but in
1951 one of the hotels the Occupation had commandeered to serve
its needs—in this instance a place of relaxation and recreation near
Tokyo.

The hotel sits carefully perched on a hillside, enveloped by its gardens and the sights of this delightful little town. Several generations of
diplomats and distinguished visitors have roamed its halls, enjoying
its hospitality. Its public rooms were cozy—not the mammoth, visually cold, boring areas that one might expect—and in the tradition of
its vintage, there was a small library available to guests. Most of the
books that remained were themselves antiques, an original edition, for
example, of Captain Brinkley's history of Japan. Each room in the old
hotel was named after a flower, and the fragrance of fresh-cut blossoms sweetened the air of one's room. There was a gracefulness about
the hotel, an air of fulfilled purpose without complacency. After we
were settled, I loitered through the gardens, attempting to see Fuji, but
dusk descended too rapidly to do so.

Rain was gently falling the next morning, and the overcast sky
made the trip to Mishima a study in grays and greens save for the
occasional field of rape, flooded with yellow flowers. Mishima was a
much smaller community then than now, and the Institute was corre

spondingly less distinguished. It occupied buildings that had once been part of a Mitsui aircraft manufacturing plant, and its facilities were limited. The library consisted largely of a collection of reprints, and the laboratories were marginally equipped. These limitations could have been fatally compromising were it not for the founding fathers themselves. The director was Oguma Kan, a cytogeneticist whose early efforts to determine the number of human chromosomes, although subsequently shown to be wrong, had gained him an international reputation. A gregarious man, Oguma projected a hearty image that countered the stultifying circumstances under which he and his colleagues worked. Among the other distinguished members of the Institute, none impressed me more than Komai Taku, recently retired from his position as Professor of Zoology at Kyoto University. Komai was a soft-spoken, shallow-cheeked patrician—a friend of Prince Chichibu—whose sense of fair play and concern for others have become legendary. He and his wife accepted his young students into their home as though they were their own children, and his generosity and obvious interest in their well-being have, in turn, colored importantly the dealings of these men with their own students. Komai was in the fullest sense of the word a teacher, a *sensei*. He had been a postdoctoral fellow in Thomas Hunt Morgan's laboratories at Columbia in the late 1920s and had returned to Japan to continue his research prior to Morgan's departure from Columbia for the California Institute of Technology. Many years after Komai's death I learned from one of his Chinese contemporaries in Morgan's laboratory, a developmental biologist, of Komai's generous furtherance of genetics and experimental biology in China. We grew close over the years, but, as was to be expected, he never spoke of his many thoughtful acts on behalf of others.

When Muller and I arrived at the entrance to the Institute, he was effusively greeted and immediately launched on a tour of the facilities themselves. I tagged along at a discreet distance, as befitted my junior status. We were shown the Goldschmidt collection of reprints, which had arrived only shortly before us. Richard Goldschmidt, who had once been head of the Kaiser Wilhelm Institute in Berlin but was then a Professor of Genetics at the University of California, had long been fascinated with the Orient, particularly Japan, which he first visited in 1914. Subsequently he taught at the Imperial University in Tokyo. He had donated his collection of papers partly out of a sense of the Institute's need but, more importantly, as a measure of appreciation for the several enjoyable years he had spent in the country.

Eventually, after tea and snacks, we were ushered into the auditorium where Muller was to speak. He spoke of genetics as it then was in the United States and dwelled at length upon those recent developments that he thought were of special significance. One that he singled out—and rightly, as time was to tell—was the work of Joshua Lederberg, who had just established that bacteria, previously thought to reproduce only asexually, reproduce sexually as well. This finding, when coupled with the structure of deoxyribonucleic acid (DNA, the protein wherein is coded genetic differences) soon to be postulated by James Watson and Francis Crick, launched the molecular revolution and permitted insights into biology that still test our understanding.

Muller's presentation vibrated with his own enthusiasm, and we were swept into the web he dexterously spun. When, after an especially animated period of discussion, the seminar ended, there was a sense of loss; promises so evocatively aroused would have to wait another telling. However, the day was not yet concluded; dinner and the opportunity to socialize had been planned at a restaurant, an old Japanese inn overlooking the bay at Atami. Muller, Kodani, and I were joined by Oguma, Komai, Kihara, Shinoto, and Tanaka, five of the most distinguished members of Japanese biology. Kihara Hitoshi was still Professor of Botany at Kyoto University, and his students dominate contemporary Japanese genetics. He was, and remained through life, a man of many interests—a skillful skier who continued on the slopes into his eighties, a mountain climber of note, an enthusiastic ethnographer who led an important Japanese expedition into the Hindu Kush some years later, a practical plant breeder, a developer of the seedless watermelon—and, withal, a very endearing man. Shinoto Yoshito was a zoologist and professor at the International Christian University outside Tokyo—an able organizer whose skills proved repeatedly useful to the various professional groups to which he belonged. Tanaka Yoshimasa had been Professor of Zoology at Kyushu University and was possibly the world's authority on the genetics of the silkworm, an area of biology in which the Japanese were the unchallenged masters.

The inn to which we were taken was high on a bluff above the gently arching bay before Atami; the surf could be heard distinctly as it beat on the rocks beneath us. Japanese hospitality was unsurpassed then, something to be enjoyed, relished. It was an evening of conversation, punctuated now and then by the rattle of loose unputtied panes of glass in the windows of the inn facing the bay as they attempted to resist successive buffetings of wind. Finally, we had to leave for Numadzu and the train that was to take Komai, Kihara, Kodani, Muller, and me to Kyoto. Although I saw Komai and Kihara on numerous subsequent occasions, and we reminisced about this evening again, we never spoke about the war; there were always things of more immediate interest to discuss. As a consequence, I have no notion what their attitudes were toward Japan's militaristic gamble. Since they had both received postdoctoral training abroad, they surely knew of the relative strengths of the different national economies and their differing accesses to resources, and they must have sensed that Japan could not win a prolonged war. They were patriotic and responsible men, however, and would undoubtedly have contributed what they could to their nation's effort. Indirectly, they would speak of the difficulties of research in a wartime economy, plagued by limitless shortages.

Kyoto University, or Kyodai, as it is commonly called, is Yale to Tokyo's Harvard. Yukawa, one of its professors, was the first Japanese physicist to win a Nobel prize, and its distinction in many areas of the sciences and humanities has challenged Kyodai's more politically favored sister in Tokyo. In 1951 Kyoto University's student body was among the most radical in Japan; its more extreme members periodically erupted into senseless violence and destruction of property. The Japanese have always been, in many ways, more tolerant of the exuberances of youth than Americans. Student violence, if not condoned, was tolerated; it seemed to be viewed as a passing aberration, like the panty-raids that received so much attention in the United States in the early 1950s.

On the occasion of Muller's visit, however, there was a sense of something different; clearly the administrators at Kyodai were apprehensive about an eruption of violence whose sole purpose would be political, to embarrass the Occupation. The president obviously feared the students and was concerned not only about Muller's wellbeing but, I suspect, about the reaction of Occupation authorities if violence did break out. This was not to be. When the time came for Muller to speak, he strode purposefully into the auditorium, seemingly oblivious of the tensions. His presentation projected an experience and authority that brooked no opposition. Student objections, if they existed, were so feebly and ineffectually advanced as to be embarrassing. Afterwards, we gathered in a large room where a light luncheon was spread; if Muller was aware of the magnetism he projected, he gave no evidence of it. His conversation was always animated, but no more so than at that moment, as he inquired about the research of his colleagues in Japan, listened attentively, and pursued their activities constructively with insightful comments.

This was, for me, an unusual time. Not only was I brought into contact with a man whom I grew to respect more with each day, but through him I met Japanese investigators whom, because of my age, I could not possibly have approached on my own. One of these was a fragile cytogeneticist, Kuwada Yoshinari, whose early work on the fine structure of chromosomes—as far as such structure could be revealed by light microscopy—was already internationally recognized. Though then retired, he came to the university daily and retained the curiosity that is the hallmark of the exceptional investigator.

Mainichi Press, one of the major newspaper syndicates, had arranged a party at the Ichiriki for that evening. Japan has had many distinguished geisha houses but none with more luster than this one. It occupies a conspicuous spot in the Gion, possibly the most famous of all geisha districts in Japan. We were enthralled by the grace and the beauty of the geisha and their apprentices, the *maiko,* and amazed at their capacity to transcend the barriers of language through gesture, dance, and facial expression. Their exquisite brocaded kimono, white-powdered faces framed by jet-black hair swept into elaborate coiffures, and the delicately, presumably provocatively, exposed nape of the neck, gave them a fragile, doll-like quality. With repeated exposure over the years I became surfeited with the artificiality of the environment within which they perform, but never have I seen this environment better exploited than at the Ichiriki. Even the double entendre, at which the geisha are peerless, sparkled. When the evening ended and Muller and I were taken to the Miyako Hotel for the night, it was as though we had been freed at last from a gossamer-like web, one which held not by its physical strength but by its ephemerality. It was difficult to sleep, for the verbal interplay, the beauty of the dress and dances, washed back and forth through my mind. I was still struggling with sleep when another day confronted.

Muller and I were to go on to Kure and Hiroshima, where we

would be met by Kihara and Komai who, as Japanese, were not free to travel on the trains the Occupation forces used. It was a source of embarrassment to us that we could not travel together, but they made light of the situation, and possibly preferred a different route. In Hiroshima, Muller spoke again, this time more informally and to a smaller audience, the professional staff of the Commission. His topic was radiation-induced mutagenesis, an area that owes more to his genius than to that of any other person. Muller was clearly at ease and welcomed, I think, the opportunity to speak on a topic that had absorbed so much of his life. Later at dinner at our home with Kihara and Komai, he relaxed further and regaled us with stories of Morgan and Sturtevant and of his life in Russia. Muller was a deeply religious man, one whose temple was human dignity and intellectual liberty—I do not actually know whether he belonged to an organized church but suspect he did not.

*   *   *

The Commission was visited repeatedly by the journalists of the time, and most of the major magazines—*Life, Time, National Geographic*—sent their stars. None was more distinguished than Carl Mydans. When I met him he was already a ranking member of that group of photojournalists who had made *Life* the undisputed leader of the national weekly magazines in the United States. Mydans had covered the withdrawal of the Marines from Tsingtao in 1948 and was in Hiroshima to describe the Atomic Bomb Casualty Commission and the war's aftermath in this once-devastated city. He was a square-jawed, forthright individual whose closely cropped, graying hair gave him an almost Prussian-like appearance, a characterization he, as a liberal, would probably not find flattering. He still wore the quasi-uniform of the war correspondent and was seldom separated by more than a few feet from the canvas musette bag containing his Contaxes and their supplementary lenses. His use of the camera was almost reflexive, a trait captured in a startling photograph of him by Alfred Eisenstaedt. He was constantly searching for the photographic nub of a scene or event. I recall vividly his words to me: "Jack, a photograph worth taking is worth taking several times from different vantage points!" And "film is the cheapest item in a photographer's bag, don't use it sparingly."

When Mydans worked, his camera rose to his eye to be clicked and the film advanced as unconsciously as he breathed. Once a roll of film was completed, it was promptly rewound and placed in an empty film cassette, whose bottom was then crimped so that through inadvertence it would not be mistaken for unused. The interior of the camera was brushed and a new roll of film inserted; the tab from the film box describing the speed and type of film was placed in the flash shoe. All of this was done while a sprightly conversation occurred, as he probed for other elements in the story taking form in his mind.

The Commission's work was to be a major article in *Life,* but the beginning of the Korean war eventually relegated it to a minor position. Few of the late effects of exposure that we now recognize had as yet emerged. Cogan, Martin, and Kimura had just recently established the occurrence of radiation cataracts among some of the survivors. The first such anomaly they had seen was found in one of the Japanese waitresses, Hatsue, who served the Commission's staff, a warm, friendly young woman who subsequently died a tragic death, struck by an automobile which it is thought she could not see. Mydans, like many others since, was especially concerned about the genetic effects, the possible damage done to generations unborn. It was this concern that led him to me, as the head of the Genetics Department of the Commission and the person responsible for the program of newborn examinations that was under way. He was interested in photographing one of our home visits, if this was possible. I assured him it was, but it would take us a day to make the necessary arrangements. We would have to select several possible candidates and visit the homes ourselves to solicit the approval of the infant's parents. Koji and I set about identifying several families and drove to their homes to see if they would mind if photographs were taken of the examinations which were to occur. All graciously approved, and provisions were made to visit them on the following day. Two jeeps would be needed, one for the usual visiting team and the other to carry Koji, Mydans, and myself. We decided that Dr. Tachino and Nurse Minato would be the examiners, since both spoke some English. They were instructed to perform the standard examination, but somewhat more slowly than might normally occur so that there would be ample opportunity for Mydans to take the pictures he sought.

About mid-morning on the following day, we left the Red Cross Hospital. As we drove westward to the part of the city in which the homes we were to visit were located, we passed a number of school grounds where students were playing baseball or tennis on makeshift diamonds or courts before the dilapidated, dreary, unpainted wooden buildings that housed their classrooms. The same soft-centered, ribbed rubber ball was used to play tennis or baseball, partly for economic reasons and partly because this had been the prewar convention. Before one of the school buildings still stood the small shrine, the *hōanden,* that had contained a copy of the Emperor Meiji's Imperial Rescript on Education (the *kyōiku-chokugo*), which enjoined students to be modest, obedient, industrious, and patriotic.[1] Prior to the Occupation, several times a year, it had been customary for the principal to read this to the assembled student body, and students had to bow to the shrine housing the rescript as they passed it, in acknowledgment of the Emperor's gift of education and their indebtedness to him. Now and then one of the older students would still bow, probably reflexively, since it was no longer required. Indeed, the rescript was no longer exhibited in the shrine.

Once we reached the area of the homes, as we went from house to house, we were startled to find that tea had been prepared for us, the homes had been newly cleaned, and the infants were dressed in their best finery. We were somewhat apprehensive that with so many strangers about the infants might be upset and prone to cry. This did not happen; they seemed to relish the attention they were receiving, and Mydans worked quickly and unobtrusively. With my camera, I aped his every move, like an undisciplined apprentice, merely copying his choice rather than seeing why it was appropriate.

Visitors, singly or in groups, have winnowed their way through the corridors of the Commission since its establishment. Each has received a brief presentation of the origins, purposes, and findings of the studies and a tour of the facilities. Some have been famous, most have not. Some undoubtedly came out of curiosity, particularly the numerous students who come as a part of their seasonal school outings, others through a sense of obligation. Certainly the visits of members of the royal family would seem better described as matters of protocol. Although the emperor himself never called, many of his sons, daughters, and siblings did, especially during the Occupation.

The first member of the royal family to visit was the Crown Prince, Akihito. In April 1949, when he came, the Commission was still in temporary quarters, and he was a teen-ager filled with curiosity. He and his party arrived at the old Ujina pier by boat. At about the time that I reached Hiroshima in the summer of that year, Prince Mikasa made an official visit, and Prince Takamatsu, an urbane and courtly man, the emperor's second brother, visited the Commission shortly after we left, in November 1951. Other members of the royal family—Princess Takamatsu, Princess Chichibu, and Prince Yoshi—have also called. I was still with the Commission when, in April 1951, we were honored by the arrival of two of the emperor's daughters, Princesses Yori and Suga. They arrived on a sunlit day in an entourage of four black cars, flanked by an accidental honor guard of our blue jeeps and green, mock-wood-paneled Chevrolet station wagons. The specific vehicle in which they traveled was a lustrously polished old black limousine with its spare tires mounted in metal covers embedded in the front fenders. Both young ladies wore simple, tan, two-piece suits with pleated skirt, and they clutched brown handbags. The younger daughter trailed her elder sister by several steps, as presumably their positions demanded.

They seemed shy and somewhat ill at ease with the attention they were receiving, but possibly their demure behavior was no more than would have been expected of other young girls in Japan at that time. Given their ages, this was undoubtedly one of the first formal functions they had been obliged to fulfill as members of the royal family. They and the accompanying staff from the imperial household asked occasional, softly voiced questions as they were led through the corridors by Dr. Taylor. The visit was short, no more than an hour or so, but obviously our Japanese employees welcomed the opportunity to see the princesses at such close quarters. Prewar protocol would have denied them this chance. When the tour ended, and their names had been inscribed in the guest book, they presented to Dr. Taylor several boxes of specially prepared cigarettes bearing the imperial monogram, a chrysanthemum. These were to be distributed among the staff as small tokens of appreciation, apparently a common occurrence with all imperial visits. As their vehicles circled to leave the area before the Commission buildings, the Japanese employees who were standing in the turnabout and parking area remained rigidly at attention, facing the vehicles until they were out of view. Most of us who were not Japanese had already turned to resume our daily activities once the vehicles began to pull away. In retrospect, I was probably guilty of an unintentional breach of etiquette, for I had filmed their departure through an open, second-story window, from which, of necessity I looked down on the young princesses. My Japanese colleagues were peering, too, but more discreetly, further removed from the open window.

* * *

Japan's years of occupation were a period of acceptance and endorsement of things American, sometimes uncritically, indeed ludicrously, oftentimes not. An American import of this era, one that enjoyed a brief vogue, was the striptease. Theaters that offered only striptease shows, or a combination of movies and burlesque, opened everywhere. My colleagues and I watched for days the preparations for the opening of one such establishment in Kure, and finally a group of us, husbands and wives, decided to see what Sally Rand had bequeathed to Japan.

We had just seated ourselves when the strains of the entr'acte music from *Carmen* rose hesitantly and off-key from the orchestra pit. The curtain slowly parted, to reveal a dark backdrop of relatively little interest, but suddenly from one of the wings a young, attractive woman slowly made her way toward the center of the stage. This would not have been particularly unusual were it not for her costume—she wore only a transparent plastic raincoat through which her firm, nubile breasts, darkened areolae, and pubic hair were readily visible. In her mouth was clenched a solitary red rose, and above her head she twirled a transparent umbrella. At center stage she turned slowly, saucily thrusting her breasts at the audience several times, and then walked on toward the other wing. Just before she reached it, she turned again to the audience, took the rose from her mouth, and threw it into the seats, where it was caught by a conservatively kimonoed elderly woman who seemed startled by it all. This performance, like each subsequent one, was met with polite applause, and none of the raucous roars that greeted Sally's famed fan dance.

As we left the theater, our eyes were drawn toward a poster that advertised the next attraction, a movie and another burlesque troupe. Blazoned across the top of the poster were the English words "Scent of Riddle." What could the author of these cryptic three words have intended? For days we puzzled, individually and collectively, without a clue. Finally, like some deep revelation, it came. It had to be "Aura of Mystery." These distortions of English-language words and phrases persist in Japan. They are not as common as they once were, but a watchful eye will reveal a substantial number. The coffee additive in the United States derived from *p*owdered *cream* has the trade name Pream. In Japan, the dairies that make a similar additive reversed the words to *cream* *p*owder, and from this combination derived the word Creap, which has a connotation in English that I doubt the manufacturers intended. Similarly, the Japanese equivalent of Gatorade is known as Pocari Sweat, hardly an enticing name for a soft drink. While the Japanese proclivity to pronounce the letter *l* when it is followed by a vowel as *r* is well known, many of the humorous consequences are not. It is common to hear a cloakroom described as a "croak room," but even more entertaining to see such a room in a hotel lobby conspicuously designated "Croak."

Although the Allied occupation of Japan was benign, benignity is in the eyes of the beholder, and it is not patent that all Japanese found SCAP a nonintrusive government. While there were only a limited number of acceptable forms of self-expression, the theater was usually exempt from overly strict censorship. As a result, the Occupation, and the mannerisms and physical appearances of the occupiers, were often the good-humored but nonetheless pointed butts of many jokes. *Morgan O'Yuki*, or, as the Japanese program proclaimed, *Madame Snowflake,* opened at the Teikoku (Imperial) Theater on February 6, 1951, literally across the street from the Imperial Palace and within the shadow of the headquarters of the Supreme Commander of the Allied Powers. The play itself grew out of one of those unusual, presumably true, postscripts to history.

At the beginning of the twentieth century, Japan was visited by Dennis Morgan, a son of the famed financier J. P. Morgan. Dennis was a bit of a ne'er-do-well. He arrived in Japan, or so the story goes, in the spring, at a time when the cherry blossoms are at their finest. It is the season of the Miyako odori, the famed geisha dance festival. One of the stars at that time is O'Yuki-san. Morgan and she meet and fall in love. O'Yuki-san, however, has a childhood sweetheart, Kawashima. Eventually she marries Morgan and leaves her childhood friend distraught. Together, she and Morgan set out for the United States, via Egypt and Paris, where they agree their lives should be spent. Morgan's family raises many objections to their marriage, and his uncle implores O'Yuki-san to divorce Morgan. Her love for him is too great, and she rejects the uncle's representations. Morgan purchases an elegant house in Nice for them, but World War I erupts. Morgan, who had briefly returned to the United States, hastens to the side of O'Yuki-san, but on his way becomes ill and dies. Heartbroken, Yuki finds and plays the record she and he had recorded in memory of their honeymoon. Lonely and sad, she recalls those happy moments with her husband and from those recollections there emerges a vision of springtime in Japan, and her departed Dennis. The outline of the story should be plain—rich boy, poor girl, and fate's intervention.

The play itself was not done in the grand manner, however, but as a comic opera in thirty scenes. These included such unlikely episodes as The First Meeting, The Ceremony of the Tea, Rikisha-Man, "Tondemo-Happen" (a mixture of English and Japanese that might be translated as What never happens happened), The Balcony, The Bowler Hat, Begin the Beguine, and Fantasia. There were less than a half-dozen more-or-less straight roles, and most of the time the stage was occupied by the Nichigeki Dancing Group. These women wore as little as Japanese sensibilities would tolerate, and in a nation where mixed bathing was long the rule, these sensibilities tolerated a great deal.

The stars of the play were Furukawa Roppa, a Japanese Bert Lahr, or so he impressed me—a round-faced, bespectacled comedian who delivered lines in English as readily as Japanese—and Koshiji Fubuki, an attractive woman whose role demanded little more than a tear now and then and a sense of loving helplessness. Much of the time that Furukawa Roppa was on the stage he was delivering quick jokes in the best of the burlesque tradition so effectively pioneered by W. C. Fields, Shean and Cohan, and others. The jokes, despite the sometimes obscure accent, were humorous. Timing is the central ingredient in good burlesque, and in this respect Furukawa Roppa was as skilled as any comedian who has trod the American stage. His jokes, however, had political overtones, and it was these that made the play memorable. Morgan was more than just a simple, albeit well-intentioned, lover; he was a personification of Occupational policy, benign but not terribly well thought out. O'Yuki was a fair damsel—like Japan, too involved in her new affair to see its long-term import. The jokes were broad, the members of the chorus attractive, often stunningly so, but to the viewer who was a part of the Occupation, there was a sense of having one's leg continuously pulled. I recall a full house, liberally sprinkled with uniforms, which laughed heartily at the good-natured fun poked at the sometimes pompous ways of MacArthur's office, and prolonged applause when finally the comedy ended and Furukawa and his fellow performers took their bows. As the audience slowly departed, one heard laughter repeated here and there as a particularly apt scene or remark was recalled. *Morgan O'Yuki* has been performed on several subsequent occasions, but it is hard to believe that without the Occupation it would be as memorable.

*   *   *

My last summer in occupied Japan, the summer of 1951, saw our social club on another one of its outings, this time an overnight trip to Miyoshi on the Go River to see cormorant fishing. Clear waters in the upper reaches of many of Japan's streams abound with fish of the trout family, one of which, the ayu (*Plecoglossus altivelis,* the sweet fish, a kind of fresh-water trout), with its fine skin of dull silver, is found commonly in summer. The most intriguing and classical form of fishing for ayu is through the use of cormorants—large, long-necked, blue-brown voracious seabirds. Cormorant fishing has been pursued in Japan during the summer months for more than four hundred years. Two of the most celebrated places for cormorant fishing are the Uji River south of Kyoto and the Nagara River in Gifu Prefecture. There are other spots, however, including the Nishiki River in Yamaguchi Prefecture at Iwakuni, and the Go River in the interior of Hiroshima Prefecture, where we were bound. The Go arises some sixty or so kilometers from Hiroshima near the small commercial and industrial city of Miyoshi, a distribution center for lumber harvested from carefully husbanded forests in the area. The fishing occurs near the junction of the three small meandering rivers that combine to form the Go, which ultimately empties into the Japan Sea at Gotsu.

As dusk settled, our party assembled at the front of the inn where we were staying to walk to the river, where we were to board the boats to observe the fishing. Five wooden boats awaited us, four of which we boarded under the watchful eye of its boatman; the fifth carried not only a boatman but the fisherman and his eight or ten birds. Like a convoy of escort vessels, we pushed away from the bank after the fishing boat. Off the prow of the fishing boat hung a large torch that illumined the water, and near its stern, just ahead of the boatman who guided its course, stood a large, clothes-hamper-sized wicker creel. We presumed this was to hold the evening's catch. The birds, to each of which was attached a rein, huddled in the forward end of the boat.

Once a relatively deep, placid stretch of water was reached, the fishing began. The fisherman, who was dressed with a band tied about his head, a straw skirt, and dark top, pushed the birds into the water, one at a time. As quickly as they were in the river, they began to dive for fish. To prevent an impossible snarling of the reins, the fisherman kept passing the individual lines back and forth between his fingers and from one hand to the other, in tempo with the changing courses of the birds. It was an impressive display of dexterity. After the birds had been in the water for ten minutes or so, the fisherman shortened the reins and drew them back to the boat. As he did so, his boat and ours moved in concert toward a beaching site on a rocky bar so that we might see the fisherman recover the catch. The birds were pulled aboard the boat one at a time, grasped, and taken to where the creel was located. They were upended and their gullets milked of the fish they had caught, which tumbled into the creel. Once this was completed, the bird was released and invariably made its way to the gunwales of the boat near the torch. Here it would extend its wings and wave them slowly but steadily, as if consciously drying off. The process proceeded surprisingly rapidly. When the last of the birds had been relieved of its catch, the fisherman brought his boat close to ours so that we might see one of the birds more closely. At the base of the long neck of each was a rigid but removable ring, and to this was attached the line through which the fisherman controlled the bird. The ring, of course, made it impossible for the fish the bird caught to pass beyond the gullet into the stomach. Questions flew across the water: How long did it take to train a bird? How much would a trained one cost? How long was their useful fishing life? How many could a good fisherman handle? Each query was patiently answered. To our surprise, we learned that a good bird in the prime of its fishing years could cost several hundred dollars, a seemingly impossible sum at that time, and with luck might fish for four or five years.

Again the birds were returned to the water. This time most of us watched individual ones so that we might see how they caught the fish. Most birds held a caught fish crosswise in their beak until they could rise to the surface, when the fish was flipped into the air and swallowed head first. A skilled bird could catch several fish in the minutes that it was in the water, and as we were soon shown, they could fish rapids almost as readily as still water. Since our boat could not keep abreast of the fishing boat as it shot through the white water, once he had completed his run through, the boatman would beach the boat and the fisherman would retrieve his birds. We followed as rapidly as practical and arrived to see him relieving them again of their catch, which was as large as that taken in relatively calm water. We also observed that there was a distinct difference in the skill of the birds; some always caught fish and others usually did not. One of the latter we noticed, when placed in the water, would swim beneath the boat and rise on the darkened side and float. He did this repeatedly. Moreover, when the fisherman lifted him to remove the fish, he resolutely emitted a loud cry as if to say "No, not again!" This laggard seemed to be the clown of the gaggle, or perhaps he simply did not want to fish on this night. We suspect the former, for although the fisherman was impressively busy when his birds were in the water, it was hard to believe that if we could spot the laggard so readily that he could not do so, too.

Our fishing had begun up the river from our inn, and though we had fished for well over an hour, soon the row of inns, including our own, loomed ahead. When the boats pulled into the bank so that we could disembark, we learned that the evening's catch was included in the price we had paid to watch the cormorants fish, and provisions had been made for the ayu to be broiled over charcoal in braziers (*shichirin*) set out behind the inn.

Evening's dark had settled snugly about us, and now the mountains that surround the river were poorly etched against the sky. As we made our way across the pebble-strewn shore to the braziers, every eye was drawn to the heavens above and the multitude of twinkling stars. Only the sound of the rushing water and an occasional exclamation of surprise from one of our party as a stone turned beneath a foot breached the quiet. We drank deeply the stillness of the evening air, disturbed now and then by a light breeze sweeping up the valley. Soon we were settled about the fires, staring fixedly at the hot embers and the leap of flame as a droplet of fat from the broiling fish fell on a coal, subdued by the beauty of the moment. Conversation became more animated as the delightfully tender fish and the other accouterments that accompany a good meal at a Japanese inn were set before us. We lingered long after the last morsel of food was eaten, unwilling to admit an ending.

*   *   *

Ritual is an important support mechanism in most societies, but few have institutionalized as many aspects of life as the Japanese. An important ritual is the *omiokuri*, the "honorable send-off." Customarily, when family members, friends, or superiors leave for an extended time or trip, one goes to the airport or railway station to bid them goodbye. Special tickets can be purchased at railway stations that admit one to the arrival and departure platforms. To gain access to the latter, one must pass through a wicket where these tickets, like regular train tickets, are punched by an attendant, and the punched ticket must be relinquished as one leaves the station. Today, they can be purchased through coin-operated dispensing machines, but years ago they had to be bought at the station's ticket windows. Now the ticket costs a hundred-odd yen, then only ten.

When it came time for us to leave Japan in the late summer of 1951, we did so with a mixture of emotions. We looked forward to seeing our families and friends again, and to initiating a new, an academic, life at the University of Michigan. But we also knew we would miss the many warm friendships that had flowered in the past two years, and we were uncertain whether this new position would afford the opportunity to continue to be involved in the genetic studies in Japan. On my last day at work, I had taken leave of my colleagues in the Genetics Department and did not expect to see them again, for we were to board our train in Hiro rather than Hiroshima. It did not seem practical for them to come to the station to see us off, and we fully anticipated that only our neighbors, our housemaid, Ayako, and possibly some of the Kure staff of the Commission would be at the station.

Since I have a penchant to be at airports or railway stations early, Vicki and I were on the platform thirty minutes or so before our train was due to arrive, much less leave. We were chatting with friends when, with a tootling of horns, five of the Commission's station wagons appeared. Out tumbled all of the genetics staff, some fifteen-odd physicians and three clerks. Koji, resourceful as always, had presented their dilemma to the director of the Commission—they wanted to see us off but had no means to do so. Would it be possible to use some of the station wagons normally used to transport patients to and from the clinical facilities? Dr. Taylor, a balding man of medium stature and neatly trimmed mustache, who had succeeded Tessmer as the director, agreed, and the whole staff had left Hiroshima more than an hour earlier.

Time permitted only a few hasty words with each of the *omiokuri-nin*, the participants in the omiokuri, but this was just as well. Emotions were strained, and a heavy heart hardly inspires endless small talk. We had become a close-knit group, proud of the camaraderie that had arisen. Words could not capture the sense of what bound us together, nor could they adequately reflect the feelings each of us had toward one another. Fortunately, our train pulled into the station before all small talk was exhausted and self-consciousness arose. Our bags were handed to the porter, who placed them in our car. Vicki and I remained on the platform until the final whistle was blown announcing the imminent departure of the train and then stepped aboard. No *banzai* accompanied this leave taking—the shout itself still projected too immediately Japan's recent militarism—but eyes were misty and faces long. We stood at the door while the train pulled away from the station, waving to our friends until they could no longer be seen.

As we settled into the velveteen benches that ran along the long axis of the Pullman car to which we had been assigned and prepared for the arduous trip to Tokyo, our minds sifted, sorted, and filed the many recollections of the past two years or so. My own reflections were interrupted by Dr. Taylor, who was also on his way to Tokyo and wanted to discuss some of the personnel problems confronting the Commission. He urged that I consider returning to Japan after several months to assume the direction of the data management program and statistics. If this was impracticable, as I felt it was, he asked that I consider helping in the recruitment of suitably trained people to fill these needs, a task I willingly assumed.

At our urging, George Fukui had made arrangements for us to stay at the Marunouchi Hotel in Tokyo, the somewhat dowdy but not stodgy billet for visiting officers of all ranks in the British Commonwealth Occupation Forces. On previous trips to Tokyo, Vicki and I had enjoyed its ambience and the promise of hot, strong tea at six in the morning, served by a young waitress who invaded one's room almost surreptitiously without scarce a clatter of the several cups or the pot, carefully draped with a tea caddy.

We journeyed the next day to Yokohama to board the *President Wilson*. George, who had seen to our transportation and the completion of the departure formalities, joined us aboard for a parting drink. When finally the call came for all visitors to leave the ship, George made his way down the gangplank, to remain on the dock as the engines were started, waving cheerfully as we and all of the other passengers crowded the pier side railings of the *Wilson*. Paper streamers clutched the wharf tenuously, to drop into the water as we slowly pulled away.

Eight or so days after our departure from Yokohama, we arrived in Honolulu. At that time, Honolulu was a truly tropical city; most tourists arrived by ship, as we had done, and Waikiki was not carpeted with visitors and souvenir shops. The Moana Loa, Halekulani, and Royal Hawaiian Hotels stood almost alone on the beach; before each, muumuu-clad Hawaiians fashioned leis from a surfeit of flowers. We rented a taxi for the day to visit Diamond Head, the Pali, and other scenic spots that have become familiar to so many. Late evening saw our ship pulling slowly away from the clock-towered pier to the strains of "Aloha" and a forest of paper streamers.

Our arrival at the pier in San Francisco, and the subsequent customs and immigration formalities, passed with only one hitch. Shortly before leaving Japan, I had purchased a small traveling set for playing go, or shogi. The black and white go pieces were about the size of a pill, and in the course of packing, the box in which they rested had opened and the pieces had found their way through much of my clothing. When I opened the suitcase at the custom inspector's request, several of these pieces spilled onto the pier. Before they had stopped rolling, the agent was scrambling about trying to retrieve them. His haste hardly spoke of solicitude but rather suggested that he had caught me importing contraband drugs. When finally he had several pieces in his hand, and noted they were all plastic, he smiled broadly and asked what their use was. I dug through my suitcase until I could find the little board and distributed several pieces about to show him how the game was played. Whether from embarrassment or for other reasons, he quickly approved the remainder of our luggage, and we were free to join friends who had returned from Japan earlier and had come to the pier to meet us.

Late in the summer of 1951, at about the time of our return to San Francisco, peace formally came. Japan and her former adversaries signed a treaty officially ending the hostilities. Dismantlement of the machinery of the Occupation began immediately. The offices of the military government teams were closed; the special trains that had served the Occupation forces ceased to run; most of the previously commandeered buildings were returned to their owners; and numberless other steps, large and small, were initiated to return the governing of their islands to the Japanese. An era ended, but the American presence continued. Japan's Occupation-inspired constitution disavowed war, and in a world fraught with dissension, her national safety was only guaranteed by the presence of American forces. Their role had grown more crucial once aggression had begun on the Korean peninsula, a fact somewhat grudgingly admitted by many Japanese. However, to those of us who were a part of the Occupation, no matter how small or indirect, there remains a nostalgia, a sense of participation in an unusual happening, an enlightened administration that in many ways has contributed to the richness of life that now prevails in Japan.

One of the Protestant churches of Hiroshima as it appeared a few days after the atomic bombing. This church was situated somewhat more than a kilometer from the hypocenter. [U.S. Army Forces photograph.]

The courtyard of the Hiroshima Red Cross Hospital with several of the jeeps used in the Genetics Program. Immediately before the gate is one of the city's streetcar lines, and across the street is a portion of Hiroshima University.

The Hondori, one of Hiroshima's main shopping streets, as it was in the winter of 1949–1950. Later this section of the street was roofed over to provide a shopping mall.

Frank Poole and myself at Haneda airport in September
1949 as we await the arrival of our wives from the United
States.

A goldfish vender in Nijimura in 1950.

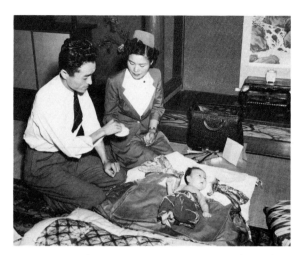

Dr. Tachino and Nurse Minato examining one of the infants enrolled in the Genetics Program. [Atomic Bomb Casualty Commission archival photograph.]

The staff of the Genetics Program in 1950. In the front row are Dr. John Wood (third from the left), myself, Dr. James V. Neel, Dr. Koji Takeshima, and Reiko Iwamoto. In the second row are Morita-san (second from the left) and Matsuda-san (second from the right). [Atomic Bomb Casualty Commission archival photograph.]

The first structure to be built in Hiroshima for the annual memorial services to those who died as a result of the atomic bombing. Note the plea for "No More Hiroshimas" on the back of the wooden stage.

A shopping street in Hiroshima at the time of the late fall sales.

A torii in Nagasaki partially destroyed by the atomic bombing but with approximately half of the cross member still precariously balanced on one of the legs of the torii.

Two young children and their mother, participants in the dedication of a new bell at Dr. Takeshima's father's temple in Saijo.

New Year's day in Kure, with families on their way to or coming from a temple.

A village scene on the island of Ninoshima in Hiroshima's harbor.

A social evening with Hermann J. Muller and senior members of the Japanese National Institute of Genetics. In the front row are Drs. Muller, Oguma, Komai, and Tanaka. In the second row are myself and Drs. Shinoto and Kihara.

The Fujiya Hotel at Miyanoshita as it appeared in 1951 when Muller visited Japan.

Hiroshima in the winter of 1954, as seen from Hijiyama. The view is northward.

The entrance to the clinical facilities of the Atomic Bomb Casualty Commission on Hijiyama at the time of the visit of Princesses Suga and Yoshi. The princesses were leaving the building on the way to their limousine.

The memorial cenotaph as it was in the winter of 1954 not long after its erection, before the beautification of the surrounding area that has since occurred.

The living room of our home at 93 Fufugawa in Nagasaki in 1960.

The buildings of Nagasaki University Medical School as they were in 1950.

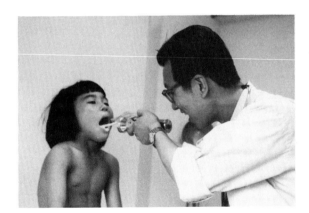

An examination of one of the little girls enrolled in the
Child Health Survey.

Dejima as represented pictorially by a Japanese woodblock artist. [This reproduction of the pre-Meiji printing was made from the original blocks, located in the Nagasaki Museum.]

阿蘭陀人圖

A Dutch merchant and his Indonesian servant as pictured by a Japanese woodblock artist. Note the pocket watch he holds and the violin in the hands of his servant. [This reproduction of the pre-Meiji printing was made from the original blocks, located in the Nagasaki Museum.]

One of the entrances to the Urakami Cathedral as it appeared in 1950.

A farm courtyard on the island of Kuroshima in 1960. Note the thatched roof and the diakon drying in the sun.

The Kuroshima parish church as seen from across the valley it faces. The farm woman is carrying nightsoil to the fields.

A Buddhist temple and the Catholic church in Hirado in 1964.

II

JAPAN'S
RESURGENCE

5

NEW AGENDAS

*J*apan in January 1954 had changed. Its resurgence was abloom, if not in full flower, and only a year later, in 1955, a little heralded event, but one with huge portent for the United States, would occur. The Toyota Motor Corporation would introduce its Toyopet Crown, which, along with such subsequent diminutive sedans as the Publica and Corolla, would initiate a revolution in transportation, placing in the reach of a growing number of Japanese, and persons elsewhere, the prospect of a car of their own. Upon the demand for these cars and the innovative assembly procedures that produced them a huge corporation was built, a challenge to the stodgy management in the American automotive industry.

The years that had intervened since my last visit had also seen the development and emergence of the Ministry of International Trade and Industry (the all-powerful MITI). The fostering of links between business and this agency, a ubiquitous, paternalistic organization, has done much to orchestrate Japan's resurgence and, in the process, has erected many barriers to a truly competitive economy. Slowly but inexorably the average Japanese was educated to a pricing structure in which the cost of articles could only rise, never fall, whatever the value of the yen or the state of world markets. American business and governmental policy contributed, at least indirectly, to these marketing inequities through a failure to recognize their inherent faults. A laissez-faire attitude prevailed, because Japanese industry then represented a small fraction of the world's total productivity, one which our ostensibly enlightened merchants could ignore.

With these impressive changes, however, the disturbing self-centeredness that had alienated the Japanese from the other peoples in the Orient resurfaced. One had to wander only a block off the Ginza

to be surrounded by signs that read "Japanese Only" or, more bla-
tantly, "No Foreigners" on many of the bars and entertainment centers that lend color to this quarter of Tokyo. Too many individuals
attributed their own improving economic circumstances solely to Japanese drive and industriousness and overlooked or minimized the
countless ways in which the Occupation had opened opportunities for
recovery. Perhaps most important had been our government's decision
to seek allies in the war in Korea wherever they could be found. The
purchase of Japanese goods and services during this conflict were an
enormous boon to the Japanese economy. Two other important lega-
cies were the Occupation's efforts to dismantle the *zaibatsu*, the business combines, and to dissolve the old landlord-tenant system through
land reform. Both helped to put the Japanese on a more equal footing,
at least temporarily. Everyone had to begin again, as it were, and this
contributed to breaking down the very strong class distinctions that
had previously existed. Although the old-boy networks have been re-
vived in Japan and possibly never actually ceased to exist, out of these
developments has grown a greater awareness of human rights in the
workplace, at home, and in society generally.

My purpose in Japan at this time was to close the clinical phase of
the genetics studies, an action prompted by the unanimous recommendation of a conference of geneticists and acccepted by the National Academy of Science's Committee on Atomic Casualties, which
set policy for the Atomic Bomb Casualty Commission. I had reached
Japan and Hiroshima shortly after the New Year via Los Angeles,
Honolulu, and Wake Island. We had left Los Angeles on a Sunday
afternoon around one o'clock, and as our Stratocruiser climbed to its
cruising altitude, the stewardess served champagne and orange juice.
Flying still had some graciousness about it then. There was only one
class—first—and seats were spaced sufficiently far apart that each
could recline much farther than is now possible. It was just as well, for
at three hundred or so miles an hour, a Boeing Stratocruiser did not
gobble up distance. Some of the tedium could be offset by walking
down the circular stairs to the small bar in the belly of the plane; this
little area, which could accommodate eight or ten of the ninety or so
passengers a Stratocruiser could carry, attracted the more gregarious
travelers. Most were business people, diplomats, or missionaries on
their way to the Orient. The flight itself, Pan American's Flight 2,
would circumnavigate the globe, terminating eventually in New York

City—some twenty thousand miles and more than one hundred hours of flying later.

After our stop in Honolulu, we rose skyward again enroute to Wake, the atoll that had fallen to the Japanese early after the war's commencement. It was awesome in a way to drone steadily westward, hour after hour, without sight of anything but water. There was a sense of the vastness of the ocean that the jets have robbed—six miles or more above the waves and at speeds of six hundred miles an hour, the enormousness of the Pacific dims. It was no less impressive to realize that our target occupied only a few square miles of this huge ocean. The smallest navigational error would prove fatal to all aboard.

Shortly after leaving Wake, the magneto on one of the inboard engines failed. I watched hypnotically as the propeller slowed and came to a stop. It was unnerving to see the motionless blade. Since we had not reached the midpoint on our way to Tokyo, the captain announced that we would return to Wake. The flight back seemed endless; our air speed was less than it had been, and apprehension drew out each minute. Conversation was muted; everyone seemed absorbed in his or her own thoughts. Finally, Wake appeared beneath us and off to the right, a reef-embraced atoll surrounded by water so clear the ocean's bottom could be seen. Tension grew less palpable.

Landing proved more complicated than we assumed, for we were much above the weight the Federal Aviation Authority considered safe. Fuel had to be dumped, a potentially dangerous undertaking. We flew on beyond Wake some distance and turned back toward the atoll into the wind. I could see several tube-like objects appear on or near the trailing edge of the wing to my left. We had been cautioned not to smoke, not to turn on or off an overhead light, and not to use the call button. Warning was heaped upon warning. An open flame or a spark induced through the opening or closing of an electrical circuit could be catastrophic. It would take about fifteen minutes or so to rid ourselves of the excess fuel, and once divestiture began our pilot would no longer be in communication with us.

I don't know what I expected—maybe something like simply turning on a faucet. This wasn't what happened. The moment we began to dump fuel the edge of the wing was wreathed in a watery, fog-like cloud that projected tailward some distance. Rather than a perceptible stream, the fuel rapidly aerosoled as it emerged from the wing tanks.

Presumably this was the consequence of our forward motion, the volatility of high octane gasoline, the temperature, and the lessened air pressure at the elevation at which we were flying. At last, it was over and we were into the final approach. Gooney birds, dressed in black, gray, and white, accompanied us down and onto the runway—graceful in the air, they are incredibly inept when they land—tumbling head over tail, veering to left and right as they attempt to put their feet down and come to a halt. We were on the ground not more than three quarters of an hour. A fifteen-cent part had cost us several hours and Pan American Airways a tidy bill in wasted fuel. Although the remainder of the flight was uneventful, it was less jovial; our confidence had been shaken, and the frailty of our craft still drummed through our minds.

Although the Commission continued to maintain an office in Tokyo and each arriving consultant was met, this trip to Haneda on the fourth of January was different from my first in a number of respects. It was drizzling on our arrival, and rain fell intermittently through the first week of my return. Arrangements had been made for me to be at the Imperial Hotel. Perhaps it was annoyance at my inability, because of my rank, to stay there on earlier occasions that made me want to be there now, but I would rather think that it was the intriguing history of the building and its architectural conception. It had, after all, been one of the few major buildings to withstand the 1923 earthquake that devastated Tokyo. At that time Tokyo had few structures that would have satisfied construction codes in any American or European city. Fire leveled more of the city than the earth's turmoil.

One approached the hotel from the west; taxis drove through an overhanging supported portico. Inside, the space was open yet limited. Porous laval rock mixed with brick provided a textural quality that was warming, inviting to the touch, and structurally effective. Individual rooms were another matter. Save possibly for the most expensive, which I never saw, they were small and disappointing. If the intent was to encourage the resident to be elsewhere as much of the day as possible, Frank Lloyd Wright had succeeded admirably. These were functional rooms; they provided a comfortable bed and adequate space but little more, certainly no enticement to dawdle the day away.

The train trip to Hiroshima was clouded with apprehension. The thought of terminating the employment of old friends weighed heavily, and it bothered me to contribute to greater difficulties for them. I knew that I would stay in the city rather than Hiro or Kure, for the Commission's quarters on Hijiyama for consultants and unmarried personnel had opened late in the spring of 1953, after the inevitable delays. The building itself is a rectangular, prize-winning structure—designed by Maekawa Kunio to offer as much flexibility in use as could be foreseen—that discreetly follows the hill-top terrain. The entrance area is dominated by a huge, circular fireplace that has never drawn well and a gracefully curved suspended staircase rising to the second floor. It contains five apartments of two or three bedrooms and a dozen or so single or double rooms, some separated by accordion doors so that they can be used jointly. It is simply but adequately furnished. Each room, whether northward or southward facing, has huge windows that rise from thirty inches or so above the floor to virtually the ceiling; these contribute to an airy, open feeling and marvelous views of the city beneath. A price is paid for these luxuries. The northward facing rooms groan under the winter winds and are often so cold that the occupants speak with frosted breath, whereas the southward ones are insufferably hot in summer.

As a result of the peace treaty, the Commission's status had changed. It was no longer attached to a military occupation; through a diplomatic "note verbale" between Japan and the United States on October 22, 1952, it had become an agency of the United States attached to the Embassy in Tokyo. It thus retained a fair degree of autonomy of operation and was freed from the legal and accounting restrictions governing most Japanese research institutions. Individual American employees were issued governmental identification documents that made it possible to use the facilities of the military bases, but we did not enjoy the diplomatic status of the regular members of the Embassy nor did we travel with diplomatic passports. Salaries were paid in dollars, rather than yen, and housing for the married staff was arranged through U.S. military housing offices. Our status was somewhere between that of the military forces assigned to the defense of Japan and members of the diplomatic corps. We were still issued special rather than standard passports, and these, like diplomatic ones, are void once one's occupational status changes. They remained valid only so long as we were Commission employees. How enthusiastically the Japanese government accepted this arrangement, although diplomatically endorsing it, is debatable. It certainly could be seen as an infringement on national sovereignty. Undoubtedly they felt ill-positioned to refuse; the stalemate continued in Korea and the purchase of supplies and equipment in Japan in the prosecution of the fighting had done much to stimulate a damaged economy.

Other less obvious changes in the Commission were soon apparent. Most of the employees, the medical professionals excepted, now belonged to a union, one affiliated with Sōhyō, a national confederation of unions. There were undoubtedly grievances, some real, some fancied, that had precipitated this, but to some extent it was merely the tenor of the times. Considerable pressure had been exerted, apparently, to get everyone to join, but the physicians were persuaded that unionization was neither in their best interests nor compatible with their view of their profession. Retrospectively, this may have been unfortunate, for as a consequence the union's officers have necessarily been drawn from the Commission's nonprofessional staff—the clerks, drivers, guards, janitorial workers, or technicians. The section chiefs and higher administrative personnel were specifically excepted. Women, while dues-paying members of the union, have had a relatively minor role; some, such as the clerically trained, have performed similar functions for the union, but usually they have been called upon only at the annual displays of solidarity—May Day or in support of a socialist candidate for the Diet. While often not willing volunteers, they are drafted nonetheless to march, sometimes tearfully, wearing the headband symbolic of the protester, the *hachimaki,* emblazoned with perfunctory political slogans.

The union's policies at the Commission have been schizophrenic. When endorsing the leftist political stance of the national confederation, the motto is "Yankee go home," but when bargaining, the position has commonly been to demand greater job security. How these incompatible aspirations were to be achieved was never discussed, nor apparently was the inherent contradiction in the two positions clear to the union's officers. Bargaining sessions were tedious and frequently characterized by snide remarks. As would be expected, the complexion of the union changed with its leadership; it was sometimes radical, oftentimes not. The agenda usually focused on pay raises, fringe benefits, holidays, and working conditions, rarely on such issues as equal opportunities for women. Indeed, while ostensibly espousing equality, the union tended to support the status quo in this regard. However, equality in Japan has a special ring; it means that, irrespective of abilities, motivation, and the like, no one is to be different. It is as though the purpose of a foot race was to see that all runners crossed the finish line together. Merit increases were anathema, therefore, for they smacked of a difference incompatible with this special notion of equality. Changes in salary were either mandated by a percentile formula or accompanied a change in job title. Commonly, too, the leaders advocated the retraining of personnel only tardily and failed to see the need to restructure the institution to adapt to the changing face of science.

Fortunately, over the years only one strike, lasting several weeks, occurred, but as a result of the threat of physical violence and the ugly, intemperate notes plastered over the buildings, it left a schism between the American and Japanese staff that was slow to heal. While it may have hastened the more equitable binationalization of the institution, I doubt that the workers actually profited in the long run, for as time passed and the Ministry of Health and Welfare came to play a progressively larger administrative role, the bargaining freedom of the union was curtailed. Now wages and ancillary perquisites are tied to the scales established for national workers generally, in which the local unions have little voice. Even the option to strike has been compromised, for it would take on an unpatriotic air, weakening the union's position within the communities the present binational Foundation serves.

Duncan McDonald, who had succeeded me as head of the Genetics Program, and Mick Rappaport, now one of the Commission's business administrators, met me at the railway station and took me to the new quarters. Duncan was soon to leave. Once he learned of the purpose of my coming and expected date of arrival, he had made his travel arrangements. Mick was an Australian national, although originally a Polish refugee to that country, of limitless tolerance and quiet certitude. An affable, balding engineer, he spoke and understood Japanese well and so by virtue of language competence and position became a member of the Commission's negotiating team with the union. He was constantly tested in this role, for commonly the disparaging remarks were directed toward him, since he was the only one of the Commission's representatives—aside from the official interpreter, who could be counted upon not to translate them—who sensed their derogatory nature.

Most of the other members of the American staff had changed. Each new cast of players sought to learn its entrances and exits, some successfully, some not. One of these was Earle Reynolds, a physical anthropologist interested in growth and development, who gained fame, or notoriety at least, at a later date when he attempted to sail his boat into the weapons-testing area in the Pacific. In 1954, however, he was inadvertently sowing bitter seeds we still harvest today. Largely at his instigation, the participants in the pediatric growth program were photographed in the nude in order to measure the possible effects of exposure on sexual maturation. Unlike statural or bodily growth and development, there are no simple quantitative measures of sexual maturity. Traditionally, it has been assessed by determining the age at which axillary and pubic hair appears, breast tissue starts to grow, menstruation begins, and the like. While he was at the Fels Institute, Earle had made an effort to develop a series of developmental landmarks that could be scored from photographs. American children and their parents were generally unconcerned about the use of this technique, but not so the Japanese, who objected to the photographs. Soon our contactors responsible for arranging appointments for the examinations brought word of increasing resistance to participation, particularly by teenaged girls. In time this aspect of the study was abandoned, but not before irreparable damage had been done. Even now, members of the Commission are occasionally pictured in the press as a group of aging voyeurs. Earle seemed oblivious to these problems.

Others of the staff made more profound impressions on me, such as William Moloney, who was head of the Department of Medicine and a Harvard hematologist. Bill is the prototypic Boston Irishman, ruddy of skin, characteristic in accent, effervescent, and perpetually charged with the excitement of science. Bill never shirked a battle and charmed, provoked, or simply challenged his staff to a pitch of biologic curiosity. Through his personal efforts and connections with pharmaceutical houses in the States, drugs that held some promise in the treatment of leukemia but were not available from Japanese sources were shipped from the United States. Under his guidance, the clinic was examining more than a thousand individuals a month.

Around Bill was a group of equally dedicated and concerned physicians. Robert Miller, a soft-spoken myopic physician with an incredible memory, directed the clinical pediatric activities and delighted us with his wry, sardonic sense of humor. Although not at the time an oncologist, he has since developed an enviable reputation in clinical cancer epidemiology, particularly as it relates to ionizing radiation.

Marvin Kastenbaum was possibly the most unlikely of the trio, a self-assured New Yorker and former muleskinner in the China-Burma-India theater who had tired of his job as a statistician with Dun and Bradstreet and sought challenges elsewhere. He brought to his job intelligence but a mind unfettered by a knowledge of biology—the kind of innate curiosity and skepticism one expects from a gifted university student.

Once or so a week, several of us, usually at Bill's instigation, would have lunch downtown rather than at the Commission's dining room, although the food there had improved immensely and was inexpensive. Our regular haunt was the Kirin Beer Hall, an unprepossessing structure at the east end of the Hondori, which had been roofed over to shelter shoppers in the years that had intervened since my last visit. Sukiyaki, prepared over gas burners, and a mug of draft beer were our usual fare. Four or five of us would gather about a single table as one of the waitresses initiated the cooking, added the appropriate amounts of salt, *mirin* (a sweet sake), and shōyu, and carefully ladled in chunks of sliced meat, vegetables, *konnyaku* (an almost transparent noodle), and bits of *takenoko* (the tender new shoots of bamboo). Raw eggs were broken into a bowl beside each of us and whipped until the yolk and white were well mixed. Into this one dipped the food drawn from the cooking pot. Inevitably the conversation turned to the studies and how these might be improved, or to a recent observation and its possible biological meaning. Bill's enthusiasm and dedication enveloped us like a reassuring intellectual blanket.

It was February before actual collection of the genetics data was suspended. A program of this scope cannot be dismantled overnight. New positions had to be found for the clerks and physicians employed in the examinations of the newborns. The registration offices had to be closed, incomplete observations completed if possible, and numerous courtesy calls made to the various agencies and persons in Hiroshima and Nagasaki who had contributed importantly to the study over the last six years. A plan of analysis of the data had to be formulated and instructions drafted for the numerous tabulations that were to occur. It was the policy then, as now, that all primary data sources were to remain in Japan. All of the information collected was on punched cards, and these had to be processed. Although our data-processing facilities were as good as any to be found in the Orient, it was still a tedious chore nonetheless. For these tasks, I had the help of

Koji and most of the same office crew who had worked with me before—Matsuda-san, Morita-san, and Sumitani-san. Alice Iwamoto had returned to the United States and been replaced by Alice Mukae; to distinguish them in conversation, one was known as Little Alice (Iwamoto) and the other as Big Alice (Mukae). Big Alice was a somewhat diffident but extremely conscientious secretary who managed the office duties with no apparent fuss and consoled herself in moments of stress with her violin.

Work went smoothly enough, but there were so many odds and ends that I had to keep a daily checklist, something I rarely do, lest one of the many tasks be overlooked. We worked from a suite of rooms on the second floor of one of the clinic buildings that had originally been designed to house small experimental animals but was never used as such. Our scurrying about was made more obvious by the untiled floors; each movement of a chair across the concrete floor scratched like chalk on a blackboard. Spirits remained high nevertheless, and everyone worked feverishly to accomplish the chores before us. All of the clerks were ultimately assigned to other departments, and the physicians who elected to remain with the Commission were transferred to the Department of Medicine, Pediatrics, or Pathology. The rest decided to open private practices; happily, all have done well.

These activities were briefly interrupted in early February by a binational meeting at the University of Tokyo of American and Japanese investigators interested in the assessment of the effects of exposure to the atomic bombings. It was not a large meeting—possibly twenty or so participants representing the various disciplines involved in the studies. John Morton, formerly Professor of Surgery at the University of Rochester, who had recently arrived to succeed Grant Taylor as the Director of the Commission, Bill Moloney, Wat Sutow, Earle Reynolds, Clara Margoles, Maki Hiroshi (the head of the Hiroshima branch of the Japanese National Institute of Health attached to ABCC), and I attended. Dr. Watanabe Susumu, a hematologist from Hiroshima University, was there, and Arthur Compton, the distinguished physicist, was also present as an observer. Earle and Wat described the pediatric studies, particularly the evidence for a growth retardation following exposure. Clara surveyed the pathological findings and the perplexing observations on kernicterus, a yellowing and degenerative lesion of the lenticular nucleus of the brain, that were emerging. Bill spoke about the recent studies on leukemia and the possible identification of a preleukemic state. The evidence was now compelling that the frequency of leukemia was higher among survivors, in rough proportion to dose. However, we still did not have estimates of individual doses and, of necessity, had to use the distance from the hypocenter of the survivors as a rough measure of exposure. Simple distances have their limitations, however, for they do not take into account the shielding that could have lessened the exposure to some degree nor do they allow for the attenuation of the radiation as it passes through the tissues of the body. These short-comings made it difficult to determine whether there was or was not a threshold, that is, a dose below which no effect of radiation occurred. Since it had been argued on theoretical grounds that a threshold was unlikely, proof of this conjecture was a matter of substantial concern in the regulation of exposures that might occur in the workplace or through accidents.

I presented the genetic studies: the preliminary analyses of the previous summer, the basis for the decision to terminate the clinical phase, and the progress that had been made with respect to the latter. Professor Komai was the Japanese discussant of my remarks, and after my presentation made a series of unsolicited but highly laudatory comments. He reiterated the importance of and need for a genetic study of the kind that we were completing, remarked on the thoughtful and dedicated management it had received, and asserted without qualification that a study of this scope would have been beyond the resources and experience of Japanese investigators. I was pleased, and not a little relieved. Dr. Komai had always been forthright, although he never failed to couch his remarks carefully and sensitively.

*   *   *

Shortly after I arrived in Hiroshima, our first hydrogen weapon was detonated over Bikini atoll. "Bravo," one bomb in this series of tests, had a yield of some fifteen megatons, a thousand times that of the uranium weapon exploded over Hiroshima. Its force exceeded expectations, and it littered the surrounding seas and islands with fallout. Highhandedly, at least in the eyes of many fishing nations in the Orient, we had, prior to the detonation, peremptorily excluded their vessels from some particularly productive fishing waters. Moreover, we had moved in a manner that did not foreclose inadvertent penetration of the prohibited area, and one ill-fortuned ship, the *Fukuryu Maru*

(Lucky Dragon), found herself fishing in the prohibited zone when Bravo was detonated.

Japanese newspapers have rarely distinguished themselves by the objectivity of their reporting. There are exceptions, such as the *Sankei Jiji Shimbun,* which has consistently tried to appeal to reason rather than to inflame emotions. However, the larger papers—*Mainichi, Asahi,* and *Yomiuri*—are locked in a perpetual struggle to outsell one another, and sensationalism is courted for the papers it sells. Bikini and the *Fukuryu Maru* incident offered unlimited opportunities to fan passions and to dissemble. Charges were thrown, to be rebuffed by countercharges. Should the *Fukuryu Maru* have been where it was? Was not this restriction of international waters illegal? Was the crew indeed sick, and, if so, were their illnesses due to exposure to fallout from the test? Were the fish that this ship carried, and others were to carry, safe to eat? How was one to know? And if the fish were contaminated, what was to become of those persons who ate them? Headlines magnified the sensational. One stated, "Fish eater's urine radioactive." A careful reader would learn that the urine of individuals who were *thought* to have eaten tuna that *may* have been contaminated was more radioactive than background soil. It wasn't stated that this is to be anticipated even in the absence of contamination; it stems from the normal concentration in urine of a naturally occurring, widely distributed radioactive isotope of potassium, known as potassium-40. This concentration has occurred for millennia in unnumbered areas of the world, wherever this isotope is found.

Over this politically and emotionally charged atmosphere presided Tsuzuki Masao, the head of Japan's nascent Institute of Radiological Sciences and former Admiral in the Imperial Naval Medical Corps, whom the Occupation had purged as it had everyone of similar rank. Although he had certainly been extremely helpful when the Genetics Program was established, there always seemed to me to be an ambivalence in his attitude toward us. This was to become clearer in the days ahead. As a result of his position and eminence, supervision of the care of the *Fukuryu Maru*'s crew fell to him and the members of his new institute.

Soon "experts" were falling all over one another, often with more embarrassment to others than themselves. Some of our staff were urged to go to Yaizu, a heretofore little known port much enjoyed by a nineteenth-century American expatriate, Lafcadio Hearn, but now, as a result of the Bikini test, a renewed focus of attention. It was to this port that the *Fukuryu Maru* returned; its proximity to Tokyo, and the national press, were presumably accidental. After cooling their heels for days, our staff members were invited to examine the crew. Only one individual was conspicuously ill, the radio operator, Kuboyama-san, who (we were to learn later) had chronic liver disease, probably of viral origin. It is debatable whether, in addition to liver disease, he did or did not have a radiation-induced illness, but the circumstances of exposure were certain to have exacerbated his prior medical problems. He subsequently died; in Japan, he is commonly thought to be a "radiation victim," but the more likely cause of his death was infectious hepatitis.

This incident illustrated to me how little the average man or woman knows of the circumstances that surround the events the press inaccurately or incompletely reports and uses to arouse emotions. None of us is immune to this appeal to prejudice. The confrontation between an independent Japan and a tolerant, but possibly insensitive, big brother, the United States, was inherently too politically rewarding to ignore. Japan's newspapers rose to the occasion. One happening will serve to illustrate the distorted perspective they fostered.

When it became apparent that the fishing trawlers returning from the area where the detonation had occurred might bring radioactively contaminated fish to their markets, the Japanese government instituted a surveillance program. The large fishing fleets, such as Taiyo, sought tuna and other similarly sized fish throughout the Pacific and returned to their fleet centers with their catches. This meant that a relatively small number of monitoring stations were needed. Shortly after surveillance began, or about that time, Merrill Eisenbud, Director of the Atomic Energy Commission's Health and Safety Laboratory and Manager of its New York operations office, came to Japan. His purposes were many, but they revolved largely around the *Fukuryu Maru* and its crew. His arrival was prominently covered in the press. At early interviews, in cautiously guarded remarks, he suggested that the risk was probably minimal. Soon thereafter, he was invited to one of the monitoring stations to see what was occurring and to comment on the procedure that was being used to assess contamination. He noted that the inspectors were merely passing the radiation counter's probe over the surface of the fish and stated that, in his opinion, this was not adequate. It would not necessarily reveal whether the fish had ingested radioactively contaminated foodstuffs. He suggested that it would be better to insert the probe into the cloaca, the common excretory chamber found in fishes. Unknown to him, the demonstration and his responses occurred on a boat whose catch of tuna was destined for the United States. The press trumpeted that this expert, who was assuring them there was little or no risk to the Japanese, was insisting on more rigid standards of surveillance for American-bound tuna.

Many other visitors, some humble and others with global reputations, still paraded through the Commission's facilities and spoke with its staff. Pearl Buck, Eleanor Roosevelt, and Norman Cousins, as well as Edwin Reischauer and Alexis Johnson, our ambassadors to Japan on different occasions, and Senator Edward Kennedy, came to listen, but then seemed to say something else. Were we unclear? Did we fail to present the emerging scientific facts comprehensibly? Scientists are not always the best communicators, for they frequently presuppose a knowledge or objectivity that does not exist and commonly qualify their remarks to the point where the main thread is lost. Or did some of these people not want to hear? Did their preconceptions or political positions interfere? Eleanor Roosevelt, for one, seemed to expect us to voice strong moral objections to our government's decision to have used atomic weapons. She heatedly reiterated the notion that the survivors were just guinea pigs, victims of our scientific curiosity. Although we had heard this charge raised by the Japanese press, it was startling to hear it from a woman of her stature and intelligence. Whatever our individual thoughts were about the morality of the use of nuclear weapons in the war, we were not openly concerned with the political implications of our findings. Our function, as we saw it, was not to publicize our individual views but to assess the biological damage that had accrued and to report it in a straightforward manner.

Numerous Japanese luminaries also visited. Yamaguchi Yoshiko (Shirley), accompanied by Isamu Noguchi, her husband at that time, later became a member of the Diet, but when she visited us she was famous as a one-time movie star and singer.

Mid-March arrived and I was still desperately trying to achieve all of the aims that I had set for myself before my scheduled departure later in the month. A respite from this madness came with the arrival of yet another group of Washington dignitaries. Our visitors were Senator Joseph Pastore and Representative Chet Holifield, the leaders of the Joint Congressional Committee on Atomic Energy. Together they provided strong, intelligent direction to congressional debate on issues surrounding the use of nuclear energy, and they supported the Commission's studies. Both had been at the weapons tests in the South Pacific and were returning through Japan so that they might familiarize themselves with another important program supported by the Atomic Energy Commission, over which they had oversight responsibilities. They toured our facilities, asked perceptive questions about the programmatic aims, community acceptance, and our future directions.

Afterwards a small cocktail party was given in their honor at Hijiyama Hall. In the course of this, it was learned that the day, March 17, was Senator Pastore's birthday, and this, too, we then celebrated. I didn't have the temerity to point out that it was also my birthday. I chose instead to enjoy the occasion vicariously, and to chat with the two congressional members, both exceptionally able men, intellectually and politically.

As each successive day passed, I avidly scanned the newspapers for possible airline disasters. It was my intention to fly home through Asia, the Middle East, and Europe. Several months before I had departed from the United States on a round-the-world ticket, British Overseas Air Corporation (BOAC) had introduced its new jet airliner, the Comet. Since it was far faster than propeller craft, I had booked myself on it from Tokyo to Manila and ultimately on to Dum Dum Airport in Calcutta, where I was to stay for several days. Not long after the Comet was introduced, it was involved in a series of tragic accidents. First one in Rome, then another in Calcutta, and still a third, if memory serves properly, all scarce weeks apart. With each of these, apprehension coursed through my breast. Should I change my reservation to a slower but safer route? Should I assume that the causes of these tragedies would be discovered and rectified before I was to leave? Would I seem too faint-hearted if I changed? Time grew shorter. Finally, unable to determine the causes, BOAC withdrew all of the Comets, never to fly them again. My problem was solved, and when at last I left Japan it was, again, on one of Pan American's Stratocruisers to Hong Kong, Bangkok, Rangoon, and Calcutta.

Almost four more years elapsed before commercial jet travel truly arrived in the form of the Boeing 707, undoubtedly a safer plane as a result of the Comet's tragedies. It was ironic that the British effort failed, for the jet engine that has made possible less expensive and more rapid air travel was their innovation. Britain's loss was never recouped, and as a consequence, American-made craft command the skies. In later years, in a similar fashion, the Japanese would borrow the innovations of American technology, and, through perspicacity and careful market research, displace American entrepreneurs in markets ranging from vacuum cleaners and stereo equipment to automobiles and supercomputers.

*   *   *

International congresses of genetics normally occur every five years. Their rhythm was interrupted, however, by World War II, and the first subsequent congress took place in 1948 in Stockholm. These gatherings are awarded by the International Union of Biological Sciences, an organization to which Japan has belonged since 1923. At the Twelfth General Assembly of the International Union in Nice in 1953, it had been agreed that a symposium on genetics should be organized in Japan, and this decision was confirmed shortly thereafter. The designation of a recipient nation is a measure of peer approval, and most countries extend themselves to assure that the meetings will be intellectually and socially rewarding. Because this was the first major international conference to be held in Japan since the war, the Japanese had an added incentive to excel, and preparations for it became a matter of national pride.

When Jim Neel and I, who were at that time members of the University of Michigan's Department of Human Genetics, learned of the proposed symposium in Japan, we faced a quandary: the dates of the Japanese meeting conflicted with those of the first International Congress of Human Genetics, to be held in Copenhagen. We had just finished our definitive analysis of the data that had accumulated in Hiroshima and Nagasaki in the years of the clinical program. This information was soon to appear as a monograph tediously but accurately entitled *The Effect of Exposure to the Atomic Bomb on Pregnancy Termination in Hiroshima and Nagasaki,* and we agreed that the results should be presented in summary form at both conferences. Preliminary reports had appeared in the English and Japanese literature as quickly as was feasible, once the clinical program had ceased, and prior to their publication we had fretted over the inferences to be drawn from the data and how to present these in a manner that would claim neither more nor less than the evidence warranted. Naturally, we had differences of opinion as to how to achieve this. We had made a number of different measurements, each presumably with its own response to mutational damage. How were they to be combined into one comprehensive statement? Moreover, since we could show no clear genetic changes in offspring whose parents had been exposed to radiation, what could be said about the range of possible outcomes that the data could exclude? How were these matters to be resolved? As scientists, we are accustomed to uncertainty: we accept differences in perspective, based upon disparate views of data that may be inconclusive. But many nonscientists become confused in the face of equivocal answers, and we particularly wanted to avoid fostering further uncertainty in the mind of the lay public.

The immediate problem of who would present and where was quickly resolved: Jim would go to Copenhagen and I to Japan. Accordingly, I was again enroute to Japan in the early summer of 1956. One of my charges on this occasion was to analyze the additional observations on the frequency of male births that had accumulated since the closure of the clinical program in early 1954. Two more years of data were available, and we hoped that this added information would serve to clarify the earlier ambiguities.

The train trip from Tokyo to Hiroshima was cleaner than on earlier occasions, for the main line had been electrified further and the rolling stock was newer and faster. I spent as little time as I could in Tokyo and was met at the central station in Hiroshima by colleagues who knew of my impending arrival. As we drove to Hijiyama Hall, I was told of the major happenings in the last two years—the most important being the visit of the Francis Committee, a group charged with a reappraisal of the Commission's studies.[1] This committee, while lauding what had been achieved, nevertheless recommended a fairly drastic change in the overall organization of the research program—a movement away from the separate studies, often based on different, opportunistically gathered samples of survivors, that had heretofore characterized the research to a unified surveillance program using fixed samples of survivors. I was also apprised of the poor morale that existed among the Commission's American professionals. They felt misunderstood by the National Research Council, and they fretted under a director, Robert Holmes, whose methods and manners some of them did not like. Whether their concerns were justified is a matter of opinion, but clearly there had been a loss of direction and sense of purpose at the Commission. No new radiation effect had emerged in five years, for radiation cataracts, the increase in leukemia, and even mental retardation among the prenatally exposed had been identified by 1951. Further refinements in the distance–effect relationships had occurred, and there had been some sharpening of the study samples, but these developments were not enough to sustain scientific interest. Shortly before I left the United States, I had been told of these discontents and assigned the task of determining whether there was a substantive basis for the concerns of the staff—an onerous job, given the shortness of my anticipated stay and the specific scientific charges I carried.

Within days of my arrival I had settled into an office and initiated the tabulating requests that would provide summaries of the data I was to analyze. Once these tabulations were completed, the tempo of work accelerated. But the days and evenings were hot and sultry, and a cool spot often hard to find. Frequently I worked bathed in my own perspiration, a sodden shirt growing progressively more clammy, trying to ponder my way through a set of seemingly illusive results. As frustration mounted and the regular fare in the Commission's dining room grew tiresome, I would prevail upon one of the other bachelors to escape downtown to a movie and dinner. Several of the larger, better theaters showed American movies with Japanese subtitles, and there was always the small neighborhood movie house, with its steady fare of distinctly grade C samurai films (*chanbara*), full of swordplay and impossible feats of heroism. Wherever we went, the main feature was always preceded by the week's newsreel, generally a collection of happenings in Tokyo or elsewhere in the world. It was of interest primarily for the commentary it afforded on what the Japanese news media felt was important. The impending Genetics Congress received its share of attention, as did the more extreme elements in the student political movement.

Often, if the movie did not last long, we would stop at a bar or cabaret. These had such unusual names as The Pit, The V (pronounced "bwee" by the Japanese, for there is no v-sound in the language), and The Loup. An old stalwart was the Grand Palace, long misidentified on its sign and by its bevy of bosomy young hostesses as the Gland Palace. Each bar or cabaret tended to have its own clientele. We frequented The V or The Pit, which was owned and operated by a

professor at Hiroshima University. Most of us drank beer, usually locally brewed, listened to the recorded music, and passed the time in
small talk with the hostesses and the proprietress, who was the professor's mistress. Like the managers of most such bars, she never volunteered her name but encouraged us and her other customers to address
her as "mama-san." She was an attractive, fine-featured woman with
lustrous eyes, who wore her small-patterned, often white silk kimonos
with exceptional grace and spoke a Japanese studded with more honorifics and more elegance than we encountered in the usual bar.
Overtly, we were as warmly greeted by the staff and mama-san as
other patrons when we came, but now and then a Japanese customer
would appear who looked upon us with askance, mumbled under his
breath or to his tablemates, and harangued the proprietress. He
seemed to imply that we were not welcome in places where he drank,
but since we were more numerous and spent substantially more than
the average patron—a nontrivial consideration in a culture that has
long placed business above moral values—he was hushed. But this
was always done in a manner that neither offended him nor jeopar-
dized our patronage. After all, who is to know what the future may
bring? Times may change, and indeed they have in ways the Japanese
of those years certainly never contemplated.

As summer progressed and the heat and humidity grew more oppressive, the urge to collect something, a particularly virulent malady
Japan prompts in me, surged upward like a rising barometer before
an approaching high pressure area. I had long been interested, pass-
ingly at least, in Japanese wood-block printing, but now the need to
know more seemed irresistible. Although Vicki and I, on our first visit,
had collected a few insipid prints of Paul Jacquollet, a long-since deceased French expatriate who had tried to combine the traditional
methods with the impressionist view of the world, and several inexpensive, poorly reproduced copies of some of the prints of Hiroshige
(his scenes along the Tokaido), the beauty and potential of this medium had slowly grown. Wood-block printing has an old history in
Japan. It was the medium in which books were printed prior to the
introduction of movable type and, from about the time of the introduction of Buddhism in the eighth century or so, a means to represent
religious art that was financially accessible to the average Japanese.
Virtually every serious Western student of art now has some familiarity with such names as Hokusai, Hiroshige, Sharaku, Utamaro, and

other greats of the seventeenth through the early nineteenth centuries, the Edo period, possibly the apogee of Japanese art. Astute observers of the foibles and excesses of their time, these artists immortalized the theater, the Yoshiwara (Tokyo's famed pleasure area), and the scenic beauty of the country before industrialization took its toll. Hokusai, with his Hogarthian view of the world, is especially appealing, and his famous *Sketchbook* is an incredibly vivid account of the unadorned commonplaces of life under the Tokugawa family's regime.

It was not these artists, however, who had aroused my curiosity but the generation of contemporary printers in the *hanga* movement. Seeking a revival of the form, they were experimenting with new techniques and new subjects; they deviated from the past not only in carving their own blocks but doing their own printing. Traditionally, a print was the culmination of the work of several individuals: the artist who made the original drawing, the carver who translated the drawing onto well-aged slabs of wood, generally cherry, and finally the printer. This convention of skills was restrictive, and the newer generation sought to remove these limitations. Each strove to do so in a slightly different way.

Munakata Shiko drew his inspiration from the simple, religiously intent style of early Buddhist art. He was a bushy haired, extremely talented man whose vision was so poor that as he carved his eyes were anchored within inches of the board on which he worked. This physical limitation notwithstanding, his chisels, driven by his intensity and creativity, scarred and gouged the wood as chips flew into his garments, his hair, and onto the floor. His themes were commonly religious, albeit often sensuously Indian, and his techniques unique. Frequently he formed his prints on thin paper onto the back of which he brushed watercolors which leaked through onto the printed image and gave it a three-dimensional quality. Generously endowed women, heads askew, flaunting their breasts and pubic hair, colored his image of the afterworld, as though sexuality were the distinguishing feature of our existence. His views of the 53 stations of the Tokaido, the route over which Japan's feudal lords passed in their periodic pilgrimages to Tokyo, clashed so markedly with those of Hiroshige as to be singularly arresting.

Watanabe Sadao, a contemporary of Munakata's, was a deeply religious man, a Catholic, and his themes were inevitably biblical, often drawn from the Old Testament. He commonly used a particular Japanese *washi,* a handmade paper, with which he could achieve a mosaic-like effect and the almost three-dimensional perspective that characterized early European art. Patently, he had been much affected by the emotional depth of Roman and Byzantine mosaics and presumably the early illustrated religious manuscripts that achieve dimensionality through texture or the sheer power of the imagery, without the technical artifices of Renaissance and post-Renaissance Western European art. There is an ecumenical Buddhist quality to his prints that identifies his origins.

Saito Kiyoshi experimented continually, and the themes that attracted him were not conspicuously driven by a personal view of the world nor a conscious religiosity. He was an eclectic. Some of his prints were angular, geometric; others were warmly, emotionally subtle. He was one of the first, if not the first, to use plywood and to deliberately incorporate the strong texture of the grain of the wood into his printing process. He also combined oils with more conventional methods, as for example, in his somber dark print, "The Assumption," an outgrowth of a visit to Nagasaki. His choice of subjects has varied constantly: scenes at the University of Michigan, Christian religous topics, and the architecturally wondrous buildings that grew out of the politically stultifying but artistically creative period when the Tokugawa ruled.

There were many others: Azechi, with his simple evocative figures of northern Japan; Hatsuyama, whose whimsical animals that adorn children's books and prints such as "Naite iru ki" (The Crying Tree) or the ostrich-like character he labeled "The Sightseer" are immediately intelligible to the young but not to their parents; Sekino Junichiro, fascinated by sumo wrestlers, castles, and children; Mori Yoshitoshi, whose treatment of feudal themes bedazzle with their strong colors and larger-than life imagery; and Onchi, one of the more difficult to characterize founders of the movement itself. His print of the poet Hagiwara captures the man so forcefully one feels obliged to straighten the disheveled hair and to sympathize with a life so indelibly etched in his face. Generally, however, Onchi's prints, particularly the later ones, eschew representation; they are abstractions in which design, composition, and the use of space take precedence over the pictorial.

Hiroshima had few galleries, and virtually no place that one could find copies of the works of these men. Many weekends found me in

Kyoto at the Red Lantern in Shinmonzen or, if circumstances permitted, at Yoseido in Tokyo. Most editions of these prints were small, for money had yet to compromise artistic ends; Hatsuyama's were often ten copies or less, but those of some of the others—Sekino or Saito, who were possibly quicker to see their economic potential—ran to a hundred or so. Each print carried the size and the number of the specific print in the edition. One intuitively presumed that there were differences between differently numbered prints, but what algorithm should one use in deciding which print to purchase, when differences in quality were so subtle? I soon learned that with the larger printings, one sought, if possible, either the first copy or one between the 10th and the 20th or so. The first copy, although often not the best, had an intrinsic appeal, much like a first-edition book, but in the second instance, the argument was more rational. Since the artist did his own printing there was necessarily some experimenting with the blocks and inks to ascertain how the specific effect he sought could be achieved, but once this was done the challenge was gone, interest waned and printing became perfunctory. Individual prints were inexpensive, a few thousand yen at a time when there were 360 yen to the dollar, and consequently one could afford to buy the work of established as well as unknown but appealing artists, and indulge one's taste, however eccentric it might be. Now, unless one is drawn to young but promising artists and trusts to inherent judgment, print collecting can be an expensive pastime. Prints of Mori, Sekino, Saito, and many other nationally and internationally acclaimed names can easily cost a thousand dollars, and a print of Munakata's will command five to ten thousand.

*   *   *

August 6th, 1956. Strangely, prior to this date, I had never participated in nor even seen one of the memorial services to those who died in the atomic bombing of Hiroshima. In 1954 I had not been in Japan at the appropriate time. And in the earlier years, although Commission members were not specifically prohibited from attending, we were certainly not encouraged to do so. I had lived in either Kure or Nijimura, and since the Commission's usual transportation to the city was not running because of the national holiday, it was not easy to attend the ceremonies.

They were much simpler back then, carried out on a covered wooden stage with the gaunt dome of the Industrial and Commercial Exhibition Building as a backdrop. Participation was largely confined to the survivors (the *hibakusha*) and local citizens, with a few official participants from the Occupation; the national and international movement to ban atomic weapons was still aborning. By 1956, however, the memorial cenotaph and the atomic-bomb museum had been constructed, and the park in which they stand had come into existence. The cenotaph had been unveiled on August 6, 1952, at which time it contained the names of sixty thousand people who had died in or immediately subsequent to the bombing, and to this list tens of thousands more have since been added.

In September 1955 the organization known as Gensuikyō, or more completely as the Japan Council against Atomic and Hydrogen Bombs, was formed. Initially, it was a group of concerned citizens who were interested only in the peace movement and the welfare of the survivors themselves. Its existence undoubtedly had an impact on Japan's basic law concerning atomic energy, a law which states that Japanese nuclear research, development, and utilization will be for peaceful purposes only. Because the causes Gensuikyō espoused were newsworthy, organized political movements soon gained ascendancy in its management, and it drifted progressively leftwards. This drift apparently was not rapid enough to satisfy all of the organization's adherents. Soon, it splintered; the first group to break away, the Kakkin Kaigi, did so in 1961, but still another, the Gensuikin, broke lose in 1965. Eventually even the extreme right sought participation. This fractionation ill-served everyone; it precipitated arguments, confrontations, even fights.

The streets of Hiroshima on every August 6th became literally a battleground, with each faction attempting to hold its own marches and ceremonies without regard to the others; jostling and shoving became a common occurrence. Eventually, at the urging of survivors, who saw their efforts to remember their dead suborned by political organizations directed from Tokyo, the city authorities stepped in to restore some semblance of order and propriety. Today, on the occasion of the anniversary, the area in which the official ceremonies are held is surrounded with barricades, and only those with invitations are permitted to attend. The various organizations are still allowed to march, and they do so with enthusiasm, often beginning several days before the actual ceremony. The central area of the city witnesses a constant parade of red banners and hears echoes to the sound of loudspeakers protesting this or that. These activities are carefully monitored, however, and nastiness is largely avoided. Because these movements have been unable to speak collectively, they have not been able to recapture the initiative they once had in the 1950s and 1960s.

As I walked down the hill from our quarters to the Peace Park on that early morning in 1956, the Peace Boulevard was filled with people moving steadily toward the park, where the city had begun to collect the various memorials that had been previously set in many places about the community. One of these has seemed to me especially poignant. When I first saw it in 1950, it stood before the Funairi Girls School, a memorial to classmates who had been on work details near the present park and had lost their lives. On its face is carved the images of three young girls in student uniforms. The leftmost one is holding a garland of flowers in her right hand, the rightmost one a dove in her left hand, and the central winged figure, on whose head the other two are placing a coronet of flowers, holds a book, across whose front is carved the equation $E = mc^2$, the fundamental energy relationship discovered by Einstein which states that the energy in a body is its mass times the square of the speed of light. On the back of the memorial, in cursively carved form, are the dedicatory remarks to the effect that those who died will never be forgotten. Surely no stonesmith, however skillful, would have thought of so subtle a representation of the intellectual feats that culminated in an atomic weapon; I have been told, but never sought to verify, that the suggestion came from one of the teachers at the school.

The ceremony itself, which was not exceptionally long, was moving and thought provoking. The typical eulogies were delivered by notables, but it was not these that made the occasion solemn and memorable. It was the tear-streaked faces upon which personal tragedies were etched. Eventually I focused on one elderly woman, neatly but somewhat poorly dressed, whose hands steadily fingered her Buddhist prayer beads, as her lips silently articulated the prayers. For whom did she pray—a husband, a son or daughter, an aged parent, or all of those whose lives were snuffed prematurely? At the propitious moment, temple bells began to peal and doves symbolic of the aspirations for peace were released to fly in ever-widening circles above the people gathered near the cenotaph. Slowly the crowd dissolved, and I walked back to Hijiyama ill-at-ease with myself, disturbed by the events that had made such a ceremony necessary.

*   *   *

This was my first international meeting, and I had been asked to speak at a plenary session. As the time neared to go to Tokyo, I grew progressively more excited. My speech was written, but slides had to be prepared, and here the resources of the Commission came to my aid. The head of the graphics department was Geoffrey Day, another friend of long-standing. A Melbournian who had completed his training in graphics design only to be called to service, he had been a cartographer in the Australian Army, but this barely touched the depths of his skills. Geoff had that uncanny eye that recognizes instantly, or so it seems, how best to prepare graphical material. He had a memory for detail and a sense of color, of perspective, and workmanship that the true artist takes for granted. He sketched, photographed, and collected books, prints, stamps, and coins with an appetite that only the person attuned to the extraordinary can appreciate. Although he has lived in Japan for more than forty years, he still speaks little Japanese—a quirk perhaps of his cultural heritage and independence, certainly not capability. Nevertheless, his sense of the culture and its creativity knows no limitation. As a consequence, when I left for Tokyo, I had slides that I knew would be well received.

Reservations had been made for me at the Marunouchi Hotel. Although the selection was not as great as now, Japanese hotels, even in Tokyo, were still inexpensive. A single room at the Marunouchi was somewhat less than 2,500 yen a night, that is, about seven dollars, and even the Imperial Hotel was only slightly more costly. Unfortunately, the Marunouchi Hotel is close to the newspaper district, and once it was known that the first major presentation of the genetics data from Hiroshima and Nagasaki was to occur, the phone in my room never stopped ringing. Representatives of *Asahi, Mainichi,* and *Yomiuri* called and came, but the most pleasant and entertaining of the correspondents was the science reporter for the *Sankei Jiji Shimbun,* a gazette sometimes compared with the *Wall Street Journal.* Unlike the others, Niwa-san, for that was his name, had been trained as a biochemist, and had come to the newspaper from a minor academic position at the University of Tokyo. We spent many hours together, along with a representative of the University of Tokyo's student newspaper. Like many students, he was a doctrinaire leftist, possibly even a member of the Zengakuren, the most radical of the university student groups. Despite our ideological differences, we got along famously, for more than anything else he was an earnest, well-meaning young man.

Other members of the press and the newsreels constantly strove to provoke an incident or to suggest a confrontation between the Soviet delegation at the meeting and myself and Harold Plough, a distinguished professor at Amherst College who was then the ranking geneticist at the Atomic Energy Commission's Washington Office. Glushchenko, Kushner, Sukhov, and the other Russians and we shook hands together for numerous photographers. Against this background of journalism, it was with some trepidation that I rose to speak, for I was fearful that my remarks, however carefully couched, could and would be misrepresented. As indeed, they subsequently were. The Tokyo University Students' Press, at my request, had submitted the questions they wanted answered in writing, and I had in turn written my answers. However, when their article appeared on September 10, not only did it misinterpret my formal presentation before the symposium, but the questions they had submitted, and my answers, had been edited to the point of obscurity. This troubled me greatly, for I had enjoyed my associations with the young student representative. Had I misplaced my trust, or was he too the victim of an editorial office?

Attendance at the symposium read like a contemporary *Who's Who* in genetics. J. B. S. Haldane and C. H. Waddington were there from England, Hans Nachtsheim from Germany, Arne Muntzing, Erik Essen-Moller, and Øjvind Winge from the Scandinavian countries, and George Beadle, Ralph Brink, Ralph Cleland, Mrko Demerec, Harold Plough, Marcus Rhoades, Ledyard Stebbins, and Curt Stern from the United States; and of course the Russians—I. E. Glushchenko, H. F. Kushner, K. S. Sukhov, A. A. Imshenezkii and at least one more, I. N. Kiselev. The Soviet delegation tended to move about the meeting as a tight little group, cordial but reserved, and as a consequence we, or at least I, never reached a first-name relationship with any of them. Glushchenko, apparently a plant breeder, was a Ukrainian and generally wore Ukrainian shirts, with their delicate needlework later affected by Khruschchev. His shiny bald head made him conspicuous in any crowd, but so did his high cheekbones and glistening complexion, undoubtedly a heritage of the Scythians who peopled the steppes of southern Russia. Kushner, a gregarious, loud-voiced, bushy black-haired individual, appeared to have rediscovered the inheritance of environmentally acquired characteristics; his Lamarckian presentation on some immunological properties of the chicken he purported to have discovered was unintelligible to Western immunogeneticists. Sukhov, a specialist in pathogenic plant viruses, was far more retiring than either Glushchenko or Kushner; possibly he was the KGB agent assigned to forestall defections or embarrassing situations. All three were ostensibly members of the Institute of Genetics of the Academy of Sciences, as was Imshenezkii. Although the reinstitution of Western genetics in the Soviet Union had begun, all of these men were probably Lysenkoists at heart. As a young aspiring academic, I felt privileged to be included as a major speaker among a group of this international distinction, even when I wondered about the scientific credentials of some.

Arrivals were spread over days prior to the opening sessions, as attendees sought to combine some sightseeing with their participation in the meetings. Haldane, who upon his retirement from the University of London had accepted a position at the Indian Statistical Institute, arrived wrapped in the robes of an Indian guru. This startled his conservative Japanese hosts, and their discomfiture was further aggravated when he and his wife, Helen Spurway, insisted on stopping for a beer enroute to their accommodations. He made matters worse when, with little prior warning, he informed his hosts that he did not want to give the opening address for which he had been scheduled. George Beadle, a Nobel laureate-to-be, who was to have given the closing address, graciously stepped into the breech.

Haldane's colorful behavior (to put it in its best light) continued, for at the final session where he did speak, he appeared on the stage in guru costume. He had a disconcerting habit of wandering away from the microphone periodically to scribble something on a blackboard in the tight little illegible hand of his generation in England; each time he did this, it was impossible to hear him. Finally, after one such episode, from the rear of the audience a rasping voice shouted, "Wireless, wireless please. We can't hear you." We all turned to see who had mustered this courage—it was his wife.

As part of the social events that accompanied the symposium, all of the participants were invited to a special performance of kabuki at the Meijiza, the second of Tokyo's kabuki theaters. On the way to our seats, each of us was given the playbill in Japanese, but inserted into the booklet were several pages of English translations of the events that had preceded and would occur in the course of the scenes we were to see. Even in the abbreviated form in which it is now presented, a kabuki performance lasts four or five hours. There are generally two performances a day, one beginning usually at 11:30 and the other about 4:30. The midday bill is not the same as the evening one, and about mid-month the bills are reversed. Normally a playbill consists of four or five vignettes from as many different plays that may or may not be separated by comic interludes, called *kyōgen*. On this specific day, the performance consisted of scenes from *Kamakura Sandaiki*, a trilogy about the struggle for power between the Toyotomi and the Tokugawa families in the early seventeenth century. Originally written for the puppet theater by Chikamatsu, it contains one of the most difficult roles of the *ohime-sama* type, the high-born princess. I remember enjoying myself immensely, enthralled by the professionalism of the actors and the sheer theatricality of the staging and settings. Kabuki is an impressive visual art that can be enjoyed without comprehension of the language.

For decades now, it has flourished only in Osaka and Tokyo, although there have been brief seasons in Kyoto, and touring companies perform in the lesser cities such as Fukuoka and Hiroshima on occasion. My first encounter with kabuki had occurred in Osaka, which seems fitting, for this city is closer to Kyoto and its river, the Kamogawa, where ostensibly kabuki had its beginnings with the performances of the temple dancer, O'Kuni, in the last half of the sixteenth century. Moreover, the Osaka theater has had a style different from that in Tokyo. This is not to imply that it has been more faithful to the original, for kabuki has undergone many changes in the four centuries or so of its existence.

Originally, the troupes included men and women, but this ceased to be so in the first years of the Tokugawa shogunate. As a result of the immorality and license that characterized early kabuki, women were banished from the stage and in their stead arose a class of actors known as the *onnagata,* who perform only female parts. Today, two onnagata dominate the kabuki stage, Onoe Baiko and Nakamura Utaemon. These men are as dissimilar as could be imagined in the roles they assume and the styles they affect. Baiko is at his finest when he plays earthy women, the fishmonger's wife and the like. Utaemon is the epitome of the star-crossed courtesan or the ill-used noble-woman. He is especially noted for the skill of his dancing; it is this aspect of his art, more than any other, that has etched kabuki so indelibly in my mind.

It is customary for kabuki performers to enter the theater at a very young age and to serve a long apprenticeship, one which, in a sense, never ends. A performer is a member of a theatrical family to which he may, but need not necessarily, be biologically related. He adopts the family name, and with time may supersede in importance all of his foster relatives, but he remains a member of the family. Recognition is marked by a succession of given names, each somewhat more lustrous in the family's history until finally, he may stand at the peak, a Danjuro, for example, if he comes from the family Ichikawa.

When kabuki began to develop, in the middle and late sixteenth century, Japan was a nation at war with itself, an era known as the Sengoku Jidai—The Country at War. It was a time of a weak shogunate and a powerless court nobility, ingredients that favored dissension, intrigue, and turmoil. Out of this period, and the restoration of order and centralization of government that immediately followed, kabuki draws the bulk of its plots. Many of these are so convoluted as to be virtually inaccessible to the observer unfamiliar with the niceties of Japanese history, but others are almost embarrassingly simplistic. One of the latter I especially relish, for it marks one of my rare moments of real comprehension of this delightful art form.

As the scene opens, one sees several members of a family, who, from their dress and language, appear to be cultured, but from their furtive and frightened manner appear to be fleeing some threat. We learn that they are a family of former substance and standing who now flee an oppressive lord who seeks to kill the father. Like many before them in this time, they have taken to the mountains, where it is easier to hide, but their plight is not necessarily made better, for the mountains are infested with bandits and masterless samurai, who survive off of the plunder they derive from hapless travelers.

Briefly, the family finds refuge in a vacant house, where they intend to gather strength and to determine their future plan of action. Unfortunately, they soon hear the sounds of pursuit and, presuming these are the minions of the lord, prepare to flee again. Before they can do so, the house is surrounded and it would seem that the end has come. However, instead of the samurai they had expected, they are confronted by a group of bandits. Little comfort can be taken in this, and they prepare to lose their paltry possessions and possibly their lives or freedom. Happily, it turns out that the bandit chieftain had once been befriended by the head of the family, and so rejoicing replaces fear. This more salubrious state of affairs comes to a quick end, for the samurai of the evil lord arrive. Now, the family has the assistance of the bandits, who soon kill the intruders. Lest this make their fortunes worse, since there is already a price on the head of the bandits, the bandit chieftain proposes to shoulder the blame himself. He will leave a note in a conspicuous spot so that when the bodies are found, it will be known that it was he and his accomplices who killed the samurai and not the members of the fleeing family. To this end he turns to the family and asks them to bring him a brush and paper. This is quickly done and he seats himself on a log to write his note.

As the family watches expectantly, nothing happens—brush is not put to paper. Finally, one of the less timorous members asks the bandit what is wrong. He says that he is merely collecting his thoughts and trying to recall how to write a specific ideogram. This seems reasonable enough, and so another period of watching follows. As time passes, the family's apprehension mounts again, and they ask the bandit what kanji it is that he cannot recall. After a moment or two of hesitation, as if recalled from deep concentration, he says *hitotsu*. This is the Japanese word for the number 1; it is the simplest of all kanji— a single horizontal line. Their benefactor is illiterate!

At this unexpected revelation, the audience, most of whom knew the story well, roared (figuratively, for Japanese audiences are more restrained than our own). For once, I too participated in their enjoyment, for I had followed the full exchange between family and bandit, from the call for a brush and paper to the final unexpected hitotsu. Obviously this event heightened my own enjoyment and sense of participation in this most unusual of theaters.

It was in the kabuki theater that the revolving stage evolved, as well as the *hanamichi* or "flowerwalk," a projection of the stage from the proscenium through the orchestra seats to the rear of the theater. The former makes possible more impressive sets, for they need not be dismantled between scenes, merely rotated to be replaced by another, and through the hanamichi a sense of dimensionality occurs that our theater does not project. Processions can be mounted literally through the audience, and the separation in space which the flowerwalk affords lends credence to an event that is often lacking in the more two-dimensional aspect that the single stage projects.

Kabuki owes much of its classical repertory as well as some of its more impressive stage gestures to the *bunraku,* Japan's fabulous puppet theater. Chikamatsu, the Japanese Shakespeare, wrote exclusively for the puppets, but almost all of his plays are now included in the kabuki repertoire, as are those of other puppet playwrights, such as Seami. Of the gestures that have been borrowed, none is more impressive than the *mie,* wherein the actor assumes an heroic posture, opens his eyes as widely as possible, then slowly, and dramatically crosses them to the accompaniment of off-stage beating of clappers, the *hyōshigi.* Its utility in the puppets as a device to heighten tension is easily seen, for although the Japanese puppet is a marvel of ingenuity (the mouth opens and closes; the eyes open, close, and can be crossed; the eyebrows can be raised or lowered; and of course the head turns), there are limits to the puppetmaker's skills. As a consequence, the stylized crossing of the puppet's eyes signified intense anger or adamancy or some other overpowering emotion. With proper make-up to heighten the eyes, this gesture can be easily translated to live actors and the kabuki theater. Other aspects of the puppet stage translate less well, such as Coxinga's removal of one of his eyes to demonstrate his fealty to his lord, or his battle with a tiger. As Chikamatsu obviously appreciated, a puppet man and a puppet tiger have a common, albeit unreal, dimension that lends credibility to their actions, particularly when these are punctuated by the rise and fall of emotion in the voice of the narrator or chanter, the *jōruri,* but not a real man and a mock tiger.

The season of 1956 at the Kabukiza, Tokyo's most famous kabuki theater, did not begin auspiciously, if one is guided by the impressions of the professional critics. They did not like the selection of scenes that were offered and seemed generally unimpressed by the performances themselves. There was an outstanding exception, which I saw during my stay in Tokyo. This involved Nakamura Utaemon, only the sixth in this illustrious family, originally from the Kyoto-Osaka area, to bear this designation. Utaemon made his appearance relatively late in the bill, as I recall. The role he was to play is touching. It is the story of a famous and beautiful courtesan who, in a fit of madness precipitated by an unhappy love affair, kills herself. The scene opens in a cemetery, at the tomb of the courtesan. Two young children come to place flowers at the grave and to pray. They fall asleep, their heads resting on the grave, and the sequence which follows is their dream. Gracefully but hauntingly from behind the tombstone, Utaemon rises, dressed as the young courtesan would have been in the happier moments of her life. The minutes that follow, encapsulate all of the tragic events which culminate in her death. The dance begins happily but in a measured manner, the expression of a beautiful and marvelously attired young woman at peace with her world. Slowly, inexorably, her fate, as in a Greek drama, is lugubriously spelled out. Costume changes actually occur on stage through divesting one kimono for another previously worn beneath. This is done with the aid of a stage assistant, clothed completely in black, whom one is presumably not to see. The changes are so beautifully integrated into the dance and the evolution of the character that there is no loss in imagery. Each successive costume evokes a sense of deeper despair and gathering doom, accentuated by the mounting frantic pace of the dance. Exquisitely coiffured hair becomes disheveled and unkempt, and one literally sees the march of madness as it overcomes this ill-fated woman. Finally, the pace can no longer continue to mount in frenzy, and the dancer collapses. Death ensues. Throughout this dazzling, spellbinding performance, one is totally unaware that the dancer himself, Utaemon, has a limp, admittedly barely perceptible but nonetheless present. As age has advanced, it has grown more apparent and he no longer dances roles of this kind, but his capacity to evoke womanliness has continued to grow. The speech, the mincing gait affected by the courtesan, the regal bearing of the noblewoman, all of these traits are now so deeply embedded in his performances that one is never conscious that this is a man being a woman.

*   *   *

In spare moments immediately before and during the Congress, I wandered about downtown Tokyo to see how it had changed from my last visit. As elsewhere in Japan, here the first glimmering signs of growing affluence were perceptible; people were better dressed, and the stores had more merchandise to tempt the shopper. More and better restaurants were available. One of these, known as Tsukasa, that specialized in Tosa no ryori, soon became a favorite. Tosa was the name of one of the feudal fiefdoms on Shikoku, and its food is somewhat more highly seasoned than other regional forms of cooking. A particularly good dish was the lightly braised, marinated fish known as *katsuo* (bonito), served with sliced garlic cloves. Coffee houses had arrived, clone after clone of them. Most were small and modest, but some were not. One, not far from the Ginza, was seven stories high, each floor a coffee house; they differed one from another only in their decor and the recorded music played to entertain their guests (some specialized in jazz, others in opera, still others in symphonic music). The coffee was relatively expensive, a dollar or more a cup, but exceptionally well prepared and an eminently suitable accompaniment to the cakes, tortes, and ice cream that were sold.

Night life in Japan (the *mizushōbai*, the water trade) seems unreal to the foreign visitor.[2] It is ubiquitous, large, and possibly involves as a source of employment several million individuals. Some of the places in which they work may be so small as to serve no more than three or four persons at a time; others entertain hundreds simultaneously. Irrespective of its size, each tends to have its own faithful clientele; this is not to say that a patron (usually a man) goes to only a single bar or nightclub but rather that he tends to go to the same groups of bars or clubs when he is out. Expense accounts pay for most of this entertaining.

Few of Japan's many such establishments have had the color of the Showboat. Located in Tokyo's central entertainment area, the Showboat gained its first international recognition when it served as a backdrop for a number of scenes in a once popular movie of the Korean War, *The Bridges of Tokori*. Its other-worldliness started at the door, where, as each party entered, a huge gong mounted in the entrance way boomed forth as many times as there were persons in the party. Almost instantaneously the appropriate number of hostesses materialized, all clothed in evening gowns and wearing little name tags. Close scrutiny revealed each to be the name of a city. Literally hundreds of hostesses were employed, and on the bosom of at least one must have been pinned the name of any middle- to large-sized city in the Orient one might think of, and most places elsewhere as well. Each patron was led through a nautical-appearing interior, up gangplanks, and across decks to a series of tables and chairs. The decks surrounded a large, central well in which a two-decked, hydraulically operated elevator rose and fell. Each deck had a band, although only one played at a time. As the elevator went up and down, the music would be below the listener, then even, and finally above. The effect this produced was bizarre, much like the sound of a trumpet, first muted and then opened, muted and opened, and on and on. Perhaps a more apt analogy is the sound of an approaching and departing train, the classic illustration of what physicists term the Doppler effect.

On the occasion that I am about to describe, the hostess who poured my drinks, offered to dance, and provided small talk was Miss Harbin; I soon learned that she spoke little English and was Korean rather than Japanese but had spent virtually all her life in Tokyo. We were a party of eight or ten, all participants in the International Genetics Symposium. Once we were seated, we were given a series of tickets, each worth 400 yen; everything served—a bottle of beer or a tokkuri of sake, peanuts, and the like—cost exactly the same amount. As new bottles of beer or other refreshments came, the sailor-suited waiter would tear off the requisite number of tickets. One always knew what the bill was, for the successive tickets were marked 400, 800, 1,200, and so forth. The amount owed was the smallest number showing. The waiters themselves were stationed at locations from which they could presumably see a group of tables and clients and identify the raised finger or hand indicating a call for further drinks. The lighting was generally so subdued that the signal could be missed, and a hostess would strike a match and hold the flaming tip in the air above the table. This gave the hall the semblance of a convention of fireflies—a glow here, a glow there. Obviously, these locations at which the waiters congregated had to have their stocks of beer, whiskey, and food replenished frequently; this was done by toy-like electrified trains that carried a case of beer on each shuttle. One kept pinching oneself to make sure that all of this was real and not merely an overly vivid dream.

More numerous than nightclubs like the Showboat are small bars, many known as stand bars, presumably because they are so small that one cannot sit down at the bar. Each has its own ambience. One in Fukuoka, for example, is called the Magic Bar. Its proprietors are all magicians who entertain their clientele with feats of sleight of hand performed within inches of one's eyes, literally daring one to see how the trick is performed. Others, particularly of late, encourage their patrons to sing, and to this end have elaborate collections of records and tapes to accompany the singer. These karaoke seek to give the uninhibited their moment on the stage. So numerous are these bars, and so varied are their individual themes, that it would take a millennia of lifetimes to explore them all.

Most of Japan's night life comes to a halt at midnight precisely, and, Cinderella-like the transformations begin. The hostesses are now anxious to send the tardy drinker homeward so that they too can leave. A traffic jam inevitably occurs in front of most of the major night spots, as taxis weave in and out seeking fares on their way home. The air is filled with shouts of "Oyasumi" (Sleep well), "Odaiji ni nasai" (Be careful), and "Mata Dozo" (Come again). Within thirty minutes or so, calm has returned and all one sees is the most tardy of convivial types weaving their way singly or in groups down the road in search of some late-closing stand bar. Japanese drunks seem less argumentative than our own; I do not remember an occasion in a bar or night club where voices were raised in obvious anger, fists clenched, and trouble threatened. I am sure these occur, but they are rare. Rather, Japanese drunks want to put their arms around you and, if you are foreign, volunteer their goodwill in limited English.

Participants at the congress were entertained at receptions given by the Ministry of Foreign Affairs, the Ministry of Education, the Governor of Tokyo, the President of the University of Tokyo, and the President of the Japan Science Council. The prospect of still another reception, magnificent though they were, eventually became too much. A couple of my colleagues wanted to visit a classical teahouse and approached me to see if this would be possible. It seemed like a marvelous opportunity to return to the Ichiriki in Kyoto, where a half dozen years earlier I had had a memorable evening with Muller, Komai, and Kihara. I should explain that this congress was one of the first of the peripatetic variety. It began in Tokyo, moved to the still young Institute of Genetics in Mishima with a brief stop at the National Institute for Sericulture, and finally ended in Kyoto. I warned my colleagues that a night at the Ichiriki might be expensive, perhaps as much as fifty dollars a person—an enormous sum in those days—but that the Ichiriki was in a class of its own. The possible cost disturbed no one, so I set about making the arrangements. This is not a teahouse to which one can go without an introduction. So I turned to an old friend on the University of Kyoto faculty, Yamashita Kosuke. He insisted there would be no problem; he was known at the Ichiriki and would call on our behalf. With this assurance, I returned to my colleagues and told them that everything had been arranged for the evening.

It was a short walk from the auditorium where the closing session was held to the Ichiriki, and so at a reasonable hour prior to our appointment at the teahouse we set out. Once there, we were met at the

door by a young woman, a maiko, who was obviously surprised to
see three *gaijins* standing in the genkan. I told her as politely as my
Japanese permitted that a call had been made on our behalf and that
we had come to eat and drink. After pondering this for several moments, she asked us to wait and hastened down a corridor. Several
minutes later she returned, accompanied by an elderly woman with a
disturbingly self-confident manner. After scrutinizing the three of us,
she carefully asked me what it was that we wanted. I repeated my
routine, again with all of the honorifics I could muster, and this time
said Yamashita Kosuke-sensei had called to make the arrangements.
She said abruptly, and with no pretense of politeness, "Dare" (who). I
repeated what I had said. She looked at us imperiously, and said "Ko-
nokata shirimasen" (I don't know that man!) The interview was obviously over; she turned and left, with the maiko in tow. It was clear
that we could not stand in the entrance to the teahouse interminably,
and it was unlikely that I could effect a change in our welcome. With
substantial embarrassment, particularly to me, we left. Instead of an
evening of teahouse entertainment, we ate at the Karafune, a well-
known restaurant, and sampled their tempura ice-cream, more a curiosity than a gourmet experience. Did Yamashita actually call? Or had
I, misguided by what I interpreted to be his Westernness, unintentionally placed him in a position in which he could not say no to my request. Or was the prospect of trying to entertain three foreigners un-
accompanied by Japanese friends merely too much for the sedate old
teahouse?

As soon as practicable after the last scientific session, I hastened
back to Hiroshima to participate in an informal review of the genetic
activities of the Commission. To this review had been invited several
of the more distinguished symposium participants. The meeting had
two aims: first, to determine what further genetic studies, if any, could
be done in Hiroshima and Nagasaki and, second, to explore ways and
means by which Japanese geneticists might contribute more importantly to the evaluation of the genetic effects of exposure to ionizing
radiation. Neither of these objectives was especially well served, since,
save for myself, none of the other participants, all of whom were ex-
perimentalists, had a grasp of the research milieu in Japan nor an appreciation of the complexities of epidemiological studies. Possibly the
only constructive suggestion to emerge was the recommendation that
young Japanese geneticists, especially those interested in human ge-

netics, be encouraged to seek further training in the United States. Without sources of funds to support such training, however, even this recommendation had little influence.

My last day in Tokyo was dreary. Showers fell intermittently, growing in frequency and intensity as evening approached. I knew that a typhoon was nearing Japan's coast and was concerned that the flight I was taking to Honolulu on my way home might be canceled. When I checked in at Haneda, I was informed that departure would be on time and that because of prevailing winds we would fly nonstop to Hawaii. As we headed for the Stratocruiser, rain probed the runway like the wrath of God. Urgency propelled us across the space from the terminal to the waiting stairs of the plane but not quickly enough to avoid a thorough drenching from the sheets of rain ricocheting in the wind. Soggy pants dripped their way to my assigned location near the plane's rear. I settled into a generous seat and drew out Oliver Stadler's *Modern Japanese Prints,* published a scant three months earlier. It had aroused my interest in hanga and influenced enormously my recent purchases of contemporary Japanese prints. Although my tastes are generally eclectic, like others I suppose they need reinforcement, and this was an opportunity to learn whether I had done well or poorly at the galleries I had patronized. I leafed my way through the book as the pilot revved each motor in turn. We taxied to the runway beneath an ever-darkening sky disgorging torrents of water. As our plane gathered speed and slowly rose, the engines groaned under their labor. We inched our way into the lower layer of clouds apprehensively, as if expecting the worst. It was not long in coming. The pilot informed the cabin crew and passengers to tighten our seatbelts, for although we would be skirting the approaching storm, it would be a turbulent forty-five minutes or so.

He had hardly completed his announcement when the tempo of the engines increased and we dropped several hundred feet. Conversations grew subdued, and clench-lipped, blanched faces stared fixedly forward, as though intent on discerning some premonitory sign of the next lurch. We wallowed, pitched, and bounced through the night as though our errant bird was attempting to prune its own tail. Airsickness set in, and those of us fortunate enough to have sturdier senses of equilibrium searched for the airsickness bags just in case. Slowly, almost imperceptibly, the clouds grew less dense, the wing-tip lights clearer, unfiltered by drops of suspended water, and finally, like a tardy moth emerging from its chrysalis, we edged into a clear sky and the setting sun to our west. Color returned to whitened cheeks, and the volume of conversation rose. Only the gathering line of individuals awaiting the toilet or a seat in the small bar in the underbelly of the plane spoke to the recency of our concerns. Normalcy returned to cabin service, but few stomachs seemed prepared, as yet, to test their strength. When some twelve or so hours later we glided into the old terminal in Honolulu, only a few recalcitrant neurons still recorded our brush with nature's fury.

6

NAGASAKI, 1959

*M*rs. Yamamoto was dying. In August 1959, shortly after my arrival once again in Japan, I called her home to say that I was in Hiroshima and hoped to pay my respects. I had brought an autographed copy of the book describing the genetic study to which she and her associates had contributed so much to present to her. One of the other elderly midwives in the association answered the call and informed us of her perilous health and the probable imminence of her death. I restated that I wanted to visit as soon as possible, for I would shortly go to Nagasaki and not return to Hiroshima for some time.

At Mrs. Yamamoto's, I was welcomed by a former officer of the Midwives' Association who led me to the room in which my friend rested. As my eyes adjusted to the darkness, I saw her. If possible, this tiny woman seemed even tinier, and she emerged into the world we peopled from the one into which illness had led her erratically, like an aged butterfly exhausted by its efforts to penetrate the light. Eyes once full of intelligence, eyes that never dissimulated, were now glazed and scarcely comprehending. Her appearance was disquieting for many reasons, only a few of which I could possibly articulate. Though I knew she understood little if any English and my Japanese, inept as always, was compromised even more by the emotion that welled, I tried to identify myself. As I knelt beside her pallet, her fevered hand sought mine and her lips formed the words "Schull-sensei." A moment of lucidness prevailed but was soon gone; she had withdrawn again into a world where none of us could join her.

As I left the house, memories crowded about. Each brought its joys and its pains, the latter because of the impending loss and the sudden, penetrating realization of how little I actually knew about her. Oh, I knew the Mrs. Yamamoto who headed the Midwives' Association, but not the woman whose hand had sought mine. Had she been married? Surely she must have been. But if so, had the marriage been childless, for there were no children present to warm her final waking moments, or had the war robbed her of them? Several days later, the Midwives' Association phoned the Commission to inform me of her death.

I had returned to Japan as Director of the Nagasaki phase of The Child Health Survey, a study that had been initiated in Hiroshima in the summer of 1958.[1] It comprised a major effort to determine, through as many metrics as possible, the biological consequences of consanguineous marriages—marriages among close relatives. These had been common in Japan until very recently for a variety of reasons, largely socioeconomic. Biological experience suggested that the inbreeding which results from such marriages has deleterious genetic consequences, but no contemporary human data had been collected to quantify the problem. Our aim was to measure the biological and, to a lesser extent, social ramifications of inbreeding. Beyond the inherent interest of such information, it also had immediate pertinence for our efforts to evaluate the impact of ionizing radiation on future offspring. Again, it had been necessary to assemble a team of American and Japanese investigators, to agree on a research strategy, and to implement it. To further these ends, three young physicians, Fujiki Norio, Ohkura Koji, and Yanase Toshiyuki, had been brought to Ann Arbor for a year of planning prior to my arrival back in Japan.

The Child Health Survey was largely autonomous in its recruitment of personnel and in its scientific management, but it received its funding mainly from the Atomic Energy Commission of the United States, and we were dependent for space and data processing on the Atomic Bomb Casualty Commission, which was at that time broadening its own staff and research efforts; indeed, it was to grow again to over a thousand employees. Implementation of the recommendations of the 1955 report of the Francis Committee under the leadership of a new director, George Darling, had instilled a stronger sense of direction and purpose to the Commission's activities. The program of clinical examinations of the survivors was growing apace, the pathology study had been broadened and strengthened, the fixed surveillance samples were largely defined, and the efforts to develop individual dose estimates had redoubled.

To accommodate our needs, an expansion of the Commission's facilities was necessary, and this entailed further construction both in Hiroshima and Nagasaki. A new building, similar in style to all of the others, was added to the array already on Hijiyama in Hiroshima. So simple a solution had not been practicable in Nagasaki, however; the space the Commission occupied was too limited. It was possible to add, adjacent to the Kaikan, a small building to support the clinical examinations and to rent and rehabilitate an old wooden, two-storied building beside the small shōyu factory across the street for offices and other administrative uses.

These tasks required time, and as a consequence, my first month in Japan was involved in closing the Hiroshima phase of the Child Health Survey and awaiting the completion of the construction in Nagasaki so that we might begin our work there. While the closure was occurring, Vicki and I set up temporary living quarters in one of the double rooms at Hijiyama Hall and waited eagerly for the move to Nagasaki. Although Hiroshima was changing, it was not as yet an attractive place to live; the river edges were squalid collections of squatters' huts and to the southwest Miyajima, looming like a two-dimensional, three-humped dragon rising slowly from the sea, was continually encased in a sultry summer haze from which there was little relief.

Activities at the Commission were bustling nevertheless. Staff had grown substantially and was still growing, which was fortunate, since those members of the Child Health Survey in Hiroshima who did not want to follow us to Nagasaki could be easily placed in other positions with the Commission. Many of the Americans had only recently moved to Hiroshima from the military compound at Nijimura, which had heretofore been one of the few available housing options. As the city grew, more houses had become available for rental or lease, and the decision had been made to move all of the families still residing in Hiro to Hiroshima or one of its suburban communities. This was a boon long overdue. It made them members of the community, which they had not been previously, and it shortened substantially the time lost in coming to and going from work. Moreover, it made possible social interaction with colleagues that had been extremely difficult over such long distances. Senior medical staff and young physicians from the Public Health Service bustled busily through the clinics and laboratories, aiding and supervising their Japanese associates. Numerous new research initiatives were under way or planned, both as a means to define better the biological consequences of exposure to ionizing radiation and to provide the training experience that would sharpen the clinical skills of our examiners.

Since our earlier clinical studies suggested a possible elevation in mortality among the offspring of the survivors, one of my first tasks was to define and identify a cohort of children suitable for long-term mortality surveillance, and in this undertaking I worked closely with Kato Hiroo, a young epidemiologist who had recently joined the Commission as a member of the Branch Laboratory of the Japanese National Institute of Health. Over several weeks, we developed a roster of about 54,000 individuals born alive after May 1946 but prior to January 1, 1959. These were equally divided among three age, sex, and city-matched groups—one group had been born to parents one or both of whom were within 2,000 meters of the hypocenter in Hiroshima or Nagasaki, a second group to parents one or both of whom were exposed but at distances of 2,500 meters or more, and a third group to parents of whom neither had been in these cities at the time of the bombing. Procedures had to be developed to follow the life status of these individuals and to obtain copies of their death certificates so that the cause of death could be determined.

For convenience and because the procedures seemed sound to us, we elected to imitate the methods used in the surveillance of the survivors themselves. These involved periodic examination of the household records (koseki), which Japanese law obliges each family to maintain. These records are stored in special offices of the Ministry of Justice. Every change in the composition of the family, such as a birth, marriage, or death, must be reported to this office within a specified period of time, and through perusal of this record it is possible to determine whether an individual is dead or alive, wherever in Japan they might be or have been residing.[2] Surveillance becomes essentially a clerical operation, but a time-consuming one nonetheless because of the sheer size of the cohort under review. However, all of the proposed procedures, as well as the arguments for selecting the cohort we did, had to be set out in detail in a bilingual technical report for the guidance of others who might be subsequently involved in the study. While I worked on the English version of this report, Hiroo steadily translated it into Japanese. Once this task was completed, our proposal had to be reviewed by the ABCC's Research Committee and their approval obtained before work could begin.

Eventually in 1960 everything was in place and the surveillance started. The first step was to determine which of these 54,000 individuals had died between birth and the initiation of the follow-up. Arrangements were made through the appropriate local municipal offices to examine the relevant koseki, and if the appropriate record was stored at an office outside of Hiroshima or Nagasaki, letters requesting a copy of it were sent to the authorities having custody of the record. Responses to all our inquiries were forthcoming, and surprisingly few individuals—usually persons who had migrated from the country—were lost to follow-up.

*   *   *

By 1959 Nagasaki was booming. No longer did an unsightly warren of buildings and lanes, colorful as they were, stand before the train station; they were gone, replaced by new roads that hastened traffic into the Urakami valley. Most of the signs of the atomic bombing had disappeared, even in the vicinity of the hypocenter, where now there stood a new park, a small museum, and a heroic statue, some thirty-two feet tall, dedicated to the victims, alive and dead, and to peace. Nagasaki's mammoth Mitsubishi Shipyards in Akunoura were humming; every shipway had a tanker under construction. Some of these exceeded 200,000 tons and were formidable enterprises to launch. All-in-all there was a welcomed sense of growing prosperity.

Once it was possible to move to Nagasaki, Vicki and I did so with alacrity. I was anxious to get on with the tasks that had brought me to Japan. The Child Health Survey's new quarters proved adequate if not luxurious. The old clapboard building, somewhat drearily painted, provided a small suite of offices for the supervisory staff—Robert Miller (a pediatrician), Jerry Niswander (a dentist), and myself—and space for our young physicians and contactors (people who arranged clinic appointments, called upon families to solicit their participation, and shepherded them to and from the clinical examination when the need arose). The contactors had a tedious task, for Nagasaki's hilly terrain and narrow streets made it difficult to reach many of the houses with a vehicle, and bicycles were not a suitable means of transportation. Briefly, I toyed with the thought of purchasing or leasing for the staff a fleet of small motor bikes—Rabbits or Doves—but this came to naught after I myself tried to navigate one up the narrow roads in our neighborhood and saw how potentially dangerous they were not only to the rider but to pedestrians as well. Finally, we simply provided our contactors with books of taxi tickets and trusted that they could walk the final yards necessary to visit a family.

As time passed and we grew more accustomed to our surroundings, we felt privileged not to be in the main building, with its overcrowding and hurly-burly. Bob and Jerry had to spend more time there than did I, for they were involved in the pediatric appraisals and the dental examinations. My duties were largely administrative. Thirty to forty children were seen most days, and since we provided their transportation, a constant stream of taxis arrived and left, bringing or returning the participants. Each of the taxis flew on its right front fender a small green pennant that identified our study and served as a continuous reminder of the survey to those who saw our vehicles.

The Commission's professional staff in Nagasaki was not as large as that in Hiroshima, since only one-third of the survivors under scrutiny were in Nagasaki. Nagai Isamu was director of the branch laboratory of the Japanese National Institute of Health and the ranking member of the Japanese professional staff. Shortly after we arrived, several more young American physicians were assigned to the Nagasaki branch of the Commission. Most of the daily examinations of survivors were performed by these young people, who were imbued with a sense of purpose that one expects from youth.

August 1959 saw Vicki and me staying at Matsuda Hall briefly while the renovation of our house was being completed. Shortly after the war Matsuda Hall had been commandeered by the military government to serve as Bachelor Officer's Quarters and after the Occupation ended, it was leased by the Commission for the same purpose until the early 1960s. Through the years that the house was administered by the Commission, it rarely had more than three or four residents, usually members of the Hiroshima staff temporarily in Nagasaki. However, it was so close to the Commission's clinical facilities in the Kaikan that the hall's dining room served the luncheon needs of a far larger number of employees.

The Matsuda family has been one of the principal forces in the economic policies of the Ju-hachi Ginko (the Eighteenth Bank), which is the dominant local bank in Nagasaki Prefecture. Shortly after the Meiji restoration in 1868, Japan's banking structure was overhauled. Prior to that date, money had been borrowed and repaid, but the fiscal and economic system in use under the feudal regime was inadequate to modern needs. The new system differed from the old in a number of respects. One was the establishment of a central bank, the Bank of Japan. This institution carries out the government's monetary policy and controls credit through the issuance of paper money; it serves as the bankers' and government's bank. The banking system made provision for ordinary banks, however, that could handle deposits and money transfers, make loans, and undertake investments on behalf of its depositors. As in most countries, these monetary establishments operate under a series of conditions set forth in a succession of Bank Acts. Many of the first banks chartered under the new act bore no name, merely a number. Few of these numbered banks remain, but at least one, the Ju-hachi Ginko, still flourishes.

It was inevitable that the Matsudas, as one of the more wealthy local families, would have a large home, and perhaps equally inevitable that when the Occupation occurred, the size, construction, and location of that home would lead to its being commandeered. On the evening of August 14, as Vicki and I passed the small cemetery not far from Matsuda Hall, we were surprised to see the area ablaze with candles and lanterns and thronged with people. Curiosity led us up the stairs and into the grounds. Before us was a pavilion. Within it could be seen some of the Matsudas, whom we recognized but really did not know. They recognized us, although our names were equally unknown to them. We stopped briefly to extend our greetings and found ourselves sipping Orange Fanta.

The occasion was OBon, August 15, the Buddhist Memorial Day, and Nagasaki makes more of this season than do many other cities. One feature is a large parade of boat-shaped floats, memorials to those who have died. These are fabricated in the streets, in the yards of homes, wherever there is space to assemble them. Some are small—testimonies to high regard but little money—and others are huge, literally tens of feet long, testimonies as much to prestige and affluence as respect. This was the year of Hatsu-bon for the Matsudas, the first year after the death of someone, when custom demands the boats of remembrance. Vicki and I had seen their *fune* (boat) under construction before the hall but had not known its purpose.

Ordinarily, on the day prior to OBon, families go to the cemeteries to decorate the graves of their dead and to welcome visitors and friends. The public display—the parade of the boats through the streets of the city—is accompanied by fireworks, sake drinking, and general noise making. It covers a well-established route through Nagasaki's center and ends at or near the prefectural offices. As the parade passes over its route, it is steadily augmented by more and more floats. It starts as soon as darkness begins to settle and continues until midnight, attracting most of the city's inhabitants into the streets. It is not an occasion of sadness but one of joy and sharing; indeed, it parallels the celebration of All Saints Day in Spanish countries.

Over the years, we saw many such celebrations in Nagasaki, but none had the sense of personal involvement of this, our first. Subsequent ones were longer, the boats bigger, the noise of firecrackers greater, the smell of cordite (or whatever firecrackers are made of) more pronounced, and the crowd more numerous. The particulars of this festival, the "festival of lanterns," have changed surprisingly little with time, and the description of it by Carl Peter Thunberg, written over two hundred years ago, remains apt. He wrote, "It lasts three days; but the greatest solemnity is on the evening and night of the second day: it has been established in honor of the dead who return, they say, to visit their parents and their friends . . . To receive them, on all of the graves are placed bamboo poles from which are suspended lanterns with candles . . . They construct small boats of straw with lanterns and lighted candles. At midnight they carry these boats in a procession to the sound of music and with much pomp to the sea; there they abandon them to the winds and the waves, which are not tardy in engulfing them; sometimes they are consumed by fire before water penetrates them."[3]

*　　*　　*

Centuries before Frank Lloyd Wright and Le Corbusier "discovered" the economies and beauty of modular construction, it was an essential feature of the architecture of Japan. The origins of the modular unit, the *tatami*, are obscure; it was already an integral part of the dwellings of the Kamakura period (1192–1333), as judged by the scrolls of the time, and may have arisen still earlier. The tatami determines the dimensions of a room, the width of sliding doors, columns, verandas, and so forth. Allegedly, it represents the smallest area in which a man can sit, work, rest, and sleep. Though the tatami's dimensions vary slightly in different parts of Japan, the length hovers around six *shaku* (roughly six feet) and the width around three shaku. It is made of tightly woven rice straw covered with rush; its edges are bound with cloth, usually black in more modest homes. Rooms are, conventionally, 2, 3, 4, 4.5, 6, 8, 10, and 12 mats in size. With little hesitation, the average Japanese adult can readily describe the size of every room in his or her home in terms of mats. And the number of arrangements that culminate in a square or rectangle with each different number of mats can titillate the most jaded set-theorist.

After entering a Japanese home through a foyer, where one removes one's shoes and puts on soft-soled slippers, one usually proceeds down a wood-floored corridor (the *rōka*), which runs along one or more of the outside walls, and into the guest or drawing room. This is set off from the rōka by the *shōji*, sliding doors of translucent paper mounted on wood. A Japanese home may have one or more drawing rooms, where visitors are received and entertained. A feature of this room is an alcove, the *tokonoma*, a place of honor where a hanging scroll and a vase of arranged flowers, or some other artistic ornament, is displayed. Its purposes are aesthetic, to encourage tranquility of mind. Usually the limit of the tokonoma is a wooden pillar, different from the other pillars in the house, called the *tokobashira*. Another common decorative touch is a series of staggered shelves built into the wall adjacent to the tokonoma; these shelves, of different lengths, are known as the *chigaidana*. Often above them across the wall is an enclosed shelf or cupboard, the *fukurotodana*, literally the bag closet. The drawing room is separated from adjacent rooms, and they one from another, by *fusuma*, sliding interior panels of heavy, decorative paper mounted on a lattice-like frame, edged usually with lacquered wood.

Ninety-three Fufugawa machi, where Vicki and I made our home in Nagasaki, is a mixture of Japanese and Western architecture, with the good and the bad of each. It is not a house that speaks of wealth nor of conspicuous charm. Its eight rooms are distributed over three levels. Three steps above the main level is a wood-floored room of about six mats, which we used as the master bedroom but which may have originally been a storeroom. On the main level at one end of the house is a hardwood-floored room of eight mats that became our maid's room. Next to this is the living room, an eight-mat area with a floor of tatami and a tokonoma, tokobashira, and chigaidana. The fukurotodana makes up the south wall. The west wall is largely of glass, with a view out over the city. The east wall is shōji. The north wall is fusuma, which separate the living room from a tatami-floored room of six mats. The fusuma are of a cream-colored paper with a motif of mountains in black and gold, a simple but effective pattern. Abutted to these rooms is a 4.5 mat room with tatami floor that served us as a small den. Aside from our bedroom, the den was the only heated room in the winter, and only when we were present. The municipal gas pressure was so erratic that the heater would often go off; and, if no one was present to relight it when the pressure mounted again, gas streamed into the room. Two other small rooms jointly provide all of the facilities of a bath. One contains a wash basin with toilet and the other a wash basin with a sunken bath. Both have large picture windows, at a height that does not compromise privacy. Beneath the main floor is the lower level of the house, which is not really a basement, for only one wall is set into the ground. Here is a dining room, a pantry, and a kitchen, all in the Western manner. They communicate with the floor above through two stairways, one connecting the rōka with the dining room and the other the rōka with the kitchen. The lower floor has windows only in the west wall; these look out on the second-stories of the houses below.

As can be gathered, the house is set into a hillside lot. Entrance is gained from the uppermost reaches of the lot through a small but pleasant garden of pines, azaleas, yucca, and agapanthus. A prominent feature is the heavy stones that make up the walk from the front gate to the genkan. Throughout the interior of the house the walls, when we were there, were painted with a very light, seafoam green paint; the ceilings where they were not wooden were painted white. Both baths had sea-foam-colored tiled floors and were tiled to a height of perhaps five feet on the walls. The main floor was furnished in part Western and part Japanese style. In the den there was a small, black, four-foot-long love seat, an upholstered green chair, bookcase, lamps, and the like. In the larger Japanese room were a low central table of lacquered wood, several zabuton (cushions) in shades of gold, a ko-tatsu, and a variety of decor pieces, including a *biwa* with *bachi* (a Japanese lute with wooden plectrum). The other Japanese room had only one major piece of furniture, a chest of rosewood.

A new day at 93 Fufugawa began when I reluctantly rolled out from beneath an electric blanket, if it was winter, and, after turning on the small gas space heater, headed toward the unheated bathroom to shower and shave. Often cold but now awake, I would rush back to the bedroom to dress and then hurry down to the dining room, the only warm spot in the house save that beneath the electric blanket. Jeannie, our maid, would have fetched the morning paper, a day-old version of the Tokyo-printed *Japan Times,* which inevitably reported an improving economic situation in Japan and a worsening state of affairs in southeast Asia. Hastening back upstairs to my briefcase, and throwing a "See you later" in the general direction of my sleeping wife, I would then pick my way up our garden walk to the front gate.

Hardly would I reach the gate before a window in our neighbors' home would slide open and their little daughter, Kusumoto Chiaki, would press her head out. A smile would spread across her face at seeing me, and I would be greeted with a stream of "Oji-san" (Uncle), which continued as I made my way through the gate and long after I had acknowledged her greeting. The first sight to greet my eyes when I pulled open the gate was a small cemetery. The *inumaki* (*Podocarpus macrophylla*), a shrubby yew, along the ledge that connected our garden with the main pathway served to focus attention on a steel-gray headstone bearing in inlaid gold lettering the name Yamaguchi. While there is much to be said for neighbors as quiet as Yamaguchi, it is unnerving to be greeted each morning by an unseen presence as one departs for work. This never-ending testimony to the transient nature of life wears on one after a few days.

I hesitate to call the main pathway I followed to work a road or a street because it was not wide enough to accommodate most three- or four-wheeled vehicles, and the steps that jutted across it at irregular intervals denied one even the use of smaller motorized conveyances such as a motorbike. Nonetheless it was more than a mere concrete-covered path because up and down it flowed foodstuffs, construction material, everything to maintain the sizable community of persons living on the side of the mountain with us. The route led gently downward for a distance of perhaps 200 feet, turned left for 50 feet, and then to the right for another 70. This brought one to the lip of a small cliff, along which there was an unprotected walkway. The holes at periodic intervals suggested that at one time there had been a guard rail, probably of iron, most likely sacrificed during World War II for the greater good of Japan. From this unprotected walk one passed into a temple courtyard. On the left of the courtyard was a rust-colored gate leading to the temple itself, Shuntokuji, which stands where once a church, Todos os Santos (All Saints), stood. Along the right was a waist-high stone wall terminating at a magnificent camphor tree. The wall of the courtyard was head high. Parallel to and a foot in front of it was a row of six Buddhist figures, the *rokujizō* (Jizo of the six states of existence). Of all the gods in the Buddhist pantheon, none is more lovable than Jizō. He is simultaneously the patron of pregnant women and guardian deity of children, and possesses some of the physical attributes of both. Time and children had left their marks on the rokujizō before Shuntokuji. But these were softened by growths of moss and small ferns.

The camphor tree was four or five feet thick and leaned gently away from the hill at an angle that has tempted generations of youngsters into its branches. It is certainly the last living thing to have heard the evening angelus rung from Todos os Santos. Its crown, delicately greened in spring, is carefully trimmed every year or so, lest some gust topple it from its precarious footing, and its fine bark, almost golden in the light of a setting sun, has resisted the probing of more than one child's finger or knife, or merely the curious interested in its odor. My path led on down some thirty cobblestone steps that curve about and down past this camphor giant. Nestled near the bottom of the first flight of stairs, close to the roots of the tree, was another foot-high Jizō. Before him stood a small trough of water flanked by bamboo vases in which fresh flowers appeared daily. I never encountered the child or pregnant woman who so faithfully tended to Jizō's wants.

My progress toward work could easily be traced by the cries of "harro" and "Amerika-jin" from children that followed me almost from the moment our front door closed until I escaped into my office. At the hour of the day when I made my way to work I met scores of children of different ages, sex, and size on their way to school. Within the immediate area there was a primary school, a middle school, and a high school, as well as a portion of the prefectural women's junior college. Six primary-school students functioned as traffic cadets near our offices. Armed with whistles, red and white flags, characteristic caps, and stools on which to stand, they carefully guarded their fellow schoolmates across a major thoroughfare enroute to the Irabayashi School. The flourish with which they used their flags, and the unison with which their whistles were blown, were manifestations of the importance they attached to their task. One of the features of Japanese life that never ceased to amaze me was the pride in job one constantly encountered. It takes many forms—from the rhythmic beat with which the streetcar conductor punches a ticket to the performance of these children.

The path was so narrow that I could sometimes touch buildings on both sides of me; but, strangely, I had no feeling of physical confinement as I wandered through this maze each morning. I was embarrassed, however, by unintentionally participating in the private activities of so many households—the morning yawn, the sound of water being drawn into a basin, the gurgle of a flushed toilet, the clatter of dishes, and a host of other intimate household sounds. To the American used to space, there is no privacy in Japan, and my embarrassment was not so much for myself as for those who would never know privacy in their lifetimes.

One quickly adjusted to the other local sounds—the plaintive early morning note of the tofu-maker's horn announcing the availability of his newly made bean curds, the clap of the *hyōshigi* (the blocks of wood struck together by the night fire-watch to announce that all was well), the rub of a sliding door as a neighbor left for work, the persistent call of a mother attempting to awaken a sleeping child or husband, or the rattle of a loose chain on a passing bicycle. These noises, unlike the intimate ones, never gave me a sense of intruding; rather there was a feeling of being a part of a community, a neighborhood, a living and caring collection of human beings. Through Jeannie we were a party to the deaths and births, the joys and tribulations that beset our neighbors; perhaps out of a sense of our strangeness and a cultural uncertainty, they shared the happy moments of life with us more willingly than the sad ones. And the children—their openness and exuberance bridged the gaps of age, income, language, origin, religion, and social status. Would that we could all retain these qualities; the social and cultural veneer that follows time and age blunts our capacity to see one another as equally transitory members of a fragile globe.

Shortly after Vicki and I were settled, Jeannie suggested we make a formal call on each of our neighbors, since this was expected of any Japanese who might move into a neighborhood. She also advised us to join the *tonarigumi*, the neighborhood group, if possible. These were not tasks to be taken lightly. Our formal introduction into the neighborhood could not be a casual dropping-in, for the neighbor would be expected to serve tea and some light refreshment—cookies or a soybean-filled sweet—or something equally appropriate. Appointments had to be made, and Vicki had to have suitable gifts for each household. These could not be so overly expensive as to place an onerous burden on the neighbor, who would feel obligated to reciprocate in kind, nor could it be so inexpensive as to be unfitting to ourselves and the position our neighbors assumed we filled by virtue of education and occupation. Since we did not know either of these bounds, Vicki was in a quandary.

After much consultation, gift-wrapped scented bars of soap, still a small luxury, emerged as the most appropriate token. Packages were purchased, and finally the day came. Vicki and Jeannie visited each of a dozen or so neighbors to introduce themselves and exchange pleasantries. A day apprehensively begun culminated, however, in an experience that warmed subsequent relationships and made intelligible behavior and events that otherwise would not have been so. We came to understand how spatially limited the average Japanese home is, the constraints this placed upon individuals, and some of the reasons for the inculcation of the "group psyche." We also observed the Japanese penchant for souvenirs, often tawdry and at variance with the architectural understatement that characterizes the interior of Japanese homes, even modest ones.

When the forerunners of the tonarigumi—the neighborhood groups—came into existence is clouded in the mist of time, but it is commonly thought that their immediate antecedents, the *gonin-gumi* (literally five-man group), were modeled upon a system of group organization employed in the Tang dynasty in China. Over three centuries ago, the gonin-gumi existed in Nagasaki, which—the name notwithstanding—even then did not consist of five men but five households defined by contiguity. A rich farmer might find himself grouped with his poorest tenant, or a wealthy merchant with a blacksmith or a day laborer. Such grouping served as a leavener in a society otherwise rigidly proscriptive. For once the Tokugawa shogunate was firmly entrenched, it issued a series of sumptuary laws governing the various social classes. These defined most, if not all, of the perquisites of one's *bungen* or status and contributed significantly to the codification of gift-giving.

No aspect of life escaped these regulations. Farmers, although ranked immediately beneath the samurai in the social scale, were wretchedly poor, burdened with onerous rice taxes and small, miserably productive landholdings, yet the size of a farmer's home was prescribed and its design defined (no parlor nor a tiled roof was permitted). Moreover, the gifts a farmer might give a daughter on the occasion of her marriage were enumerated, the guests who could be invited specified, and the parting gifts of food (the *hikidemono*) to be taken by the wedding guests itemized. A farmer and his family were never permitted to wear silk clothing, perquisites of other classes. Even the nature of the gifts to be given by grandparents on the birth of a first child were set out, and directions given as to appropriate gifts to boys and girls on their respective festival days. An ordinary laborer was permitted the use of an umbrella only under the direst of circumstances; he had to content himself with a straw raincoat.

Whether the gonin-gumi, and subsequently the tonarigumi, had their origins in a need for collective defense in troubled times or were merely a surveillance mechanism, imposed from above to counter dissent and social injustice, is uncertain. However, they have certainly seen service in both capacities. In the feudal era, the gonin-gumi was both an insurance company and a charitable organization; the neighbors joined in reestablishing a household consumed by fire, took care of foundlings on a prorated basis, adjudicated conflicts within the group, and shared in the culpability of any of its members guilty of wrong-doing. Under the military regimes that flourished in Japan in the 1930s and early 1940s, the tonarigumi, which usually consisted of ten to fifteen households, was the smallest unit of general mobilization. Participation was compulsory. Each unit was collectively responsible for circulating notices from the central and local governments, fire-fighting activities, civil defense, public health, and so forth. The thought-control police used the tonarigumi as an instrument of coercion and to stifle independent thought. These were excesses of the past, however, when we moved to Fufugawa machi; in fact, the tonarigumi had been officially abolished in 1947, although they lingered on without official sanction.

The head of our group, our *tonarigumi gashira* (literally, the pillar of the tonarigumi), was a middle-aged woman, Uwaki-san, who lived several houses up the hill from us. Her husband was dead, her children grown. Startled and initially uncertain by the presence within her group of a family of foreigners, she soon became to us, a pillar. She insisted upon treating us as just another family, but for us she was more than just a neighbor. On our frequent trips, she saw that our house was protected from intruders and safe from fire. These were obligations to her, not just courtesies, and she expected no reward. Occasionally she would accept a small jar of coffee, but too overt an expression of our appreciation was embarrassing to her.

Japan has been, and remains, a land in which a calling card (a *meishi*) is an indispensable part of social and business life. Calling cards were surely the custom of Europeans when they arrived on these shores in substantial numbers more than a century ago, but they had also been a custom in China of some antiquity. Matteo Ricci, for example, commented on the calling cards he received from the Chinese literati who called upon him. Meishi vary in their elegance and the information they contain. Some are handmade paper, often from Echigo, part of the modern prefecture of Fukui; others are simpler. A well-established artist's meishi may contain little more than his professional name, cursively written, whereas those of lesser renown will have their names, titles, home and business address, and phone numbers. In the early years, the home phone number was not necessarily one's own but that of the nearest available phone. Indeed, owning a phone bore a social responsibility to acknowledge calls to one's neighbors.

My meishi was bilingual, English on one side and my kana-transliterated name and address on the other. If the meishi was to be used in formal calls on local dignitaries, such as the mayor, added to it, often in red, was the statement that this was a formal greeting card.

Elaborate courtesies attend the use of these cards. Proper respect must be exhibited. One bows as one proffers the card, which is held in the two hands like a gift, and accepts the inevitable return of another's card with an equal display of appreciation. Frequently, after the latter is received, again in the two hands, it is raised slightly toward the head, as would be done with any other gift. The existence of these cards has generated an industry of its own—printers who specialize in their printing, boxes in which to file the cards, carrying cases, and the like. For most Americans in Japan, the exchange of calling cards is a welcome practice, for it removes the onus of trying to remember the name of everyone when group introductions occur.

A personal stamp (a *han*) is no less important in Japanese society than a calling card. Stamps are the only legally acceptable form of written endorsement of a transaction. Even workers, on reaching their place of employment, will enter their arrival time and endorse the entry with their stamp. One's stamp is customarily registered with the ward or city office. Until quite recently they were hand-carved, and on scrutiny would be as individual as a signature. Generally they are round, ovoid, or square and are indexed in some manner so that the top can be recognized from the bottom when the face is applied to paper. The ideograms etched into the stamp's face may be simple or highly ornamental. Almost any carvable material can be used—bamboo, hardwoods, jade, quartz, ivory, bone—and the sizes of stamps vary from a desktop model that might be the official stamp of an organization or business to one sufficiently small to fit into a watch-pocket. To prevent breaking, pocket-sized versions are usually stored in a little case that contains an even smaller inked pad. Prices vary as much as sizes. If one has a relatively common Japanese surname, a han can be cheaply purchased ready-made. But more exotic materials or less common names must be carved on order. Here, too, time has wrought changes, and now self-inking, spring-loaded stamps can be purchased. Vicki and I, too, found the stamps convenient, socially and professionally, but we wrestled with whether our names should be etched in roman, or written phonetically using one of the syllabaries, or even rendered in ideograms if the pronunciation was self-evident and approximated our names. Since several ideograms have the same pronunciation, it became a matter of selecting those that conjured the best image—our surname was usually written with characters that imply a pretty picture and was read as *sharu,* a sound as close to Schull as we could get.

Vicki, whose appreciation of some aspects of Japanese art had grown more rapidly than my own, had, in her spare moments, become a student of ink painting (*sumi-e* or *suiboku*), an art form begun in China but developed to new levels of astringency in Japan.

Her teacher, Obeiya-sensei, was an aging, delightful man whose gentle lined face was usually creased with a smile. He visited our home weekly with his teaching materials wrapped in a tattered furoshiki, often with Takenaka-san, the jeweler, if he was in town, who was not only his friend but ours. Obeiya would unroll his delicate paper on the floor, prepare his ink and brushes, and, with the nonchalant skill of the master, hastily fill it with the figures—bamboo leaves, a chrysanthemum, an iris, plum—that students use as practice models. As he watched Vicki emulate his models, and the wet and dry techniques, he sketched autumnal flowers, branches of loquats, or whatever else struck his fancy. Generally these pictorial musings, after he had placed his stamp in the corner, were left with her for further practice.

We saved them all, and, as roll after roll accumulated, I sought to have several that I found especially appealing mounted in the hanging scroll manner, as *kakemono*, but did not know to whom to turn for this. I asked Obeiya. He sent me to a *hyōgusa*. I found this strange; they are makers of shōji and fusuma. But he told me they could mount anything that required gluing. The particular one to whom he sent me had his shop adjacent to the Nakagawa River near the main shopping street. The proprietor recognized Obeiya's work immediately, without even glancing at his mark, and laid before me several roles of brocade cloth he thought would make a suitable background. I selected one for each of the several drawings to be mounted. When I returned in a week, the results were not only a testimony to Obeiya's skills but to those of the hyōgusa. The drawings I had left bore not only the wrinkles inherent in the paper itself but the consequences of weeks of being rolled. Neither were now visible; instead, we had six or so scrolls which, while certainly not Sesshu's, are of more immediate personal value to us. They are the products of a man, now dead, whom we knew as warm and humorous, without pretensions.

*   *   *

Supporters of the Child Health Survey and the local citizenry soon learned where we were housed and called upon us directly. One of the most regular of these visitors was Dr. Hayashi Ichiro, Professor and Head of the First Department of Pathology at Nagasaki University and the editor of the *Nagasaki University Medical School Journal*. Dr. Hayashi was a medium-statured man, with shallow cheeks and a somewhat prognathic lower jaw, an upper lip decorated with a small nose-width mustache, and a disconcerting habit of seeming to have his mind elsewhere whenever I spoke to him. He always arrived, or so it appeared, at an inopportune time, his aged briefcase under his arm, bulging with manuscripts in various stages of editing and review. These were rapidly spread upon my desk and his interests explained. He had set as his editorial goal not only the raising of the scientific quality of the journal he edited but securing for it a wider international readership. To achieve the latter, he reasoned, more articles had to be written in English, or at least had to carry intelligible English summaries. I was the vehicle to these ends. Almost every English-speaking scientist is asked, sooner or later, to review an English article and to improve the author's use of the language. Over the years, I have read hundreds of such papers of Japanese colleagues, students, acquaintances, and others whom I did not know but who apparently knew of me. It is a burdensome task, particularly if one seeks to salvage as much of the author's own words as possible, for the manuscript is often incredibly obscure. Generally, however, there is little immediacy to the rewriting, and one can do it on a schedule that fits one's other activities.

I had no compunction about being of help with the medical journal, for I shared Dr. Hayashi's aspirations. But as an editor, he had deadlines to meet, and that meant that I did, too. It took us weeks to work out an effective modus vivendi, and for him to realize that I could not respond on demand, as it were; I had other responsibilities. An accommodation did come. Slowly but steadily the manuscripts came to me at the time of their submission, which meant that I could work on the English while the scientific review of the submission was occurring, and not be the final stage in the process.

Vicki, too, was engaged in English-language instruction. She regularly held classes in conversational English at the local Women's Prefectural Junior College. Vicki encouraged her students to stop at our home, and some did; others didn't. Most were in their late teens and giggly. Each wore a student's uniform—white blouse, black skirt, short socks, and tennis shoes. They were at ease with "sensei," but my arrival, especially if unexpected, was the source of endless tittering and purposeless embarrassed actions. After a time the more regular visitors grew less ill-at-ease when I was about, and some even tried to draw me out in conversation. We grew extremely close to several of these young women and still visit with many on our trips to Japan.

They were, or so they purported themselves to be, liberated women. They intended to have a larger hand in their own destinies than their mothers had had. In subsequent years, as young matrons, they have become reconciled to less liberation than they anticipated. They chaff at the constraints in which their lives are bound, but they do not openly rebel. Those who married upwardly mobile young men find themselves second wife to their husband's corporation. They have all of the trappings of success and the promise of more—a company home, a car, children in the proper schools, and enough money to indulge themselves in travel, designer clothes, cosmetics, and the like. But their husbands are rarely home before nine in the evening and then are often too disgruntled from pressures at work to cope lightly with neighborhood affairs and the children's problems. The children are either in bed or studying when he arrives home on weekdays. His corporation insists that he project the image of the thoughtful, caring father, however, and so Sundays are spent with the children.

Corporate America may be wrapped in an overly evaluated sense of importance, but matters are far worse in Japan. The salaryman is an automaton who marches lockstep to any tune his corporation calls. It is a life devoid of opportunities for independence, originality, creativity, or even the sense of extended family the corporate life purportedly seeks to further. Proper dress, proper language, proper attitude, and proper credentials—the right schools and courses—are the essentials of success. Even so, a life at the corporation's dictation may bring no more than obligatory retirement in one's fifties.

Some of these young matrons and others we knew have directed all of their energies toward securing a successful future for their children, through striving to gain them admittance to the better schools. In Japan, moreso than in the United States, the recognition one's children receive reflects back on the family, especially the mother. Some have sought liberation through teaching crafts, at which they are particularly skillful, not so much as a means to further financial success as to achieve an identity of their own. A few Japanese women, annoyed at the stereotypic roles in which they have been cast, have sought professional careers as designers, dentists, lawyers, or physicians. Some have entered politics, if not as active candidates for political office (although the present head of the Socialist Party is a woman) then as a part of the political process of campaigning. Nevertheless, the abysmally constricted lives most Japanese women lead represents one of the less obvious costs of Japan's economic achievement and its cultural traditions. Possibly only on the farms do women still share their lives with their husbands, and even here mechanization has compromised the physical sharing of good and ill-fortune.

American men in Japan have tended to take on many of the habits of this male-dominated culture. For example, once the Child Health Survey had been launched, the male professional staff reinstituted the custom of a monthly night on the town. Generally we numbered eight or ten, but occasionally we were joined by more. Shortly after five or so in the afternoon, once the last patient was seen, we would gather in the Chinese-like lounge in Matsuda Hall for preprandials and a debate as to where we should eat. At this time, one of the younger physicians was designated as the treasurer for the evening. As we moved from place to place, it was his function to collect the statement of charges and either pay it or inform the proprietor that we would pay him the next day. There was never an objection to this, and it had the advantage that we did not have to burden ourselves with carrying as much money as we anticipated spending. Bob, Jerry, and I usually made certain that we had sufficient money so that if an objection was raised we could settle the bill; however, this was seen as a group affair, not the supervised and the supervisors, and we tried not to intrude on the arrangements our young colleagues had made. The following morning the statements were totaled and each person was asked to contribute, but only for that portion of the evening's expenditures that were incurred while he was actually present, for sometimes one or two people would leave early.

Once adequately fortified, we would leave the hall to walk to dinner and, at eight or nine in the evening, to a club. The Oranda, in the entertainment quarter of Nagasaki, was one of the largest, but by no means the most expensive, of the city's nightclubs that appealed to the wallets of our younger colleagues. The Ginbasha and Junibankan were much more costly. Mostly we talked, often about cases seen in the course of the day, drank beer, listened to the music, or danced with the hostesses. We were certainly not as boisterous as most of the business types—Japanese and foreign—that frequented the club. Indeed, the hostesses that attended our needs had a relatively simple task; this dawned on us forcefully when we found that the mama-san who managed the hostesses kept staffing our table with slightly inebriated young women who had grown tired of fending off their more lecherous customers. We apparently served as a form of rest and recuperation. As each new one joined us, her ensuing conversation with me followed a banal, well-worn path. Where did I live? How long had I been in Japan? Where had I learned Japanese? My answers were invariably rewarded with an "ojōzu ne" (skillful), which I knew to be blatant flattery, for my command of the language certainly does not warrant this description. Inevitably, I would be asked how old I was. The Japanese seem to be preoccupied with one another's age, and this interest extends to foreigners in Japan as well. They are unable, or at least pretend to be unable, to judge how old we might be; they do not recognize the hallmarks of our advancing years—the graying and loss of hair, coarsening of features, a skin that no longer fits, a memory more faithful to the past than the present, the occurrence of pigmented spots in the skin, and on the list goes.

About eleven or so, tired of the noise, we would wander off to a smaller, more intimate club to a second party. Invariably, some of our members left us at this juncture, especially those who lived near the outskirts of the city, but enough remained to warrant some further beer and natter.

Midnight brought a halt to our carousing but not an end to the evening. Warmer nights found us at a small restaurant, usually eating *nori-chazuke*, a dish of rice with seaweed, sesame seeds, and minute crackers over which is poured green tea. This may not sound especially appetizing, but it is very filling. In winter and early spring, if the night was cool, we would seek out one of the numerous small *sushi-ya*. These were often little more than the traditional sushi bar and possibly a small four-mat alcove with two tables, but since Nagasaki is a major port, the fish was always fresh. We would seat ourselves at the bar and clutch the steaming cup of green tea set before us. Sushi-ya use extra large cups, commonly five or so inches high and possibly four in diameter; they are large enough that one can conveniently wrap both hands about the cup, while the heat permeates fingers numbed by the cold. The bar itself is generally a dressed plank of cypress two or more inches thick, and perhaps ten feet long; opposite the stools on which one sits are refrigerated glass containers, accessible to the sushi maker, in which the various raw fish, roe, and so on are exhibited. Sliced pickled ginger along with a small mound of ground horse-radish (*wasabi*) is set either on the small wooden pallet on which the sushi are served or in a flat dish adjacent to the pallet. Beside the sushimaker, who was aproned with a jauntily and characteristically tied white hachimaki on his head, stood a large container of vinegared rice. An appropriate amount of this would be gathered into his hand and worked into an egg-shaped ball, wasabi applied to the top, and upon this placed a slice of the food chosen. Each such piece was dipped into shōyu, to which one might add the wasabi on the pallet. Each of us had his favorite seafood, some traditionally prized, like tuna (*maguro*), others not. I enjoyed the fresh shrimp, often still crawling lethargically in the cold. The head and legs were quickly snapped off, and the shell over the carapace and tail removed. Initially, it was disconcerting to see the fluid still circulating in the veins, but the queasiness this aroused was easily suppressed. Sea bream and the eggs of the sea urchin were also my favorites. But I have never been able to conquer my upbringing with regard to another Nagasaki delicacy, raw *iwashi,* small sardines. These are set before the diner still flipping pathetically on a plate.

Other forms of sushi, *maki-sushi,* were rolled in dried seaweed with the aid of a small square of bamboo (a *makisu*) and sliced before being placed upon the pallet. Often these contained an especially crisp cucumber in the center and were called *kappa maki.* It was a matter of some concern to me how the sushi maker kept track of the number of pieces each patron ate, until I was told that his system was simple. On his side of the bar, behind the refrigerated containers, he placed a grain of rice arrayed according to price each time he prepared a piece of sushi for a particular person, and to reckon the bill all he needed to do was count the grains of rice before the customer.

*   *   *

Shibui—this word, though not exactly translatable, implies a tastefulness above the ordinary. More than this, it suggests a simple, quiet, almost austere quality. A vase may be shibui, a design may be shibui, and, I presume, even a funeral may be shibui. Rarely, however, is the latter so, for the Japanese perception of the nuances of color and harmony of design seems to flee in the face of death. This was forcefully brought home to me when I was obliged to attend the funeral of one of our employees, a young man whose untimely death was made grimmer by the bleakness of the ceremony marking his demise.

It could be said that he came from a humble, albeit proud, family. Their home was in Isahaya in an area noted for neither its poverty nor its wealth, a commonplace district. Accompanied by his section chief (his kachō), I arrived at the home where the funeral was to be held shortly after mid-day. Following a somewhat strained greeting, I was shown into a room with two hibachi, several zabuton, and a soiled table on which were sake bottles, cups, and the remnants of a half-eaten meal. Adjacent to this room was another one in which, through the open fusuma, could be seen the family altar. Before it sake, rice, and fruit were placed. Above these and within the embracing doors of the altar was a recent photograph of the young man. In front of the food, incense was burning, and beside the incense burner (kōro), the incense money (kōden) was neatly stacked. On either side of the altar (the *butsudan*) stood a large bouquet of paper flowers, imitations of the carnations and chrysanthemums then in season. Their colors ranged from white through yellow to a brilliant red. The flowers were so obviously artificial as to be sordid under the circumstances. Above the bouquet on the right a white paper dove hovered. Before each bouquet, thrust into a bamboo vase, were gilded leaves and flowers of the lotus. Their falseness jarred and desecrated the sanctity of the altar. To the right of the butsudan was a long tokonoma, within which hung a scroll. Overall, the scene was one of shabby gentility.

Outside, the sky was clear and the sun was shining, but within the house there was a cold dampness not dispelled by the heat from the hibachi. While we huddled in our thoughts for the arrival of the Shinshu priest, tea, cookies, and fruit were placed before us. The food failed to invade our consciousnesses and went untouched. We had waited only a few moments when the priest and his assistant arrived. While the former greeted and consoled the family, the latter was ushered into the room with the altar to perform several duties. Among these was the writing of certain particulars concerning the young man on small wooden memorial tablets (the *ihai*). It was winter, and the most pervasive sound was the sniffle. The assistant began to unpack the paraphernalia of his office from a furoshiki—cymbals, bell, and robe. He requested an inkwell with water, and from the voluminous sleeves of his kimono he drew ink and a Japanese writing brush. Slowly but methodically he mixed the ink and the water. When he had achieved the color he sought, he carefully pointed his brush and, drawing from his sleeve a piece of paper, vigorously rubbed the surfaces of the ihai. This completed, he wiped his nose and shoved the paper back into his sleeve. A brief discussion with the young man's brother ensued over the appropriate kanji to be used in writing the dead youth's name. Satisfied that he understood, the assistant inscribed the memorial tablets.

Almost at the moment this duty was discharged, we were joined by the priest, who introduced himself. The introductions over, he said a short prayer before the altar and prepared to robe himself for the ceremony. He stripped off his black outer kimono; underneath was a white one tied about the waist with a narrow black sash, an obi. By this time, his assistant was holding a gossamer-thin salmon-colored crested kimono into which the priest hastily stepped. Around his waist was tied a purple and gold brocaded sash, and over his left shoulder a cape-like garment of the same material was draped. The latter was held in place by a tie passing under the right arm. Thus attired, he sat on a zabuton immediately before the altar. The assistant seated himself a step to the rear and left of the priest and placed before himself his cymbals and bell.

In a monotone the priest began to chant one of the sutra. Occasionally he was joined by his assistant. The chanting was periodically punctuated by a churning and clashing of the cymbals, a tinkling of the bell, or more frequently by the refrain, "Namu Amida Butsu," repeated several times over. The only overt act of the priest, aside from the chanting, was to bless a boat-like wooden vessel containing incense, which he ignited. This was placed in front of the deceased's mother and his grief-stricken wife, both clothed in the black kimono of mourning, and subsequently before each of us. To this vessel we carefully contributed three pinches of a sawdust-like incense. As each pinch fell, a small puff of smoke rose slowly from the vessel but was quickly lost in the drafts of the house. The ceremony seemed endless but actually lasted only thirty minutes and culminated with the reading of a letter of condolence from George Darling, Director of the Atomic Bomb Casualty Commission, the young man's employer. The letter, which had been placed among the kōden, was carefully opened before the altar by the brother-in-law of the deceased, who bowed to the altar before he began to read. He addressed himself not so much to the family as to the deceased, which gave to the reading the sense of a valedictory rather than a letter of sympathy. This act completed, it was clear that we were free to leave.

Amid bows, reiterated condolences, and a sense of inadequacy because of my inability to communicate to the family the empathy I felt, I bid them farewell and moved toward the genkan and my shoes. As I drew the door closed behind me, the tear-reddened eyes and puffy faces of the mother and her daughter-in-law, who had followed me to the entrance, were driven deeply into my mind.

Life is full of mysteries, but the greatest of these is death. It is rarely beautiful, rarely opportune; more frequently there is about it merely an air of inevitability. This death was depressing. Tears welled in my eyes, to be subdued only with effort. Did the depression that so deeply troubled me stem from the fear that I too might have life snuffed out prematurely? Was I, in searching for a reason for his existence, searching for one for my own? Must this young man who in life was an average individual, a *bonjin*, be also accounted a bonjin in death? He did give to his wife a child, and to that child life. But what other solace can the widow and her infant daughter find? Was happiness for his wife infrequent? When it was present, did she and he always appreciate the value of living? Was there warmth in their love, or merely urgency? How would those moments of shared gladness be kept alive in the weeks and years of loneliness she faced? What would she tell her infant daughter in later years about her husband, the child's father? Would the child's only recognition of him be an aging, slowly deteriorating photograph mounted above the household altar?

*   *   *

On New Year's (Oshōgatsu) in Japan, most businesses are closed and only essential services are available. Rail lines and flights are overburdened with people on the move. This is the season to visit one's family, to go to one's temple or shrine, to exchange gifts, and to enjoy a variety of special dishes. Preparations for the New Year begin weeks in advance of its actual arrival, and one of the more common of the pre-holiday festivities is the *bōnenkai,* the year-end party. It is an occasion for employer and employee to review the past year and their respective roles in it and to identify sources of friction so that ill-feeling will not carry over into the new year.

Christmas and New Year's comprise the largest shopping season by far in Japan, aided by the year-end bonus, which can amount to an additional two or three months' salary. In 1959, the two major stores that competed for these extra yen in Nagasaki were Hamaya and Okamasa. Of the two, Hamaya was the larger, but Okamasa had a new wing under construction. These two *depato,* as they are called, engaged in advertising antics worthy of Americans. On sale days, the stores could be located from almost any spot in Nagasaki by the large balloons lazing in the sky above them. Each store offered to its customers a version of the Christmas savings club to which one contributes 500 yen monthly. This money could be withdrawn at the end of the year, applied to the purchase of a major household appliance, or used in a variety of other ways. One's funds drew no interest; instead, several times a year the store sponsored for its club members the appearance of a kabuki troupe, a Noh performance, or some similar form of entertainment not generally available in Nagasaki. The audience at one of these affairs had all of the earmarks of a convention of garden clubs; the buzz of busy voices clashed incongruously with the slow, deliberate movements of a Noh drama, the oldest of Japan's extant professional theaters, a form of musical dance-drama using a simple, undecorated stage and largely poetic language.

Okamasa's organ gave that establishment the nod in the seasonal competition for a shopper's yen. At regular intervals from the tower atop the store came a wondrous tootling. The sound resembled that of an old steam calliope as it drifted up to our home in Fufugawa from the valley below. As entertaining as the sound was, the selection of music was even more intriguing. Everyday except Wednesday, when the store was closed, at nine in the morning and three in the afternoon, Stephen Foster's "Home Sweet Home" sallied forth over the city. On Wednesday, "Home Sweet Home" was replaced by Schubert's "Wild Rose." At six in the evening, at a time when Nagasakians were on their way home from work, their footsteps were cheered by the Largo, "Going Home," from Dvorak's "New World Symphony." At nine in the evening we are informed, with Stephen Foster's help again, that "there is no place like home." These are the only four tunes this electric monster was programmed to play, apparently. There must be a story behind this odd repertoire, but modest efforts on my part have failed to reveal who selected the tunes or why.

On January 1, 1960, the New Year of which I write, Christmas had not yet become a large occasion, although all of the premonitory signs were there. Nagasaki's major shopping area rang with the sounds of Christmas carols, and images of Santa and his elfin helpers were to be found in both of the department stores. Across the face of the Ginbasha (Silver Horse Carriage) night club, fifteen feet or so above the street, a large Santa and his reindeers raced. Many small stores had added their iotas to the seasonal spirit. Before others stood the shochikubai.

With the approach of the holidays, Vicki and I, as well as other friends, decided to welcome the New Year together, but not at a private party; our celebration would be with the city. It started with dinner at a restaurant, the Ginrei (Silver bells), locally known for its steaks and good beer and the eclectic collections of watches, steins, and other memorabilia of its owner. At about ten o'clock, surfeited with food, we went next door to the Bon Soir, a small cocktail lounge run by the owners of the Ginrei. Upon entering, each of us was presented with a small gift, a pottery ashtray with the names Ginrei and Bon Soir baked into the lip. We talked, danced unenthusiastically to recorded music, and waited. It was not our intention to greet the New

Year at the Bon Soir; we had plans for a more lively greeting, at the Junibankan, the Number 12 Club. We had ample time and decided to walk the half dozen or so short blocks from the one place to the other. We were too numerous to walk the narrow sidewalks together, and so, strung out in twos and threes, we emerged onto the street running before the Bon Soir through the district known as Teramachi. Although it was not cold, there was an autumnal quality to the air, crisp and invigorating, and just the faintest appearance of one's breath as we strolled through the city's shopping area. The streets were crowded but no more so than on any other evening; there were as yet no signs of New Year's revelry. Most of the stores had closed, but those few still open seemed to be doing a brisk business—late shoppers intent on completing their purchases for the impending holidays. As the last shopper left, the lights within the store were snuffed and the tallying of the day's receipts hastily begun.

It was not yet eleven when we reached our destination. We were welcomed by hostesses who were unaccustomed to seeing a group of husbands and wives but seemed to understand the special occasion. We danced some, enjoyed one another's company, and listened to the music. Perez Prado's "Cherry Pink and Apple Blossom White" was a favorite, but it really did not matter what the band played, the beat was always the same—not wholly objectionable, not correct either. Previous New Year's Eves drifted through our minds as we swayed to the music. Finally, at the stroke of twelve, we wished one another the best for the New Year, and a wave of nostalgia broke across us as we joined in the singing of "Auld Lang Syne"; thoughts of home and loved ones crowded out the present.

* * *

Summer of 1960 was enlivened by an American teen-ager, one barely so. Patty, a daughter of close friends in Ann Arbor and a surrogate niece, was to spend a portion of her summer vacation with us. In our anxiety over the prospect of entertaining a teenager for sixteen hours a day, we had attempted to rent or lease a television set, but to no avail. Jeannie told our neighbor, Kusumoto-san, of our problem, and within a day of his learning of our unsuccessful efforts a new television arrived with a tasseled brocaded bit of cloth to drape over its sightless eye when not in use. It had come from one of the stores that had re-

buffed our inquiries; obviously his position meant more than our en-
treaties.

We had written Patty's parents to say that I would be in Tokyo from
May 30 through June 9 to participate in the Thirty-second Session of
the International Statistical Institute and it would be especially convenient if Patty could arrive at that time. This would give her several
days in the capital and an opportunity to participate in the functions
planned for Congress members. I was to speak on the morning of June
1st, the third day of the meeting, but would be free to accompany her
on subsequent days on some of the planned activities and to take her
to see the stores and places she might otherwise not see. Among the
planned social functions was a gagaku concert, a kabuki performance,
trips to Nikko and Kanazawa, and a reception banquet sponsored by
Prime Minister Kishi, to be attended by the Crown Prince and Princess.

I arrived in Tokyo by train from Nagasaki on the eve of the Congress, and Patty arrived at Haneda the next day. As a part of the registration formalities for the Congress, one was asked to complete a brief
questionnaire. Its principal purpose was to identify oneself, provide a
suitable mailing address for the Congress authorities, and indicate
whether one was or was not accompanied by someone—a wife, son,
or daughter, for example. I hastily answered the questions about myself but faced a quandary when confronted with the items on accompanying individuals. It was simple enough to enter Patty's name, but
how was I to describe our relationship? Finally, I entered the word
niece. While this was incorrect, it did suggest the bond we shared. All
this perplexity was for naught. When Patty received her invitations to
the social functions and her identification badge, all were clearly and
precisely labeled—Mrs. Niece Schull. No end of explanation was
needed when others saw her name or wondered who the attractive
and personable young woman was whom I had in tow.

I was still mulling these events the next morning as I stood on the
upper level of the terminal at Haneda, watching Patty's aircraft taxi to
a stop. Top-loading had not yet reached Japan; one walked down the
stairs wheeled to the plane and across several hundred feet of tarmac
to the airport. As I squinted into the morning sun trying to identify
her, a tall girl, taller than her age warranted, began to wave. It was
Patty. As her steps brought her closer to the spot beneath which I
stood, it was clear she was no longer the youngster we had left almost

a year earlier but a young woman. Within moments, we were together and she was pouring out news from home, exuding excitement, and obviously trying to still the jet lag that tempered her curiosity.

Congresses of the magnitude of the periodic meetings of the International Institute of Statistics routinely have one or more honorary presidents—usually established investigators in the later years of their scientific careers. Sir Ronald Fisher was one of those who presided over this meeting. He was an exceptionally gifted man, holder of the prestigious position in mathematics of Wrangler at Cambridge at a very early age; his contributions to contemporary statistics were enormous and certainly warranted this recognition. Indeed, he had the respect of everyone interested in statistics and genetics, but strangely, he, like many of his English scientific peers, including J. B. S. Haldane, had a need for adulation, or at least continued recognition of his standing, that could take unusual twists, even thoughtless, indefensible ones, to my thinking. These actions reflected not on his scientific gifts but on his sensitivity. One such episode occurred at the session on statistical methods for biologists at which I spoke.

Approximately half of the speakers at this session were young Japanese investigators who were ill-at-ease with English. But since the official languages of the International Institute of Statistics are English and French, they were obliged to present their results in one or the other of these languages. Most chose English and read their manuscripts hesitatingly, not attempting to present them extemporaneously. Fisher, a bearded, bespectacled man, drifted in and out of most of the various scientific sessions like an eminence grise. He was conspicuously present, seated near the rear of the auditorium, at ours. Normally, the individual presentations are printed in advance and the manuscripts are available to participants to peruse as the speaker presents. Copies of the presentations had been spread on a table within but near the entrance to the auditorium. Presumably the intent was to make them accessible as one entered the auditorium. Midway through a presentation by one of the young Japanese, Fisher rose from his chair to obtain a copy of the paper; however, instead of walking about the rear of the auditorium to the table, as he could easily have done, he strode down the central aisle and before the podium, acknowledging friends distributed through the audience, as he walked along. This tactic would have unnerved all but the most experienced, and the young man was certainly not this. Almost from the moment of Fisher's rising, the speaker grew tense, promptly lost his place in his manuscript, and became conspicuously flustered as he searched to regain his equanimity and the appropriate line in his manuscript. It seemed an unpardonable demand for attention, gained at the expense of an inexperienced speaker.

I was the last speaker in the session, and when my time came I rose with apprehension, uncertain what to expect from Fisher. I need not have been worried; he listened intently and at the end asked a very perceptive question, one possible only if he was following my remarks carefully. Chary of what to expect, I answered in a more elliptical and guarded way than I needed to. Afterwards, as I reassembled the pages of my manuscript, Fisher stopped to express his interest in what I had said and chided me good-naturedly, eyes atwinkle, for the less than direct answer that I had given him. Clearly he was a complex man who could be exceptionally gracious one moment and thoughtless the next. Most of his students have spoken glowingly of his willingness to be of help as they wrestled with their dissertations, and possibly this episode was merely the behavior of an aging man too acutely aware that his productive years were behind him.

At least his insensitivity was not as objectionable as that of Haldane, if credence can be placed in a commonly told story. Allegedly, as a result of injuries sustained in research during World War II, Haldane had difficulty sitting for prolonged periods on seats as uncomfortable as those usually found in an auditorium. When sitting was necessary, he carried an inflatable rubber doughnut to use as a cushion. He would, or so the story goes, invariably arrive in the auditorium after the speaker had begun, make his way ponderously to the first row of seats, where one was reserved for him, and then, turning his back to the speaker, inflate his doughnut while still standing, gazing fixedly at the audience. Such a performance was guaranteed to upstage any speaker, however interesting his remarks.

Some months before the Congress, the romance and wedding of the young Crown Prince, Akihito, and his pretty, ashen-faced princess, Michiko, had attracted the attention of the international press, and as a consequence their names were well known, especially to the young. It was with more than usual excitement, therefore, that Patty looked forward to the banquet at the Imperial Hotel at which they would preside. The fact that the Prime Minister, Kishi Nobusuke, would also be present meant substantially less to her. It surprised me, however, that he would be present at all, for there was trouble in the streets. His announced intention of renewing the defense pact with the United States had aroused substantial opposition from the nation's leftists, particularly groups such as the Zengakuren. Student demonstrations were mounted daily, and deliberate efforts to provoke Japan's well-trained riot police occurred hourly. Snake dancing, arms-linked, was a common tactic designed to snarl traffic and further confrontation. It generally took place in the vicinity of railway stations or along the Ginza or wherever inconvenience and delay might ensue. Given this state of affairs, we presumed Kishi would send an alternate, but he did not. Indeed, he was a garrulous man, well aware that his stand would undoubtedly cost him his party's leadership. When it did—although in customary Japanese political fashion he retained his Diet seat and the faction that supported him—he was succeeded by his brother, Satoh Eisaku. As children, they had been separated by adoption, a common practice in Japan among sonless families, and hence the difference in family names. Kishi smoked and chatted freely and comfortably with the Congress delegates.

Soon we were called to dinner. Protocol demanded that we be in place when the Prince and his Princess entered. We were seated close to the head table, not more than eight or ten places away. Patty's eyes were at full-stop as we stood to greet their arrival. Once they were ushered to their seats, dinner began. It was hard to keep Patty engaged in conversation, for her eyes kept straying to our young hosts. Dinner ended quickly, and the customary addresses were brief. On cue, we all rose as their Highnesses left. Patty had hoped, I suspect, to get their autographs but settled for Kishi's instead. He graciously signed her dinner menu. It is ironic that the Crown Prince had to wait in the wings for so long; his father's reign extended well beyond a half century. Kishi, an important architect in Japan's postwar recovery, has completed his life.

7

MERCHANTS, MISSIONARIES, AND MARRIAGES

*E*uropeans in the fifteenth, sixteenth, and seventeenth centuries were driven by economic circumstances and by scientific and technological advances into a period of exploration unique in the history of mankind. It was in a sense inevitable, once a sea route to Asia had been charted, that they would ultimately find their way to Japan. Marco Polo, in his book describing his travels through the Far East, had spoken of a land of gold—an island empire called Zipangu—situated on the eastern confines of Asia. The picture of Japan he painted was more than enough to interest the Europeans, already trade-conscious, and by the early sixteenth century European merchantmen reached the Far East. In 1542 a Portuguese vessel, driven by a storm, ran ashore at Tanegashima, an island just south of Kyushu.

In these same years, Europe experienced a religious upheaval of a magnitude heretofore unseen. The new enlightenment of the Renaissance and the Reformation of the Catholic Church, precipitated by the moral and theological challenges of Martin Luther and others, gave rise not only to different Christian sects but to new religious orders within the Roman Church. Many of these, imbued with a spirit of renewal, saw among their challenges the dissemination of the faith to peoples everywhere. No orders were more important in this regard than the Jesuits and the Franciscans.

On August 15, 1549, Saint Francis Xavier landed in Kagoshima, accompanied by two missionaries and three Japanese Christians from the Asian mainland. Francis Xavier—a man of unusual abilities and drive, and a founder along with Ignatius Loyola of the Jesuit order—made numerous converts in the years following 1550, some through political persuasion and hope of economic gain, but many for less calloused reasons. When Xavier sailed from Japan in 1551, he left behind him four or five congregations with an aggregate of from 1,500 to 2,000 Christians. In the evangelical period from 1549 to 1640, it has been estimated (over optimistically perhaps) that several hundreds of thousands of Japanese became Christians. The major portion of this growth occurred before 1614.

At the time of the advent of Christianity, Japan was a loose federation of feudal districts ruled by a powerless god-emperor, Go-Nara, and the once strong but then-tottering family, the Ashikaga, his military lieutenants.[1] The inability of the Ashikaga to control the feudal lords, and the constant wars waged by them among themselves, made Japan ripe for rebellion. This came when the Ashikaga shogun was overthrown and Nobunaga came to power, shortly after the arrival of Christianity. He fostered and encouraged the work of the Christian missionaries as a means of counterbalancing the power of some of the Buddhist priests and monasteries who had become involved in political manipulation. It was not uncommon, under the Ashikaga, for the more militant Buddhist monasteries to maintain large bodies of troops and to be a law unto themselves. This was obviously not consistent with Nobunaga's attempt to unify Japan, and by fostering the spread of Christianity, he was able to divert the activities of some of these groups while strengthening his own position.

Upon Nobunaga's death through assassination in 1582, Hideyoshi, one of his generals, seized control of the country. For a few years he continued the policy toward Christianity that Nobunaga had advanced, but by 1587 the political picture had changed. Undoubtedly motivated by the need to offset the growing power and questionable allegiance of some of the Christian nobles of southern Japan, Hideyoshi published in 1587 an edict that banned the Jesuit missionaries from Japan. There were perhaps other motivations as well. The long civil wars and the misery that accompanied them had impoverished the peasantry, and a trafficking in slaves arose. An unknown number of Japanese were bought, sold, and transported to the Philippines and southeast Asia, aided and abetted by Portuguese and other European adventurers. Hideyoshi repeatedly published decrees threatening slave traders and even the purchasers of slaves with death. So overt, apparently, was the trafficking that even the Catholic bishops and priests inveighed against it. Whether slave trafficking by Europeans figured in Hideyoshi's decision to ban the Christian church from Japan is not known, but unquestionably the slave trade impressed him and other Japanese unfavorably.

The young Christian church went underground for a few years until this opposition seemed to pass, though shortly after its reappearance, the first inklings of the suppression to follow became apparent. In 1597, at Nishizaka in Nagasaki city, Hideyoshi had twenty-six Spanish and Japanese priests and brothers crucified. Hideyoshi himself died the following year. The regency he created to ensure the succession of his son did not do so, and eventually Tokugawa Ieyasu became shogun.

A period of relative prosperity for the church followed, since it was critical to the security of Ieyasu himself and his family that he encourage continued trade with Europe. Initially, much of this commerce had focused on Nagasaki, which had been a small fishing village until the middle of the sixteenth century, when its sheltered harbor made it an important market for foreign trade, particularly with Portugal and Spain as well as China, the Philippines, and Thailand. The years 1600–1625 brought the first efforts of the Dutch and English to establish viable mercantile centers in Japan, but on the island of Hirado rather than in Nagasaki. By early in the 1600s, as many as 1,000 of the 8,000 inhabitants of this island immediately to the west of Kyushu were said to be Dutch, English, Portuguese, Spanish, or Chinese.

Ordinarily the Dutch sent one or two ships a year to Japan. These left Batavia (now Jakarta), the headquarters of the Dutch East Indies Company, in June laden with sugar, lead, rouge, raw silk, and medicines, and returned at the end of the year carrying copper, burnt camphor, lacquerware, porcelains, and silks. Sake, rice, and soy sauce were also bought but on speculation, since they had little assured commercial importance. The Dutch became quite successful on Hirado, but the English met with several bouts of ill fortune and by 1623 relinquished their trading post in Japan.

In 1614, with his own family's supremacy better established, Tokugawa Ieyasu promulgated an edict that banned all Catholic missionaries from Japan, without exception. This was tantamount to an open declaration of a war of extermination on the Christian minority. Although he had issued other perfunctory bans following his rise to power, these had not been drastically enforced. They were in a sense warnings: Religion would be tolerated so long as it did not interfere with his government. However, tiring of the constant conniving of one European community against another, and especially when he learned of the malfeasance of one of his officials on behalf of a Christian lord and, through his spies, of the ambitions of some of the European monarchs, he seized upon an alleged conspiracy between some of his vassals and the foreigners, and acted.[2] In the following twenty years, Ieyasu, his son Hidetada, and his grandson Iemitsu succeeded in destroying all overt signs of the presence of Christian communities in Japan. The last major act in this destruction was the suppression of the uprising in Shimabara on the peninsula to the east of Nagasaki in 1637–38, in which it has been estimated that as many as 40,000 Christian and non-Christian peasants, who had the temerity to rebel against the oppressive measures of their feudal lords, were killed. The Dutch, zealous to control the Japanese trade and without interest in or concern over the rightness of the peasant cause, aided the Tokugawa in this enterprise; one of their warships was sent to Shimabara to shell the castle the rebels had seized.

Shortly after Iemitsu became shogun in 1623, following the abdication of his father, Hidetada, he ordered the deportation of the Spanish and prohibited the Portuguese from residing permanently in Japan; but he did allow the Portuguese to continue to trade. This prohibition on residence was not vigorously enforced, and some of the merchants and one of the captains-general, who participated as an envoy in the annual delegation from the Senate of Macao to the shogun's court, tended to reside in Nagasaki the year round. This was a matter of convenience. The trade ships from Macao usually sailed from Nagasaki in November or December, but the annual trip to Tokyo occurred in January or February when the prevailing winds upon which the ships were dependent were no longer as favorable. Finally, in 1638, Iemitsu ousted all of the Portuguese merchants from Japan, presumably because of their complicity with the peasants in the Shimabara rebellion. Now only the Dutch and the Chinese were allowed to trade there, but surely unexpectedly to the Dutch, they were soon to be confined to a small island, Dejima, in Nagasaki bay.

Dejima (or Deshima) literally means the "out or fore island"; its original fan-like outline is still to be seen in Nagasaki's dock area, but it is no longer the separate island wrestled from the shallows of Nagasaki Bay that it once was. Dejima had been built for the Portuguese in the years 1634–1636 with funds from some twenty-five wealthy Japanese traders, but after the Portuguese were ousted in 1638, the Dutch, who had been in Hirado, were obliged to transfer their residences and factory to this new location in 1641, rented at an exorbitant price, over 6,500 taels a year (about 9,000 ounces of silver). It has been alleged that the shogunal government, awed by the naval aid offered by the Dutch in the Shimabara rebellion, did not trust them any more than other Westerners but allowed them to stay because of their artillery and gun-casting skills, which were deemed essential to the Tokugawa family's dominion over Japan. Whatever the logic behind the Tokugawa decision, it was clear to the Japanese administrative authorities that isolation on Dejima would inhibit possible mischief-making but still provide access to armaments and trade.

Scarcely half a fathom above high tide, the island itself was a scant 130 acres, framed in a wall designed to incarcerate and stifle exchanges between the townspeople and the Dutch. Warehouses, homes for the factory's employees, and an office for the Japanese interpreters comprised the island's only structures. Some pastureland for oxen, sheep, and pigs, and limited garden space, were available. A single narrow bridge, the Dejimabashi, was the sole access. The ships that came from Batavia stood out in the water and discharged their cargoes into lighters, to be rowed to the warehouse wharves; or, if circumstances permitted, moved into the bay side of the island, where two doors in the wall facilitated the lading and unlading of cargo.

Some of the factors (or *opperhoofd,* as they were termed in Dutch) were men of exceptional ability, but most were merely merchants who must have found their several years in Nagasaki tedious. Usually less than fifteen men, generally unmarried employees of the opperhoofd and sailors, made up the factory's complement. One of these men was characteristically a physician, often not Dutch but recruited from Germany. All were members of Jan Compagnie, the Batavia-based center of the Dutch East Indies Company. Much of the first century of Dutch occupancy was punctuated by harsh feelings between themselves and their Japanese interpreters and the *bugyō,* the official of Nagasaki under whose jurisdiction the island fell. With the passage of time, some amelioration of their condition occurred, and now and then wives were permitted to join their husbands in their service to the company.

Life, even for those men whose wives had accompanied them, consisted of long periods of boredom broken only by the continuous haggling with the bugyō's lieutenants. When, finally, each year the ships arrived from Batavia, mercantile efforts took precedence over all others, save possibly the refitting of the ships for their return to the Indies and Europe. News of loved ones, or reassignment to more hospitable circumstances, must have brightened the ship's arrival, but undoubtedly could be enjoyed only after the new merchandise was stored in the godowns and arrangements had been made to load the outbound cargo.

All through the period of exclusion (1636–1854), foreign learning filtered into Japan through Dejima and Nagasaki.[3] In this era, Dejima was briefly the home of Engelbert Kaempfer, a German physician with the Dutch East Indies Company whose *History of Japan: Giving an Account of the Antient and Present State and Government of that Empire* is the earliest authoritative description of the country in a European language. Carl Peter Thunberg, the peripatetic Swedish botanist and friend of the naturalist Carl Linnaeus, also served for a time (1775–1776) as the factory's physician, as did Philip Franz von Siebold, a German physician and surgeon who did much to introduce Western medicine into Japan. Events on Dejima intrigued the Japanese, for it was the one window on the Western world still open to them. A whole genre of woodblock printing arose, the Nagasaki-e, which sought to communicate to others the drollery of the Dutch— their dress, the appearance of their wives and slaves, the games they played, their ships.

As these prints disclose, life on Dejima was languorous, if not indolent. Thunberg wrote disparagingly of "the custom of passing the evening at the home of the chief of the factory, after having made a tour of the island and a short promenade in the two streets. Their visit lasts six hours until eleven or so; this makes for a very monotonous style of life, good only for automatons."[4] He wrote, too, of prostitution and the ease with which the Dutch could find a companion "to relieve the tedium of their solitude."[5] These unfortunates were drawn from impoverished families with too many daughters who sold them at the age of four or so into the service of Japanese officials, who, in turn, engineered the arrangements with the foreigners on Dejima. In childhood, the girls were little more than slaves performing menial tasks for established prostitutes, whose ranks they ultimately joined. Few, if any of these liaisons with the Dutch were permanent, although some lasted for weeks or months, but Thunberg said that it was necessary for the Japanese women to stay at least three days. He notes, too, that rarely did the liaisons give rise to children, but describes a girl of about six who, he stated, much resembled her European father. He was assured that such children were sent to Batavia when they were fifteen or so, but he seemed skeptical.[6]

Although Thunberg does not accuse the Dutch of being pederasts, he noted disapprovingly that homosexuality was common among the samurai and Buddhist priests, a practice he saw as bestial and depraved, diminishing to the young boys drafted into the service of their social superiors. Obviously, only the most motivated foreigner could make an intellectually rewarding experience out of his stay in a country in which, Thunberg smugly observes, "continence is not a favorite virtue."[7]

In 1709 the legendary Captain Lemuel Gulliver visited Japan and condemned the Dutch for their submission to the *efumi*,[8] the trampling on Christian religious figures instituted by the shogunate in 1619 and continued until 1859, when, as a part of a treaty with France, it was officially abolished.[9] This New Year ritual was a part of a general inquisition aimed at disclosing Christian missionaries and believers and forcing them to apostasize under pain of death. Some, both lay and clergy, did; but many, possibly thousands, did not and stoically accepted their fate. Within a few years after the initiation of the persecutions, certainly by the middle of the seventeenth century, to all outward appearances Christianity had been stamped out. But the shogunal government remained vigilant, for the Catholic Church continued to try to penetrate Japan. The last such effort occurred in 1708, when Father Sidotti landed at Yakushima, an island immediately to the south of Kyushu. He was quickly arrested and sent to Tokyo for trial, where he died in 1715.

Many sites in or near Nagasaki are hallowed by the steadfastness of the early Christians. In the harbor stands the rocky islet, known as Papenberg, the isle of the Papists, or Takayama to the Japanese, where numerous Christians were hurled to their deaths in the early years of persecution. And at Unzen are the sulfurous volcanic cauldrons, still actively bubbling, into which Christians were suspended, head downwards, to be slowly scalded to death if they did not recant, and quickly enough to forestall death.

*   *   *

In the summer of 1853 Matthew Perry and an armada of American vessels of war appeared in Tokyo Bay requesting, indeed virtually demanding, a treaty of friendship and commerce between the United States and Japan. This development apparently caught the shogunal government, the Bakufu, unawares, divided in the appropriate response to the American request, and wont to temporize. Perry did not press for immediate action but announced that he would return the next year, in 1854, for the government's statement. This he did, on February 11, again with a flotilla. Reluctantly, but seeing no other alternative, the shogunal government approved the proposed treaty. Soon other countries—England, France, and Russia—were demanding similar trade privileges. A flood had begun, one a bureaucracy woefully out of touch with the remainder of the world could not stem. The more conservative members of the warrior class, the samurai, were repelled by the actions of the Bakufu, which they interpreted as weakness, and despised the thought of foreigners in Japan. The seeds of rebellion were sown, and in 1868, following a brief civil war, a group of powerful lords, the *tozama,* who were not vassals of the Tokugawa, succeeded in overthrowing the shogun Yoshinobu and restoring to the Emperor Meiji his legitimate rights.

Once Japan was reopened to trade, foreign merchants flocked to Nagasaki, Yokohama, and Hakodate (in Hokkaido), the three cities initially designated to receive them. In 1864 a French missionary in Nagasaki, Bernard-Thaddée Petitjean, erected a Catholic church to serve the foreign residents of his city. Petitjean had been born in 1829 in the village of Blanzy (Saône-et-Loire) and had arrived in Japan in 1860. His church, dedicated to the twenty-six martyrs of 1597, was not built without difficulty. He complained bitterly to his superiors in France of the numerous problems raised by the contractor, Koyama Hidenoshin, over costs and deadlines, and his frequent threats to abandon construction of the church. Providence intervened. Petitjean was asked by the governor to accept a chair as Professor of French at the local college, but he demurred until his church was completed. When the governor's agents inquired when this might be, he unburdened himself of his difficulties but expressed hope that construction might be completed by January 1, 1865. He was assured that there need be no problem meeting his wishes.[10]

Less than three months after its consecration, the church was the scene of an event Petitjean saw as miraculous, for to his knowledge Christianity had been erased in the years of persecution. Shortly after noon on Friday, March 17, 1865, a group of twelve to fifteen Japanese—men, women, and children—appeared before the church door. He opened it to them, thinking they were motivated by curiosity about Westerners. To his surprise, they genuflected before the tabernacle, and one of the Japanese women, speaking in a lowered voice as though the walls could hear, said their sentiments were as his. When he asked where they had come from, he was told that all were from Urakami and that many others of similar faith lived there as well. As he subsequently wrote, an older woman asked him, "Sancta Maria no go zo wa doko?" (Where is the image of the Virgin Mary?) With this, he had no further doubts; he was surely in the presence of descendants of the ancient Christians of Japan whom the Europeans had long since assumed were annihilated. They asked him other questions and told him this was the season of the year they knew as *kanashimi no setsu* (the time of sadness—Lent) and wondered if he too celebrated this solemnity. He said that he did, and pointed out that this was the seventeenth day of Lent.

Soon other secret communities of Christians were identified in the Goto, at Imamura, on Hirado and Kuroshima, and in the Amakusa. They had survived the persecutions through geographic isolation and worshipping in secret. Although the rituals they followed varied somewhat from village to village, overall they were surprisingly similar, and the exceptions that existed were often curious. At Imamura, it was said that "on certain days, they gather together for prayer. They have retained in Latin the Our Father and the Hail Mary, repeat often the Kyrie Eleison and the act of contrition, and finally, curiously, they recite 63 Hail Marys in honor of the 63 years the Holy Virgin lived on earth." [11] There were also differences in the days of abstinence. In some villages, only the last two days of the week were involved, whereas in others, abstinence occurred on Wednesday, Friday, and Saturday. Those believers that adhered to the former regime were known as the Paterneshû, and the others Kirishitanshû.

These communities had evolved a hierarchical structure to serve their religious needs, and legends circulated that prophesied the return of the priests in black ships, who would be known by their celibacy, their devotion to Mary, and their acknowledgment of the supremacy of Rome. The members of these communities knew Japanese versions of prayers such as the Confiteor, Ave Maria, Pater Noster, and Credo, and were familiar with the Commandments. In addition some possessed a copy of a book, the *Konturisan* (Contrition), published in 1603, and a catechism, published in Nagasaki about 1607, in one of the oldest versions of romanized Japanese.[12] Religious medals and rosaries were common, and the major religious festivals were Christmas and the period of Lent.

These facts, in the minds of the nineteenth-century Catholic clergy, sufficed to identify the communities with the Christians of the period 1549–1640, and an association with the church in Nagasaki began. Petitjean was circumspect about his discovery, fearing a new wave of persecution, since freedom of conscience did not exist. However, word of these communities seeped out. Renewed oppression began in 1866, in Nagasaki, and continued until world opinion forced a halt in 1873. At that time, although freedom of conscience was not granted and the only religions officially endorsed by the Emperor Meiji and his government were Buddhism and Shintoism, a toleration of Christianity was proposed, but not before another three dozen of Urakami's population of Catholics met a tragic end at Tsuwano in Yamaguchi Prefecture at the hands of the Kamei family. Many were incarcerated for weeks in barred wooden boxes little more than a yard square; others were beaten or immersed naked in frigid waters in an effort to force them to apostasize. Petitjean vigorously pleaded their cause and ultimately played a significant role in achieving their freedom, physically and religiously.

Finally, a new constitution, promulgated on February 11, 1889, granted complete freedom of worship, and this has continued more or less undisturbed to the present time. Some measure of suspicion was directed at the clergy during the period of intense nationalism preceding and lasting through World War II, and again, the basic issues were the same: the irreconcilable positions of Christianity and the state religion, Shintoism, and the spiritual allegiance of the Catholic Church to the Pope.

The extent of the isolation of the Christian communities during the Tokugawa period is speculative. It is known that they divorced themselves as much as possible from other Japanese. This isolation, geographical or social, could not be overly obvious for fear of bringing on persecution and defeating its purpose. If the faithfulness to prayers, baptism, and other observances is an indication of the steadfastness of this group in all matters of Catholic teaching, then it is reasonable to assume that there was relatively little mixing of Christian and non-Christian. Furthermore, from the beginning of the persecution under the Tokugawas to the promulgation of freedom of conscience, there was a monetary value—100 pieces of silver—placed upon informing against a Christian living under the same roof as the informant, with higher bounties placed upon clergymen and minor church officials. This would seem to produce either a tendency of Christian to marry Christian to avoid exposure or, in the case of the marriage of a Christian to a non-Chrisitan, a tendency of the Christian to separate himself from the religious community, something that would limit marriages of the active Christians to others adhering to the precepts of their religion. But if these communities were small, as we in the Child Health Study believed them to be, this suggested that within a Christian village or hamlet consanguineous marriages might be common, and we were particularly interested in the biological effects of such marriages.

At the discovery of these communities by churchmen in the nineteenth century, the Christians numbered at least 10,000 individuals (some estimates run as high as 50,000). It is doubtful that they would have increased in number through conversion in the period of persecution, but they may have increased through growth of the population. At a time when infanticide was an acceptable means of controlling family size among most Japanese, its abhorrence among Christians could have accounted for a disproportionate increase in their number.

*   *   *

When the atomic bomb fell on Nagasaki, the city's Catholic community—most of whom resided within the sounds of the Urakami Cathedral's bells, Japan's largest Catholic church and the center of its largest parish—suffered disproportionately in dead and injured. One of the survivors, Dr. Nagai Takashi, a Catholic physician and author of *Nagasaki no Kane* (Bells of Nagasaki), *Kono Ko o Nokoshite* (Leaving this Child Behind), *Watashitachi wa Nagasaki ni ita* (We of Nagasaki), and numerous other books, was living closeby in 1945. He was dying of leukemia, possibly occupationally induced, since he had been chief of radiology at the university. Prior to the bombing, he had received many lengthy courses of x-ray treatment, seemingly to no avail, and his staff and colleagues had abandoned hope of recovery or amelioration of the course of his disease. On August 9 he was standing before a window at the medical school when the atomic bomb fell. He managed to escape death from flying and falling debris but was so irradiated that for several years he seemed fully cured of his leukemia. Subsequently, there was a relapse, but he did survive for six years following the bombing.

Another of Nagasaki's Catholic survivors was a man we call Gus. He was christened Hitoshi Augustine, for he was born on August 28th, Saint Augustine's day. For years he has been our Japanese voice in negotiations with the community. His family are old Urakami Catholics; indeed, they are among those families whose Christian roots predate the origins of modern Japan. His uncle was a monsignor in Yokohama, and several of his aunts were nuns; and his mother hoped that he would follow in this tradition and become a priest. He is one of ten children, only five of whom were alive at the time of the atomic bombing. Gus's oldest brother was in military service, another brother, also older, was employed in the Mitsubishi factories, a married sister with children lived in the vicinity of the Nagasaki railway station, and he and a younger sister were still at home.

Gus was in junior high school and, like most young men his age, had been mobilized to work at one of the factories supporting Japan's war effort. He was employed at the Mitsubishi steel factory, located about 1,100 meters from the Nagasaki hypocenter. On August 9th, there had been an air raid warning prior to the one that terminated in the atomic bombing. He and others had gone to the shelter near the boy's school. When the all-clear sounded, Gus lingered and fell asleep. Like most growing boys, he was perpetually hungry and tired—food was inadequate, and growth competed with work for the limited calories available. Gus awakened with the atomic bombing and emerged from the shelter into a world he could not believe. Destruction was everywhere; familiar landmarks were gone. Uppermost in his mind was his family. Where could they be? Days earlier his sister and her children had returned to the family home, on the supposition that it would be safer than her residence near the railway station. The family home was only 500 meters or so from the hypocenter, near the Urakami Cathedral.

Gus set out to find his parents and siblings but wandered aimlessly, bearingless. As darkness neared, he grew more desperate and his search more frenzied, but to no avail. He could not find them. Subsequently, he learned that he had undoubtedly walked past the spot where his parents, attended by his younger sister, were dying. They had survived the blast but had succumbed to radiation and injuries. His older sister and her children died, too; the bodies of the children were recovered, but not his sister's. His brother was seriously injured but survived. Gus himself was adequately shielded and exhibited none of the symptoms associated with radiation sickness.

As quickly as practical, Gus, his younger sister, and brother built a temporary shelter where their home had been and began a new life. Hunger was ever present; the limited rations available were hardly adequate for a sixteen-year-old. He found work on a farm in the area. At that time, most fathers in the Urakami valley worked at one of the Mitsubishi enterprises, but many had small farms that their wives cultivated more or less alone save at harvest, when the husband would take leave to help. Part of Gus's pay was in the produce he helped to raise and harvest. Gradually, the economic situation improved, and mindful of his mother's aspirations for him, he entered a Jesuit-run diocesan seminary in Tokyo as a representative of the Osaka diocese.

Upon completion of his studies in Tokyo, and prior to ordination, he was chosen in 1952 to go to Montreal to enroll in a two-year course in philosophy at the Sulpician Seminary there. As this course neared its end, he was sent in 1954 to the Grand Seminary in Montreal for four years of theology. The years passed, and as ordination approached he became progressively more worried about the responsibilities he would assume as a priest and less certain that the priesthood was his calling. Finally, in the fall of 1957, he left the seminary, persuaded that he could contribute more effectively to his community in other ways. He is now extremely active in his parish and church affairs generally and is the father of three children. The oldest is an autistic child; his son, the middle one, will soon graduate from a college in Kochi, and his youngest daughter has a degree in biochemistry from Kagoshima University.

Gus does not worry much about the possible late effects of his exposure but clearly sublimates events on that horrendous day. He is especially pained by the knowledge that he could have consoled his parents as they lay dying and faults himself for his inability to have done so. Gus, like most Nagasaki Catholics, is of the mind that their travail was "the will of God," and reiterates the old remark that Nagasakians found solace in prayer, whereas the people of Hiroshima responded with anger and accusation. He credits this to Catholic beliefs, and to Dr. Nagai, whose writing undoubtedly influenced public opinion, although initially his aim may have been merely to express or reflect those beliefs.

To all of us who live to be old, there comes a time when, among the people we have known, the dead outnumber the living; to Gus and the other survivors, this moment came early.

*   *   *

Kuroshima had come to my attention some ten years earlier, when on one of my periodic trips to Nagasaki from Hiroshima I was told of an island on which there occurred communities of Buddhists and Catholics. A mental note of the place was made, but by the summer of 1959 when I returned to Nagasaki, the name of the island had long been forgotten. As soon as the Child Health Survey was running smoothly, however, I set about trying to identify the name again. It proved easier than I had imagined, and happier still, the research opportunities I had previously envisaged—to determine how frequent consanguineous marriages were in these isolated Christian groups and to begin an assessment of their biological effects—seemed still to be there.

On Saturday, November 7, 1959, Dr. Yanase, Dr. Nemoto, and I departed for Sasebo on the 10:02 semi-express. This particular train, a diesel, had only third-class accommodations, and one of the hazards of traveling third class was that there were generally more passengers than places to put them. To forestall the possibility of finding ourselves without seats, we had made our plans for the trip very carefully. We arrived at the station thirty minutes before our train was to leave and positioned ourselves as close to the gate as possible. Our strategy called for Dr. Yanase and me to carry the heavier pieces of luggage, while Dr. Nemoto, the youngest and presumably fleetest, sprinted ahead to the train. There is a real art to securing seats under the circumstances I describe, but Dr. Nemoto was an artist. Timing is all important. When the attendant appears to punch the tickets, you must carefully brace yourself and begin to push with all of your might, wedging into the narrow entrance to the train platform. It is impossible to be adept at this, if you have a tendency to hesitate to elbow a person out of the way because of sex or age. You shove your ticket at the attendant, lean backwards ever so slightly into the crowd as the ticket is punched, and relax. If the relaxation has been properly timed, the crowd behind propels you through the wicket with sufficient momentum to carry you well along in the general direction of your train. If you have been quick enough, you can sprint through the coach door with no difficulty; if not, the maneuver at the ticket gate must be repeated at the coach.

Once on the train, you pick out the best possible spot and strew objects on several adjacent seats to reserve them for less fortunate or adept traveling partners. Before the train leaves the station, you provide yourself with all of the food needed enroute. At this season, Japan's mandarin oranges were the choice of everyone, and the peelings cast haphazardly on the floor added color to an otherwise drab coach. Each bag of mikans comes with a little folded paper sack into which the peelings can be placed, but only old ladies and foreigners are so compliant as to do so. Dr. Nemoto was successful, and the three of us enjoyed fairly comfortable seats. We equipped ourselves with two bags of oranges and settled back. The trip to Sasebo seemed interminable because of the large number of stops. Some five mikans after our departure, at 12:10 to be exact, we arrived at Sasebo Station, where we were met by Ura Shuji of the Sasebo City Office.

We stopped briefly at Tamaya, a department store, to purchase the gifts needed on the island. We bought two *shō*, somewhat more than three quarts, of Hakushima's Extra Special, a middling good sake, and decided this should be ample manifestation of our interest in the islanders and testimony to our station in life. At about 2 o'clock in the afternoon we boarded a ferry, the *Kuroshima Maru*. It was a sturdy, unattractive tug-like vessel of fifty or sixty feet in length that appeared to be the major source of communication between Kuroshima and the mainland. We were deposited aboard among bags of fertilizer, bottles of medicine, shōyu, and a host of other objects destined for the people on the island. The ferry had two small passenger compartments, one forward, one aft. Instead of seats, the compartments were raised about eighteen inches above the level of the deck; one sat on tatami. When we boarded, both compartments were fully occupied and we sat on the forward end of the ship on bags of fertilizer or the gangplank.

The ferry cautiously picked its way amidst a variety of floating objects ranging from mikan peelings to American landing craft, fishing sampans, and some sizable support vessels of the U.S. fleet. Sasebo was the United States' second largest naval base in Japan, exceeded in importance only by Yokosuka near Tokyo. The water in Sasebo Bay was calm; but once we passed through the harbor mouth, the sea became choppy. There was an open stretch of eight miles of water across which we steadily moved. As we journeyed toward the island, we repeatedly heard the faint chirp of a bird. It could not be coming from the shore several miles distant nor from the occasional gull passing overhead, for the chirp was distinctly not that of a gull. Our attention was finally attracted to the pilot's cabin, where there were three small *mejiro,* the Japanese silver-eye, a canary-sized gray-black bird whose eyes (*me*) are ringed by a white (*shiroi*) band.

Once this mystery was solved, we entertained ourselves by reading the material Ura-san had given us. We learned that Kuroshima is approximately 4.5 kilometers long and that somewhat more than 2,300 individuals live there, of whom about 420 are Buddhists and the remainder are Catholics. We learned further that the Buddhists are descended from samurai families from Hirado, whereas many of the Catholics trace their ancestry to individuals who moved to the island in the early seventeenth century to escape the waves of persecution sweeping over Kyushu.

By 4 o'clock in the afternoon we had made land on the eastern end of the island and sedately churned water while a small boat put out from shore to relieve us of several passengers and some of the sundries aboard. Within minutes we were again on our way to Todohira, the island's only protected harbor. Viewed from the sea in the late afternoon, Kuroshima appears little different from many other islands studding the west coast of Japan, but we were impressed by the tidy appearance of the few farms we could see as our boat rounded the various headlands on its way to the north side. At about 4:20 the *Kuroshima Maru* glided into a small breakwater-protected cove. The ferry was to remain overnight and depart for the mainland at 7 o'clock in the morning. With this knowledge, we and the dozen or so other passengers debarked.

On the breakwater, Kawahara Nobuyuki, an employee of the local branch of the Sasebo city office, who had been notified of our trip by Ura-san, was awaiting to take us to the island's one inn for the night. A scarce few hundred yards from the harbor, this establishment had seen better days. There were two buildings, the newer of which was divided into two modest-sized rooms; the larger two-storied building, perhaps fifty to seventy mats in size, was divided roughly in half. One of the halves served as the home of the owners and the other functioned as a meeting place, a dining room, and on occasion a bedroom. We were shown into this area, provided with tea, and asked to wait while Kawahara-san went to contact the other persons on the island we were to see. Within a very short time he returned, freshly attired and accompanied by Magome-san, the officer in charge of the Kuroshima branch of the Sasebo City Hall. After the usual introductions and exchange of calling cards, we set out in his company for the church.

The road was perhaps eight feet in width, wide enough to accommodate a single bata-bata, and as we toiled up it, the peace and serenity that pervades the island settled upon us. The only sounds we heard, aside from human voices, were those of nature—a bird, a frog croaking in a rice paddy, the rustle of leaves. The one road on the island makes a figure eight, the intersection of which occurs a few hundred yards from the church and is the hub of the island. In addition to the church, nearby are the primary and middle schools, the branch office of the city hall, the cemetery, and the largest cluster of buildings on the island. While we made no precise count, I guessed that there were perhaps a dozen to as many as two dozen houses or stores within a radius of two hundred yards of the intersection.

As one rounds a sharp bend in the road, a valley suddenly appears. The structure that dominates this scene is a lovely French-provincial-style church nestling against the side of one of the higher peaks amidst a group of pine trees almost as tall as the church itself. It was begun in Meiji 33, the year 1900. The builder, Joseph Ferdinand Marmand, a French priest, lies buried in the island's cemetery. The church is constructed of red brick, and all of the windows are of stained glass; both materials were imported from France. The interior is striking. The floor is of very broad hardwood planks uncluttered with pews but pierced here and there with wooden pegs. The altar is rough but lovely. Its roughness stems from hands unskilled in the intricacies of carving altars. Overhead hang several large chandeliers. Behind the church, higher on the hill, is the parish house, reached through a small, two-tiered garden. The rectory is a Japanese structure of moderate size.

We were greeted at the door by Father Tagawa, a man about sixty-three years of age with closely cropped gray hair and sharp un-Japanese-like features who had served the parish for a decade or more. He showed us into the small, sparsely furnished room that functions as the reception room of the rectory and served us black tea with ample quantities of sugar and cookies. From this room one gazes out on the curved wall of the sacristy of the church, framed by the various plants of the garden, the partially opened shōji, and the rōka. The setting sun imparted a rosy hue that glinted off the foot-high image of Christ on the cross which hung on the church wall. Occasionally, the wind rustled the pine boughs and set into motion the branches of trees denuded of their foliage by the approaching winter. There was an air of serenity and an invitation to contemplation seldom encountered.

Father Tagawa indicated that he would assist us in our local survey of consanguineous marriages and their health consequences in any way that he could. While we were speaking, darkness descended and the overhead light was turned on, closing off our view into the garden and to the church beyond. The island has its own small generating station, which provides electricity from 5:30 until 10:30 on weekday evenings and apparently all day on Sunday. Shortly after 6 o'clock we excused ourselves and began to wend our way to the inn for supper and the meeting to occur later in the evening with the village elders. The sky was clear, the air autumnal, and overhead a half moon served to illumine our way.

Half-way through supper Kawahara, Magome, and we were joined by Iwasaki-san and somewhat later by two school teachers, Toyomura-sensei, head of the middle school, and Kawaji-sensei, the assistant principal of the primary school. For the next couple of hours we had a very interesting, more or less continuous question-and-answer period with these five people. The school teachers said there were slightly more than five hundred students on the island. Neither of the teachers were Kuroshima islanders; they had been sent from Sasebo to teach. Of the three islanders present, Kawahara-san was a Buddhist, whereas the other two were Catholics. From their conversation, one gathered that the Catholic and Buddhist communities got along satisfactorily and that there was no semblance of discrimination, other than matrimonial, between the two groups.

While we were talking, one or another of the members of the owner's family had been carefully stoking the boiler beneath the bath in the back portion of the inn. After nine our guests departed, allowing

Dr. Yanase, Dr. Nemoto, and me to get to the bath and properly attire ourselves for bed. When it came my turn to bathe, I was taken to a remote room in the inn and shown the door into the room where I was told the *ofuro* was. The tub, if it can be called that, was a giant cast-iron cauldron. On top of the water floated a wooden platform whose purpose was to keep one's feet off the heated bottom of the ofuro. One was supposed to place one's feet on the platform and slowly sink into the water. Only subsequently did I learn that this cauldron-like ofuro is known to the Japanese as a *goemon,* its name stemming from a notorious, almost legendary bandit of the pre-Tokugawa era who was captured and boiled to death in a similar contraption. Had I known this, my reluctance to step onto the wooden float would have been even greater than it was. I found the water much too hot for comfort and was trying to slowly acclimate by sticking first a toe, then a foot, and finally an ankle into the water. I had previously soaped and rinsed myself clean beside the tub; one does not actually wash in an ofuro, for its purpose is to luxuriate in the hot water. This protracted effort to immerse myself was happening while I was naked, of course. Suddenly, I felt a draft and turned to the open window beside me to close it. Imagine my surprise when just beneath me a bit, my eyes fell on an elderly, toothless woman carefully stoking the fire beneath the tub with small pieces of wood. She smiled and unaffectedly said to me, "Komban wa" (Good evening). I was so embarrassed, however, that it was minutes before I could respond.

We adjourned to our room at about 10:30, chatted for a few moments, and then found ourselves without electricity and perforce decided that sleep was the only thing left to us. There was a touch of dampness in the air, but thanks to my thermal pajamas, not to mention two heavy quilts, I was quite warm. Sometime after midnight we were awakened by a loud noise at the front door. From the generally unintelligible conversation, and the stumbling about, it was evident that two visitors into their cups were attempting to gain entrance. They rattled the door for four or five minutes and finally gave up. Instead of moving on, however, they went around to the side of the house, where there was a window that could be opened. They proceeded to open it and to raise their lamp to a height to see what might prevent their access to the house. The light fell directly upon my face, and to these intoxicated gentlemen the appearance of a white face must have been an apparition from hell. One of them emitted a loud "saaa," the window was banged shut, and there followed a stumbling effort to beat a hasty retreat down the hill.

I awakened early on Sunday morning, and at about eight the three of us commenced our walk toward Nagiri and the church. Mass was at 8:30, and I had promised myself that I would arrive in time. On the way we passed the branch office of the city hall, whose location Dr. Nemoto carefully noted. Near the church we encountered other people on their way to Mass. Adult or child, they greeted us warmly. To the children, we, and particularly me, were the objects of considerable curiosity but never exceeding the bounds of politeness. By the time we reached the church, it was almost filled. We took off our shoes, storing them in the receptacles placed beside the entrance for this purpose, and stepped inside. The interior presented a most attractive sight; the light streamed through the simple stained-glass windows, creating a spectrum of colors on the white veils with which the women covered their heads in church. Dr. Yanase, who had not previously been in a Catholic church, was impressed by the beauty of the service, and I could not help but think that many parish priests would envy Father Tagawa his flock if they could witness their devotion.

Shortly after the introductory group prayers, Dr. Nemoto returned to the offices of the local government to examine the municipal records. Midway through the Mass, Dr. Yanase and I were approached by two women who were carrying a small bench for us to use so that we would not have to squat on the pewless floor. This was a thoughtful gesture, but we grew increasingly embarrassed at towering above the congregation. As the Mass—a Tridentine one, for the reforms of the ritual inaugurated by Vatican II had yet to come—progressed and the priest intoned the offertory, the sun splintered off the golden chalice he raised skyward and glinted spectrally off the head coverings. Immersed in their own piety, row after row of parishioners rose for Communion. Most remained after the "Ite missa est" to participate with Father Tagawa in the elevation of the Blessed Sacrament as the incense rose languidly, swept upwards from the thurible the young thurifer carried.

Once free to leave, Dr. Yanase and I visited the cemetery before going to the local office of City Hall for an appointment with Kawahara-san's uncle, one of the senior citizens of Motomura, the principal Buddhist community on the island. The cemetery is at the end of a well-worn path almost immediately across from the local office of the Sasebo city government. As one strolls through it, one is impressed by several things that make it different from other Japanese cemeteries. All of the grave markers, be they flat stones superimposed on the grave or the more conventional headstones, carry a large cross. One notices, too, that the markers record for most persons two given names, a Japanese one and a baptismal name of biblical origin. A further difference is that all of the grave markers bear the dates of birth and death not only in terms of era years, the conventional Japanese way, but also in terms of the Roman calendar. Embedded in the ground before some were receptacles for flowers, and here and there fresh chrysanthemums bravely confronted the chill. Scattered between the gravestones were a number of large, handsome cherry trees. In the spring of the year when these trees bloom it must be an appealing spot, with the blossoms an adornment for the graves. The only site Dr. Yanase and I specifically sought was the grave of the Reverend Father Joseph Ferdinand Marmand, who had died in 1912 and lies among the souls he patiently shepherded while he was alive. One is inclined to imagine that missionaries are put to rest in spots soon untended, forgotten; this is not the case for Father Marmand. Though the years have pitted and eroded the granite marking his grave, there is ample testimony that he has not been forgotten. Fresh flowers unfailingly rest there, as though sprouted.

Before Dr. Yanase and I reached the local governmental offices, Dr. Nemoto had succeeded in copying onto a sheet of paper all of the names of the Buddhist families on the island in whom we were interested, for they too lived in similar isolation and could serve as a comparison group for our findings among the Catholics. He happily informed us that the household records for the island people, which we would need to identify family relationships and verify reproductive histories, were all available on the island. The elder Kawahara, a florid-faced, open countenanced farmer with a rasping, deep voice, appeared, and after a few general questions, Dr. Yanase and I left Kawahara-san to Dr. Nemoto and went to a building presently occupied by a nurse but formerly the home of a physician. There, we were chagrined to learn that through some mix-up on the part of the city authorities in Sasebo, the islanders were expecting us to perform medical examinations. Dr. Yanase took upon himself the task of seeing the two dozen or so persons who were waiting for medical attention. I helped for a half hour or so as best I could. Shortly afterward we were joined by Dr. Nemoto, and a brief council was held. It was decided that Dr. Yanase would continue single-handedly, while Dr. Nemoto and I went to the church to see what was available in the way of church records of use to us. It was approaching 11 o'clock when we reached Father Tagawa's house.

He carefully described the church records and gave us the opportunity to peruse some of them. We learned that all of the matrimonial, baptismal, and death records normally maintained in a church were available for Kuroshima, and, moreover, they were complete from the founding of the parish in the late 1870s. Collectively, these records would permit us to identify consanguineous and nonconsanguineous marriages, to determine the number of children born to a couple, and to ascertain how long each child survived. This would allow us to learn, in turn, whether a consanguineous marriage was more or less fertile than a nonconsanguineous one, and whether the children of related parents were more or less likely to survive to adulthood.

As these thoughts crossed my mind, it was fascinating to page rapidly through the book of marriages and to note that the first weddings recorded were performed on the day before Christmas in 1879, only a year after the initial wooden chapel was built in Nagiri. The earliest marriages at which Father Matrat, Marmand's predecessor, officiated, judging from the ages of the bride and groom, sanctified unions that had undoubtedly occurred decades earlier before the parish was established. Most of these entries, including those of Father Marmand, were written in the spidery handwriting so common in the late nineteenth century.

Once we had satisfied ourselves on the completeness of these records, I approached Father Tagawa about the possiblity of having access to them or, better still, being able to copy them. He indicated that we were welcome to use the records in any way convenient for us so long as no defacement was involved. In short, with a little effort on our part we could obtain copies of all of the records at the Kuroshima parish. Before we knew it, it was twelve o'clock and Dr. Nemoto and I sheepishly decided that we had left to Dr. Yanase too much of the tedious work and should return to his assistance.

As soon as the last patient was examined, we hurriedly made our way to the wharf to board what we had been told was a high-speed boat for Ainoura, a small port to the south of Sasebo. To our dismay, the boat that was to rush us to Ainoura was one of the common fishing sampans one sees in all of the waters around Japan putt-putting along at about five or six miles per hour. We were thankful the sea was calm and placid. As we boarded the boat, Magome-san suggested that if we were to plan a major study on the island, the best time for us to return would be in December or January. There are no crops to be planted or harvested then, and fishing is at a minimum.

We did return, not at the time Magome had suggested but not long thereafter. This time we were a more impressive contingent, for in addition to Drs. Yanase and Nemoto and me, we had an additional physician, Dr. Kidera, two medical technicians, two clerks, and one of the Commission's photographers to copy the church records. Kitasako, the photographer, had not only brought a motion picture camera to record our examinations but had included a projector and several small rolls of film. We surmised that if it was known that a film was to be shown, we would have a larger gathering at our meeting of the village elders that evening, for films were a rarity. This proved true, and after a few carefully rehearsed remarks by me in Japanese and the showing of the movie, Dr. Yanase explained our purpose and answered questions from the audience.

Our greater number posed logistic problems that had not arisen when only Dr. Yanase, Dr. Nemoto, and I had visited the island earlier. The Minatoya, the old inn at the harbor, was not large enough to accommodate all of us, and as a consequence the five women with us were housed in a residence in Nagiri, near the school where the examinations were to occur, that had once been an inn itself.

Each day we examined about a hundred individuals, both parents and children. After the daily examinations were completed and before supper, I would often walk from the school toward Warabe, one of the western districts of the island, to enjoy the quiet of the island and the sunset as it glinted off the stone cliff or seeped through the pines and cedars overhead. I always found myself accompanied by eight or ten school boys who delighted in cavorting before me like so many playful puppies. They were pleasant company. One encountered none of the smart-aleckiness sometimes seen in the city. When, in my halting Japanese, I asked them to identify a flower, a vine, or other greenery, they vied with one another to see who could shout the name quickest. It became a game for them and for me. I learned the names of the island's vegetation, and they learned that foreigners were not to be feared and were curious about Japan. I still recall the small leafed vine, called *issho*, that clothes so many of the rocks on the island, giving them life, to which they introduced me.

As our examinations neared their end, a curious incident arose. Dr. Yanase and I had left the school for a few minutes in the afternoon, and upon our return found that a young newspaper correspondent had not only made his way to the island but to the school where we were at work. I do not know how he was clued to our presence on the island, but someone's tongue had apparently been wagging loosely. The poor reporter, who was from Nagasaki, was undoubtedly one of the most misinformed persons that I have ever encountered. He believed that our undertaking was an ABCC affair, which it was not, and that we had gone to Kuroshima for nefarious reasons that he had been led to believe had something to do with the necessity to obtain blood for tests presumably sinister in motive. We were drawing blood, but for genetic tests to measure the biologic differences between the Buddhist and Catholic communities.

Fortunately, the young reporter was of a type to whom one could speak and who was willing to be set straight if he was wrong. Dr. Yanase satisfied him with regard to our purposes, invited him to watch our examinations, and dissuaded him completely from his previous erroneous notions. The article he subsequently wrote for the *Nagasaki Shimbun* was an eminently fair and unbiased presentation of our study. It was interesting, however, to see the consternation that developed in our group with his appearance. Dr. Kidera dashed off immediately to the dispensary so that she would not be obliged to speak to him, and everyone else made themselves as scarce as they possibly could. I found this surprising; but on other occasions since, I have been amazed at the extent to which the press in Japan seems feared by scientists. Are they fearful of misrepresentation, or do they merely seek to shun publicity? We are accustomed to sloppy or misdirected reporting in the United States, and to the embarrassing consequences of it, but certainly there is no fear of the media.

Gradually a greater and greater curiosity built in me regarding Father Marmand, who was obviously a key figure in the history of the church on Kuroshima. As one is wont to do, I had formed an image of him in my mind. He had undoubtedly been an ascetic cleric of saintly mien, cheeks thinned by the hardships of the missionary's life. Father

Tagawa, I learned, had a picture of Father Marmand, and I could hardly contain myself until I could see it—obviously to confirm my image. The photograph, however, revealed Marmand to be a stocky man with white hair and a long white beard, more Santa than saint. I was to hear from others that he had been a gentle, short, fat, low-voiced priest who had come to Kuroshima directly from France. When younger, he had had dark brown hair, but most of those who were still alive and could remember him had known him only as a white-haired man. He shepherded his island flock for over twenty years, but I never learned whether he ever visited France again—given the time, probably not. His only contact may have been through his church, which must have reminded him of many in rural France, and an occasional letter from home, and these undoubtedly grew more infrequent as he aged and his relatives and friends preceded him in death.

On our return to Nagasaki, we immediately began the collating and tabulating of the information that we had collected. As the analysis of these data proceeded, it became obvious that the children born to consanguineous marriages were more likely to die early than the children of nonconsanguineous marriages. Most of these deaths occurred in the first year following birth, but we could discern no specific cause or causes to which this greater mortality could be ascribed. However, consanguineous marriages appeared to be more fertile, and thus almost the same number of individuals survived to adulthood in these marriages as in nonconsanguineous ones. There was no simple socioeconomic explanation for this, for we had been unable to discern any significant differences in the ways of life of the Buddhists and Catholics, or, within each of these groups, between couples consanguineously and nonconsanguineously married. Did it imply a conscious compensation for the greater loss of children in consanguineous marriages? And if so, what were the genetic implications of this? As frequently happens in science, a study initiated to answer one question culminates in still more questions. These could be answered only by a larger undertaking than the one on Kuroshima.

As I now look at the notes I made so many years ago, I cannot help wondering whether the island was really as lovely as I thought. Or was this just one of those rare concatenations of events, when the weather, one's psyche, companions, and everything are attuned to a memorable occasion, and a memorable occasion would have occurred, wherever one went? Or is this thought just the cynicism that colors our lives?

Perhaps the island does in fact enjoy a special mandate. I am loathe to seek the answer.

*   *   *

Although the roles of the Dutch and Portuguese in Nagasaki's history have been well publicized, that of the Chinese has not. However, when Nagasaki began to develop as a significant port, it was not long before Chinese merchants settled there. Initially, they came when they wished, and in such numbers and with such merchandise as they pleased. However, as Kaempfer observed, "The liberty, which the Chinese for some time enjoy'd in this country, was too great to continue long without alteration, and it came quickly to a fatal end." [13] As their number and trade grew, suspicions and circumspection among the Japanese arose, and in 1688 the Chinese too were confined to a specific area of the city, known as Juzenji, a short five-minute walk from the present Dejima dock. At the time of its establishment or shortly thereafter, this compound, now bounded by Junin Machi and Kannai, was said to have housed several thousand Chinese. Few of the early buildings have weathered the ravages of time, and those that have are mere shells of their former selves. Under the Tokugawa this area was enclosed in earthen walls and strict rules were published to prevent the Chinese from living outside the settlement. Their lot could not have been an easy one, but as Kaempfer contemptuously observed, "So great is their covetousness and love of gain, that they suffer themselves to be so narrowly watch'd, and every bit so badly, if not worse, accommodated, than . . . the Dutch to be at Desima." [14] Some of his scorn, however, may have reflected his annoyance that they were not treated as vassals or hostages, as Kaempfer described the plight of the Dutch; indeed, he used the Japanese word for a hostage, *hitojichi*, although he spelled it hitozitz. The Chinese were free to buy their victuals and provisions at the very gates to their compound, without the bureaucratic Japanese intermediaries the Hollanders were obliged to support. Nor were the Chinese obligated to the onerous and expensive annual trip to the shogunal court that was the lot of the Dutch. These trips were not only physically arduous, involving travel by horseback, boat, and palanquin; they took weeks, even months, to complete. Once in Tokyo (then called Edo), the Dutch often found themselves cooling their heels until the shogunal government thought the time for their reception was auspicious. And at court, they were obliged to dance, parade about, and act the role of buffoons. It was a demeaning experience. However, out of this travel, through the eyes of Kaempfer and Thunberg, came the limited knowledge the West had of feudal Japan, its culture and people. The differences in the restrictions placed upon the Chinese and Dutch were more matters of degree than kind, for the shogunate distrusted all foreigners and sought to prevent both the Chinese and the Dutch from disseminating political and religious thoughts held to be subversive. So paranoid about these matters were the shogunal officials that the Chinese merchant vessels that called at Nagasaki were thoroughly searched for contraband literature.

The Chinese attempted to convince the Bakufu that they were Taoists and Buddhists, not Christians. They built a number of temples in Nagasaki, some of which are still active, such as the magnificent Sofukuji and the even older Kofukuji, and staged parades of their gods through the city's streets. Even now, the Chinese dragon is an integral part of the festivities which accompany Okunchi, the big Suwa Shrine festival. The merchants who were confined to Juzenji had their warehouses in an area known as Shinchi, closer to the harbor. With the restoration of the Meiji and the dissolution of the restrictions that confined the Dutch and Chinese to the Dejima and Juzenji areas, most of the merchants relocated themselves to Shinchi to be nearer their warehouses. Today, this remains Nagasaki's Chinatown.

Throughout most of the last hundred years, the Chinese community in Nagasaki has been substantial, and as a result there has inevitably been a Chinese consulate there. With the rise of Mao Tse Tung, and the expulsion of the Chiang Kai Shek government to Taiwan, the consulate was staffed with members of the Taiwanese government. Once Japan recognized the People's Republic of China, this relationship ceased and the consulate was closed, but I write of an earlier time.

The Chinese consulate as Vicki and I knew it was located at 52 Nakagawa machi, not far from our home in Fufugawa. Each year on October 10th, the consulate entertained local authorities and the members of the foreign community in Nagasaki at a reception. The occasion was the famous 10–10 celebration commemorating the overthrow of the Ching government on the 10th day of the 10th month (October) in 1912. In 1959, the consul's name was Chang Chia-kai and his vice-consul was Liu, who, like most Taiwanese consular officers, expected to be assigned to a post in an English-speaking country at some point in his career. While he spoke English reasonably well, his wife spoke very little. It was she who wanted to learn English and who offered to teach Vicki and Bernice Goldstein, the wife of one of our friends, Sanford Goldstein, a Fulbright professor at Nagasaki University, Chinese cooking in return for an afternoon of English conversation.

Mrs. Liu decided what dishes to make, and brought the necessary ingredients. It was supposed to be a meal and a social experience for the three women, but either Sanford or I also profited, for there was more than ample food. Whether he got it or I did depended upon whose home was the scene of the lesson. Normally, a couple of dishes were prepared, and Mrs. Liu tried to provide a variety of beef, pork, and chicken menus. On occasion, the women were joined by So-san, the Chinese cook who worked at the Commission's Matsuda Hall. Once a month So-san prepared a Chinese dinner, and on that evening none of us ate at home. A multicourse meal at the equivalent of a dollar was too much to miss, but it was more than economy that motivated us. So-san, an exceptionally good cook, whose own life had been especially trying, clucked about us like a mother hen, making certain that we tried each of her dishes and repeatedly so.

I have often wondered whether Mrs. Liu's English instruction has had as lasting an effect as her teaching, for almost thirty years later we are still eating the delicious dishes she taught Vicki to prepare.

8

HIRADO, 1964

*V*icki and I were in Japan again, this time with a somewhat more confused purpose. We were to spend six months at the Atomic Bomb Casualty Commission and six months with a separately funded joint research program, involving the University of Michigan and the Commission, whose purpose was to define further the effects of consanguineous marriages on the outcome of pregnancy. The first three months, late February to late May, would be in Hiroshima, then we would spend six months on the island of Hirado, in the sea to the west of Kyushu, and a final three months back in Hiroshima. Most of the time in Hiroshima would be devoted to examining heretofore unanalyzed data that had accumulated on the possible genetic effects of exposure to the atomic bombings and to planning the Hirado study. We arrived with the chill of winter but were welcomed at the railway station with the warmth that old friendships had led us to expect. Soon we were ensconced in one of the small apartments in Hijiyama Hall and swept up into the small social circle the Commission's American contingent seemed to continuously sustain.

Each day made more perceptible the changes that had occurred both in Japan and at the Commission. Life had measurably improved for both. Reinvestment of profits into plant modernization and new construction in the preceding years and the emphasis upon export trade were beginning to ameliorate hardship and better the quality of life among the Japanese. Massive governmental investment had furthered these ends. This was the year for the summer Olympics, and in preparation for the games work had begun several years earlier on the new high-speed electric rail line, the Shinkansen, that was to connect Tokyo and Osaka, and on the first stages of the Tokyo subway system. Every safety feature conceivable had been incorporated into the Shinkansen, including a wholly new rail bed elevated on concrete trestles virtually free of grade crossings and a sensing system that automatically halted the trains if the continuity in power was altered, as might happen with an earthquake or landslide. The trains had been completely redesigned to be less wind resistant and to tolerate curves at higher speed with minimum sway. New stations had been built, generally as extensions of the old in the larger cities, but many local stops had been eliminated. Two classes of trains plied the line: the Hikari—super expresses that made only two stops between Tokyo and Osaka, at Nagoya and Kyoto—and the Kodama, which stopped at smaller cities along the route. Schedules were adjusted so that residents in these small communities could begin their trips on the Kodama but at the first major station transfer to a Hikari, often from the same platform. To accommodate as many pocketbooks as practicable, two classes of travel were created, first class, the so-called "green cars," and a second class with somewhat more crowding and the option to reserve or not reserve a seat, the former at extra charge.

An almost instant success because of its speed—only three hours from Tokyo to Osaka—the Shinkansen in its early years was so profitable that it literally supported the annual loss incurred by the older, conventional line. As its fame has grown, however, other regions of the country have clamored for an extension, and the nation's politicians have acquiesced. Subsequent lengthenings have been enormously more expensive to build and have served areas with population densities too low to amortize the cost. Glutted with superfluous personnel, the system has sunk into debt.

At the Atomic Bomb Casualty Commission, George Darling, a former Yale Professor of Human Ecology, had been Director since 1957 or so. An affable, portly, deep-voiced man with a cultivated, courtly manner, a sense of drama, and a flair for binational sensitivities, he was nonetheless accepted by his more scientifically oriented associates with some reservation. Presumably they expected not only an adroit administrator but a scientific father-figure, and George was not this. He delegated that role to his successive chiefs of research and to the heads of the individual research departments. He intervened little or not at all in the Commission's research program. His function, as he saw it, was to adjudicate policy and to provide the bridge to the lay and scientific Japanese communities, a function at which he was exceptionally adept. Soon after his arrival he had organized liaison committees in Hiroshima and Nagasaki and had established a Japanese Advisory Committee that met annually, or nearly so. He chose men of distinction, either because of the office they held or the renown they had achieved through research, to serve on these bodies. George could appear more formal than they, bow deeper at important social functions, and, in his carefully tailored striped suit, seem peerless at memorial or dedicatory ceremonies.

To these same ends, he introduced the use of bilingual forms throughout the organization, a program of bilingual publications, technical and nontechnical, joint translation of staff meetings, and similar changes designed to emphasize the binationality of the Commission. He encouraged the participation of the professional staff in scientific meetings in Japan and elsewhere as a means of furthering their technical competence and spreading more widely the achievements of the institution. Almost single-handedly, he redressed the Japanese community's dim perception of our presence and activities. He was the inveterate joiner, a member of the Rotary, a member of the local Japan-America Society and the like, all of which served a purpose. This sense of his function was inevitably destined to alienate some of the professional staff, who sought a scientific colleague. Disconcertingly, he seemed to speak at us, through us, around us, but rarely to us. Nevertheless, he was an important transition figure, for without his careful orchestration the reorganization that was to come in 1975 would have been much more difficult, as Japan grew in its own perception of importance. Directorship of the Commission has always been a thankless lot, torn between a sometimes militant union and a mounting Japanese sense of self-esteem, a distant National Academy of Sciences, and the financial shortfalls in the budgets of the Atomic Energy Commission and its successors, the Energy Research and Development Administration and the Department of Energy, where a research institution in Japan had few advocates in a Congress dominated by local interests.

Elsewhere, with singular exceptions, the staff had changed. Howard Hamilton continued to preside over the clinical laboratories, as he had since 1956, and Walter Russell was still chief of the Department of Radiology. There was a new head of the Department of Medicine, Kenneth Johnson, but its routine clinical activities hinged heavily, as they had for some time, on young Public Health Service physicians who elected this option in preference to military service. Few had research credentials, but the initiation of new research was not their primary function. Other methods of recruitment, short of this unacknowledged draft, had been tried, unsuccessfully. Institutional affiliations had been sought, on the supposition that these would serve as foci not only of recruitment but of continued scientific guidance as well. Arrangements had been made with the Department of Medicine at Yale, the Department of Pathology at the University of California in Los Angeles, and with the Veterans Follow-up Agency in Washington. Only the last provided the long-term stability that was sought, and none of the continuity in direction and purpose the Genetics Program has always had, as a result of its affiliation with the University of Michigan. Academic affiliations, if not driven by a sense of the uniqueness of the study and the scientific opportunities it holds, must inevitably wither. They do not acknowledge the impermanence of the staff of departments, including the chairman, their commitments, nor the nature of the Commission's needs. Institutional commitments without overriding personal ones cannot succeed. It is not the immediate return that counts but perseverance and a belief in the long-term importance of the undertaking.

Research at the institution became increasingly bureaucratized. Each proposed project had to be carefully described, its budgetary implications identified, and a schedule for completion drawn up. These protocols were reviewed, generally within the Commission; but occasionally outside authorities were consulted. Program reviews occurred with intimidating regularity, and each encounter induced its own turmoil. Summaries of progress on existing studies had to be prepared and future activities defined and planned. Although these steps were essential to a properly managed scientific program, each took time that might have been better spent on further work. Moreover, most of us were unaccustomed to institutional research and the constraints it necessarily imposed. We were used to designing our own research, implementing the experimental strategy, describing what we had done, and then awaiting peer acceptance or disapproval. In an institutional setting, one's freedom as an investigator can be compromised by institutional needs and by advice that might be sound or might be capricious. A review panel can impose directions without accepting any of the responsibility for misguided advice. There is an oracular element to this process that is troubling. Fortunately, at ABCC the reviewers rarely imposed their judgment; indeed, their comments were generally uncritical. It was difficult, however, to know whether this signaled approbation or merely a poorly prepared group of referees. To the individual investigators, there often seemed little intellectual return for the efforts that went into preparation, but the Commission's perspective was different. These periodic reviews spread the responsibility for its activities around, and the organization could always point to its attempts to seek advice should anything ever go awry.

Time had wrought other changes; the grounds that surround the guest quarters had an attractiveness that had not earlier obtained. Dogwood, white and pink, honeysuckle, and native plants such as azalea and *tsubaki,* the Japanese camellia, bloomed steadily through the seasons. On spring evenings before summer had settled in full force one could see hawks soaring effortlessly on the updrafts created by Hijiyama, and plummeting occasionally when the tide was in to grasp an unwary fish that had risen too closely to the surface of the Kyobashigawa, one of the tributaries of the river that slices through the city. Still further across the valley to the west, Mount Suzugamine rose, wreathed of an early evening in a glorious halo of red and orange as the sun slipped slowly from view behind its mass. Clouds drifting impassively above would reflect the waning glow of the sun long after it had disappeared. There was a peacefulness heightened by the calmness of the air faintly tinged with the smell of the sea. It was hard to believe that scarcely twenty years earlier this same sun had set on a sea of desolation, scorched timbers, and charred bodies twisted in a last gasp for life.

*   *   *

In Japan every season has its flowers. Autumn is alive with chrysanthemums and *higanbana,* a striking, showy red flower of the genus *Lycoris,* a relative of amaryllis. February is graced with plum blossoms, May with azaleas. But no blooms capture so thoroughly the Japanese spirit as those of the cherry tree. Indeed, the behavior of these blossoms, cut off at the peak of their beauty, was an integral part of the way of the warrior, the Bushido. In Japan, the cherry tree comes in numerous forms; there is the mountain cherry, the *yamazakura,* one of the few whose leaves erupt with the blossoms; the ordinary cherry, with its many different varieties, each with its own name; the double-blossomed cherry, the *yae-zakura;* and the weeping cherry, the *shidare zakura.* Like a gigantic wave, the cherry blossoms sweep northward from Kagoshima, at the southern tip of Kyushu, until eventually even Hokkaido is engulfed. They first appear in southern Japan in late February or early March, and slowly spread northward in pace with winter's retreat. For most of Japan, though, April is the season of *hana-mi*, literally "flower-seeing." However, hana-mi is the occasion for more than the viewing of these lovely but odorless flowers; it is a time for parties and revelry—picnics under the blossoming trees. We at the Commission have come to know these festivities especially well, for we have been the uninvited guests at tens, possibly hundreds, of them.

In a year in which the weather has been especially propitious, Hiji-yama is carpeted with blooming cherry trees. They are at their best for only a week or so and soon are replaced by the erupting leaves, fragile and pale in their newness. During this week, white and red paper lanterns sprout amid the trees, and the citizens of the city descend en masse on the area to stroll beneath the blossoms or to enjoy a round of parties. The roads that sector the mountain are closed to traffic, though our cars are allowed to proceed cautiously to the clinical buildings or our quarters still higher on the hill.

Groups infiltrate the park like SWAT teams, each armed with their grass mats, box lunches, and ample quantities of sake and beer. An attractive spot is selected, preferably beneath the floral arms of one of the cherry trees, and the mats are spread. Little mounds of mikan, senbei, obento materialize from pockets, packets, and bags and are deposited on the mats. Everyone carefully removes his or her shoes and selects a spot on which to sit. In a slightly rising titer of conversation, the lunches are opened and the eating commences. However, food without drink is unthinkable. First one cap, then another, and then another is lifted from the bottles of cold beer, and sake materializes for those who find it more palatable. Sake is always difficult to serve at the proper temperature under the trees, and so more and more it is drunk at whatever temperature prevails. "Kompai" (Cheers) rings through the air. As the beer and sake gurgle down more throats, blushes begin to suffuse the faces of most of the drinkers. It is not embarrassment but physiology. Many Japanese do not detoxify alcohol rapidly, and a dilatation of the peripheral blood vessels occurs. As this takes place, more blood reaches the face, and reddening results. It is all the fault of an enzyme, aldehyde dehydrogenase, instrumental in the degradation of alcohol. Some 90 percent of Japanese are thought to have a variant form of the enzyme that does not convert acetalde-hyde, the first product in the destruction of alcohol, as readily as the form of the enzyme that is more prevalent among Westerners.

Alcohol has effects other than vasodilatation. Conversations rise measurably in volume, as though a progressive deafening had de-scended on everyone. Embarrassed remarks are made about the flush-ing, and more than one hand will seek a face to feel its warmth, as the expression "atsui" (hot) is heard. Virtually none of these conse-quences slow the drinking, and soon someone, generally a man, finds song irresistible. It is odds on that the words of "Sake nomu na," an old drinking ditty, will spill from his lips. As his voice strengthens, others may join in the singing or will clap in rhythm and voice the refrain, "oi! oi!" at the appropriate places. Another song follows, and soon a momentum has been achieved that will sustain itself for some time. And what is song without dance? A situation to be remedied!

A samisen appears. A circle of dancers forms on the mats and be-gins to move to the cadence of the instrument and the singing. Folk dances serve these occasions especially well. They accommodate to the space, or to dancers one to ten, and are slow-moving enough to be compatible with the mats and those seated about. Inevitably the old favorites are danced—the tanko bushi, soran bushi, awaji odori, and on. Soon the moment of decision arrives—to stop, flushed but exhila-rated, or to continue to party to an alcoholic stupor. If it is a midday picnic, the decision is usually an easy if reluctant one, since many have to return to work. If it is a late afternoon or early evening affair, that's a different matter. Beer and sake will continue to flow, the dancing will become more boisterous, erratic, and wobbly, and the singing will grow mumbled or wholly incoherent. Occasionally, the end comes with all survivors asleep on the mats, oblivious to passers-by and other parties. More commonly, there will be a loud leave-taking and an unsteady descent to the tram line at the bottom of the hill. It is mostly good fun, but Hijiyama is a mess. Few of the bottles or cans of beer and sake so laboriously carried up are toted down. Remnants of the box lunches are scattered about and only fortuitously end near the receptacles placed to receive them. Each morning a living vacuum sweeper arrives to ready the mountain for the next day's guests. Gar-bage vehicles scurry about like dung beetles tidying a pasture. While all of this happens, we at the Commission grit our teeth and prepare for another day of celebrations, consoling ourselves with the knowledge that soon it will end and calm will descend as the cherries march further northward.

*   *   *

Our earlier studies in Hiroshima and Nagasaki on the health and well-being of children born to parents who are biologically related to one another, like cousins, were incomplete in several respects. They had begun by identifying a group of children for study and determining whether their parents were or were not related and, if related, how. This approach could not tell us whether consanguineous marriages are more frequently infertile. We could learn this only by identifying potential parents and afterward their reproductive history. To do this we reasoned that an area, a geographic district, had to be identified where consanguineous marriages were common and easily recognized. For a number of reasons, it was further desirable if the district was an administrative unit that collected and maintained its own birth and death records. These considerations led us to contemplate several possible study sites in southern Japan: Hirado-jima, Fukae-shi in the Goto, and Tanegashima to the immediate south of Kyushu.

One of my first functions upon returning to Japan was to determine which of the areas we had tentatively identified would be most suitable, and to do this I consulted Dr. Yanase, whose opinion I respected and whose knowledge of the local scene figured prominently in our plans. The two of us went to Hirado to examine the local records, assess the receptivity of the islanders to a health study, and examine the resources that might be available to support the undertaking we envisaged. We knew Hirado to be surprisingly homogeneous, save in religion, and that fishing and farming accounted for the livelihoods of over half its inhabitants. Most of the farmers were small landowners, ordinarily cultivating less than one *cho* (2.45 acres) of land. With such modest holdings, most farm households were obliged to seek other sources of income, but this was true of much of Japan. Only one third of farming families derived their livelihoods solely from farming; members of most households, often the head, were engaged in unskilled seasonal labor—woodcutting or road construction. Farm income was small; the modal annual income was only 150,000 yen (then somewhat more than $400). Although scarcely 200 of the 40,000 or so residents of Hirado had attended a university or professional college, illiteracy was virtually unknown, and the residents displayed a healthy curiosity about the island's future as well as pride in its past.

The study contemplated would take nine months to a year to complete and would involve a staff of ten to fifteen people, half of whom would need living quarters. Obviously, we would need offices in one, possibly more, of the island's major communities. However, housing and office space were limited in rural and semirural areas thoughout Japan, and Hirado was no exception. Moreover, we hoped to be able to complete the health assessments themselves in the school buildings distributed over the island, and this would require the approval of the local school authorities and the various parent–teacher associations. There was also the need to solicit the approval and cooperation of the local midwives and public health nursing groups, since they figured importantly in our census plans. A program of the complexity we foresaw could not possibly be arranged in a day, and we had therefore made plans to stay at the Taguchi-ro, one of the harborside inns, for several days. Indeed, before we would move to the island, several trips were necessary, but this attention to detail was amply rewarded: only nine of the over eight thousand households involved in the census refused to participate.

Prior to going to Hirado, Dr. Yanase and I had stopped in Nagasaki to discuss our proposed study with our governmental friends there and seek their endorsement. The governor's office phoned the *shisho-chō* (mayor) of Hirado, since the city is within Nagasaki Prefecture, and the governor conveyed to the mayor his support of the study, to be called the Hirado Health Survey. As a consequnce, when we arrived on the island we were cordially received. Once we had acquainted the mayor with what we hoped to do, he suggested he communicate this to his council, for he was certain they would approve, and he further indicated that the city could undoubtedly find office space for us both in Hirado-machi itself as well as in one of the more westerly communities on the island, such as Tsuyoshi, where a branch office of the city hall existed.

At that time, Hirado-jima and a dozen or so nearby islands (only two of which are inhabited, Takashima and Takushima) constituted a uniquely Japanese administrative unit, a *shi*. The word is commonly translated as city, and in most instances this is appropriate, but in some, and Hirado is one, a shi has a different connotation than we in the United States or Europe might envisage. It does not necessarily imply a continuous, high-density area. Hirado-jima had more than sixty distinct, separate communities varying in size from villages with under 100 residents to the town at the eastern end of the island that had a population of 7,000 or 8,000. This town, called Hirado-machi, is the center of government for the administrative city.

In Hirado-machi we also paid a courtesy call on Dr. Kakizoe Shinobu, a thin, energetic man whose sole passion beyond his work was his cars. Aside from the hospital ambulance, which he owned, he had four other automobiles—various Japanese ones and a large black Mercedes, his "Benz," of which he was enormously proud. Dr. Kakizoe was the most influential physician on the island, and his hospital was the source of his importance and wealth. It, like the vast majority of hospitals with fifty beds or so, was privately owned and profit-oriented and subsidized by the Japanese income tax system: to encourage physicians to participate in the national health insurance program, over 70 percent of the income they derive from their participation is tax free. The program sets the fee structure, and the user pays very modest direct charges. But the taxes paid by the salarymen to subsidize the health care system are substantial.

*   *   *

Hirado-jima, some 106 miles in circuit, lies approximately a mile to the west of Kyushu, near the border of Nagasaki and Saga prefectures. For some eight centuries, it has been the center of the Matsuura *han,* a minor fiefdom in western Japan, and has played a role in Japanese history that considerably transcends its present limited importance. Archaeological evidence reveals human habitation of some antiquity, but fact and fiction are so intimately interwoven in early historical references to Hirado that it is not until 1225 and the permanent establishment of the Matsuura family as the island's feudal lords that a firm history begins. National prominence was achieved shortly thereafter through Hirado's part in Kublai Khan's ill-fated attempt to invade Japan in the thirteenth century. The next several hundred years brought a golden age, the island's greatest national and international importance. The Portuguese and Spanish used it as a secondary trading center from the 1550s until 1638, when they were barred from further commerce with Japan, and it was the major focus of the Dutch mercantile enterprise until they were moved to Dejima in 1641.

Even the English briefly entertained notions of a Japanese commercial effort, largely as a base from which to trade with China, and to this end Captain John Saris was sent to establish a factory on Hirado, or Firando, as it was more commonly then called, in 1613. Eight Englishmen, most of whom were to meet untimely ends—Richard Cocks, William Adams, Tempest Peacock, Richard Wickham, William Eaton, Walter Carwarden, Edward Sayers, William Nealson—three Japanese interpreters, and two servants made up the group. Housed initially in quarters rented from a Chinese merchant near the Dutch factory until more suitable and commodious accommodations could be built, the Englishmen eventually expanded their enterprise into Tokyo, Kyoto, Osaka, and Sakai. But on orders of the shogun, Tokugawa Hidetada, they were obliged to abandon these branches in 1616 and restrict their trading to Hirado. Through well-intentioned but inept management and the alleged conniving of the Dutch, who had established their trading center on a grander scale on Hirado four years earlier, the British factory met with limited success and was abandoned within a decade after its inception. Matters were not made better by a devastating typhoon that struck some three months after the establishment of the factory, on September 1, 1613, destroying parts of the Dutch and English establishments and doing extensive damage to hundreds of the island homes, including that of the Matsuura daimyō.

The trials and tribulations of this enterprise are described in the diary of Richard Cocks, the Cape or chief merchant. Almost from the outset he deplored the choice of Hirado as the focus of the factory's trading and plaintively noted that the stock he was to purvey consisted mostly of woolens, which had little market value in Japan. He complained bitterly of the constant undercutting of prices by the Dutch and the fact that the Japanese used this practice to their advantage to drive prices still lower. He also observed that "Firando is a fisher towne and a very small and badd harbor, wherein not above 8 or 10 shipps can ride at a tyme without great danger to spoile one another in stormy weather; and that which is worst, noe shipping can enter in or out of that harbour, but they must have both tide and winde as also 8 or 10 penisses[1] or barkes to toe them in and out, the current runeth soe swift that otherwaies they cannot escape running ashore."[2]

His view of the local lords, the Matsuura, whose fiefdom he valued at "six mangoca" (60,000 koku) was not more flattering. He opined,

"At Firando there is the kings hym selfe, with two of his brothers, and
3 or 4 of his uncles, besides many other noble men of his kindred; all
of which look for presentes, or else it is no living amongst them; and
that which is more, they are allwaies borowing and buying, but sildom
or neaver make payment, except it be the king hym selfe. So that it
maketh me altogether aweary to live amongst them, we not being
abell to geve and lend them as the Hollanders doe, whoe geve them
other mens goods which they neaver paid for."[3]

Cock's complaints about the apparent duplicity of the local lords
were not unique. Alejandro Valignano, an apostolic visitor to Japan
and the Superior of the Far Eastern Jesuit mission in Goa (India),
wrote scathingly of Matsuura Takanobu and his son, Shigenobu.[4] Possibly some of this invective was deserved, but most was probably not.
The clash of perceptions was undoubtedly fostered by differences in
religion and a failure of the Europeans to appreciate the political
forces at work in Japan in these years. The Matsuura fiefdom was not
a major one, and its rulers necessarily resonated to the demands of
more powerful lords in their vicinity, such as the Nabeshima and Ku-
roda, and to the growing centralization of authority with the rise of
Nobunaga, Hideyoshi, and Tokugawa Ieyasu. However, when the
persecutions of Christians began, no local lords exceeded the Matsuura in zeal. Indeed, the first of the martyrs were from this island.

Not much is known about Cocks' origins; he was apparently born
in Seighford, Staffordshire, in 1565, the third son of a yeoman. His
name appears in the charter of incorporation of the East India Company, at which time seemingly he was already a merchant. His diary,
couched in quaint language with a charming disregard for consistency
in spelling, reads like a continuous bill of sale or a litany of gift-giving
and reveals sparingly his thoughts on the country and its people, as
well as the nature of the man himself. Compared with Kaempfer's or
Thunberg's subsequent accounts of their experiences with the Dutch
factory in Dejima, Cocks seems oblivious to the Japanese culture but
sharply attuned to the vicissitudes in the life of a factor obliged to deal
constantly with drunken, whoring, dissolute sailors, disgruntled colleagues, and a bureaucracy constantly trimming its sails with each political event in Edo, as Tokyo was then called. Not given to hyperbole,
he cryptically described his audience with the shogun Hidetada, in
September 1616, as follows: "The Emperours pallis is a huge thing,
all the rums being gilded with gould, both over head and upon the

walls, except som mixture of paynting amongst of lyons, tigers, onces, panthers, eagles, and other beastes and fowles, very lyvely drawne and more esteemed [then] the gilding. Non were admitted to see the Emperour by my selfe, Mr. Eaton, and Mr. Wilson. He sat alone upon a place somthing rising with 1 step, and had a silk catabra[5] of a bright blew on his backe. He sat upon tho mattes crossleged lyke a telier; and som 3 or 4 bozes or pagon pristes on his right hand in a rum somthing lower."[6]

Cocks' Chinese goldfish and his garden, where he was ostensibly the first to raise potatoes in Japan, were matters of great satisfaction to him, and he enumerates virtually every bush, fruit tree, and the like he added, but he writes of little else in his personal life, save to note the occasional purchase of clothing and his social invitations. He seems to have had an ambiguous attitude about his Dutch counterpart, Jacob Specx, whom he alternately trusted and distrusted. Whether he had dalliances we do not learn, but many of his colleagues took Japanese wives, and he refers to the baptism of their children. The fates of these children with the departure of the English go undescribed, but as it presently exists his diary is incomplete, and possibly we would know were this not so. Since these liaisons were generally matters of convenience, it seems reasonable to presume that wives and children were left behind to fare as best they could when the factory was closed. Some, possibly most, were banished when the children of the Portuguese, Spanish, and more influential Japanese Christians were expelled under the edicts of 1624 and 1638.

Occasionally other insights into local life and customs appear in Cocks' diary, for instance, his entry on August 27, 1615, reads: "This day at night all the streetes were hanged with lantarns, and the pagons vizeted all ther temples and places of buriall with lantarns and lampes, inviting their dead frendes to com and eate with them, and so remeaned till midnight; and then each one retorned to ther howses, having left rise, wine, and other viands at the graves for dead men to banquet of in their abcense, and in their howse made the lyke banquet, leving parte on an altor for their dead frendes and kindred. This feast lasteth 3 daies; but to morrow is the solomest fast day."[7] Although the date appears wrong and may reflect either an error in entry or the use of the Julian rather than the Gregorian calendar, which had only been introduced in 1582 and was not yet universally accepted, he clearly writes of the OBon, the August festival for the dead.

Cocks was undoubtedly a dutiful man, honest to his obligations, a shrewd observer on occasion, one sensitive to the needs of others as attested by his thoughtfully penned letters on the misfortunes of some of his colleagues and concern for the children of Will Adams, after their father's death. With the demise of the Hirado factory, Cocks was shabbily treated by the company he constantly addressed as Worshipful. Held wholly accountable for its losses and errors in judgment, recalled in disgrace, broken in health, he died enroute to England. These were clearly not times for the timorous, and even the fortunes of the self-confident often went askew.

It is ironic that Cocks, seemingly an intelligent and curious man, should have left Hirado so shortly before the arrival of François Caron, one of the most gifted employees of the Dutch East India Company in its early years in Japan. Little is known about this remarkable man, a French Huguenot, prior to his appearance on Hirado in 1619 as a cook's mate. Over several decades, he became sufficiently skillful in Japanese to serve as an interpreter and rose in time to the second highest position in the company. His (and Joost Schouten's) *A True Description of the Mighty Kingdoms of Japan and Siam,* compiled in 1636 and translated into English some twenty-seven years later, was one of the first histories of Japan in a Western language. Written primarily for the Company's purposes, it was an important achievement. It deals more with issues of governance and customs, less with Caron's own personal observations. Nonetheless, as chief of the factory, he brought to Dutch–Japanese relationships a new dimension: a conspicuous interest in the culture and the language. It is said that, of all of the buildings at the Dutch factory on Hirado during his tenure, his alone was built in the Japanese manner. Caron took a Japanese wife, who bore him six children on Hirado; of these, at least one son, his namesake, apparently exhibited his father's linguistic skills and achieved some distinction with the Dutch Church in the Orient.

After the English abandoned their factory and the Portuguese and Spanish were expelled from Japan, ostensibly because of their inability to separate religion from commerce, the Dutch were left with a monopoly of the trade for which all had previously competed. But the forced transfer of the Dutch factory from Hirado to Dejima in Nagasaki Bay in 1641 marked a waning of Hirado's importance. Trade with Holland was removed from the purview of the Matsuura and brought directly under Tokugawa control. Hirado's role in Japanese history had been played; the island and its people settled into rural anonymity.

*   *   *

One approached Hirado across a mile or so strait, the Hirado kaikyo, known to early Western nagivators as Specx Strait, from Hiradoguchi, a small community on the Kyushu coast immediately opposite Hirado-jima itself. The white, two-decked ferry carried not only passengers but vehicles, private and commercial, about eight or ten at a time. Larger vehicles were customarily loaded first and smaller vehicles last. Loading proceeded from a small parking area in front of the pier and the office where tickets were purchased. To an uninitiated driver it was an emotionally taxing experience to get a car onto the boat, at least the first several times. The ferry pulled into a floating pier, generally at right angles to its long axis. The pier itself was connected to the land by an articulated bridge so that it and the pier could rise and fall with the tide. Bridge and pier were narrow; the former was 10–12 feet wide and the latter about twice the bridge's width. It was necessary to back down the bridge onto the pier and then make a sharp turn. Visibility was restricted; as a result, one backed toward a member of the boat's crew responsible for the loading of the vehicles. He invited you downwards and backwards with a steady stream of "hai . . . hai . . . hai" (yes . . . yes. . . . yes), unless you strayed to the left or right in which case he would sing out the correcting direction "hidari" (left), "migi" (right).

I was not overly familiar as yet with the van I was driving; and, to make matters worse, it was filled with household and personal effects and office supplies to the point where the rearview mirror was useless and the two mounted on the front fenders gave only a limited view. I was not comfortable with the procedure looming ahead and urged my two passengers—Vicki and Horikawa Motoko—to get out and walk down onto the ferry. I took a deep breath and started backing. In hindsight, it seems far less threatening, but I do recall that I wasn't at ease until I felt the rear tires grasp the ramp into the ferry. My hands were clammy and trembling when I got out of the car to join Vicki and Motoko at the rail as we pulled away from the pier.

The current through the straits can be stiff, especially when the tide is running. The pilot allows for this by moving along the coast within the shelter of headlands until a margin of distance is purchased to trade against the drift wrought by the tide. Once the boat leaves the shelter, the force of the current is immediately felt in the laboring of the screws as the ferry makes its way toward the island. The roadstead that lies before the main town is narrow but well protected, guarded partially by a small, uninhabited island. Two channels, one above and one below this island, lead into the harbor. Through these a steady traffic flows—fishing boats coming and going, as well as ferries bound for other islands, such as Ikitsuki, Takushima, and Iki. Each incoming boat, blinded by the small island or the turn into the channel, sounds its horn to announce its approach. Time has developed a rigid proto-col among ships—precedence is determined by whether the ship is inbound or outbound, whether its power is sail or some combustible material.

Once into the harbor's protected waters, our ferry turned toward the end of the floating pier that served all of the ferries and slowly churned to a halt. As it did so, we could see the spire of the Catholic church on the hill above and the prominent stucco-covered stone wall of the enclosure surrounding the Matsuura museum. The ramp was lowered, and since we were the last on, we were the first off. This was a much easier task. We drove up onto the quay-side road that girds the harbor, one of only two paved streets, past several aging inns and a gasoline station, to the Kakizoe Hospital, where we were to reside.

Our search for a place to live had proved unexpectedly frustrating; there were very few places that even approximated our needs. We were first shown a small, dilapidated house whose sole redeeming feature was its location overlooking the harbor, but this hardly offset the holes in the walls through which one could stick an arm, or the shōji so tattered as to expose every wooden member. Clearly this would not do, so we were taken to a boarding house where a small set of rooms could be made available to us. These rooms were on the top floor, no toilet was near, and, worse still, the stairs that led to this suite were virtually perpendicular. I could barely suppress the vision of a late-night emergency and the need to negotiate this hazard in the dark. We thanked the owners profusely but stated that we really did not want to intrude. Our options were disappearing rapidly, and more and more we were being urged to accept the Kakizoes' offer to use their guest rooms. Some years earlier when his hospital was built, Dr. Kaki-zoe had included not only the family's living quarters but a small guest suite as well. The latter consisted of a bath and an eight-mat room separated discreetly from the remainder of the hospital and living quarters. It obviously would be quite comfortable, but we were disinclined to live there, for several reasons. There was no provision for cooking, and either we would have to eat out all the time or be dependent upon the Kakizoes. Neither of these alternatives was appealing. Furthermore, if we lived in the guest quarters, there would obviously be no place for guests of the Kakizoes when they came. If our stay were to be short, this might not become troublesome, but we expected to live on the island for six months. Moreover, though the Kakizoes were very hospitable and obviously concerned about our well-being, we really did not take to the idea of being so dependent upon anyone. Finally, Dr. Kakizoe proposed a solution that met our needs beautifully.

An elevator had been installed in the hospital when it was constructed that rose through the three main floors of the building into a smallish, two-story structure that provided access to the roof and housed the elevator machinery. One could ride to this fourth floor, where a small snack bar existed, walk out onto the roof, and then climb a flight of metal stairs to two rooms, one of which was being used as a small ward; the other contained the elevator machinery. Dr. Kakizoe suggested that we convert this space into a small apartment. We could install a bottled-gas stove in the smaller room, and the larger one could be made into a combined living, dining, and sleeping area. Water was already available, but there was no toilet. This need could be solved by putting a lock on one of the toilets on the second floor normally used by patients; we would have the key. While, again, the thought of having to go down two floors to the toilet was not an overpoweringly attractive one, this was a small price to pay for the privacy and independence our little apartment would afford. One would be able to gain access to our quarters only through the hospital, whose main entrance was locked at night and was attended throughout the day. Since a phone had been installed in the larger room, one of the clerks or nurses could always call to announce our visitors or to tell Vicki that I was on my way up the stairs from work.

It was soon obvious that our style of life would be a compromise between the Japanese and the American. The Kakizoes moved two upholstered chairs up to the room for our use, and we bought a low table from which to eat. To sleep, we had a frame built of cypress which supported several tatami. Onto these of an evening we spread futon purchased in Sasebo. These arrangements took some adjustment, but after a few days we found ourselves snugly comfortable. Since our kitchen window looked out over the harbor, we could see the ferries come and go and participate, as it were, in the harbor activities. Dr. Kakizoe's patients were curious about us but respected our privacy. The same could not be said of his two younger sons, who enjoyed rapping on our door and then running off to hide. As it became increasingly obvious that we would be called on the phone if we had a visitor, we simply ignored the rapping unless it was uncommonly persistent. Like other youngsters of their age, their attention spans were short and if we did not respond quickly, their interests turned to some other activity.

As the weeks marched on, we grew more and more fond of these arrangements. We were close to food and meat stores; a laundry and dry-cleaning establishment were only a few hundred feet away; and opposite the entrance to the hospital was Yamamoto's sakeya, which on an instant's notice brought us cold beer or soda. If the evening was uncommonly sultry, we often walked the inner street, past the red neon sign advertising Asahi beer before Yamamoto's, to the one stoplight in the community, and we then turned to the quay-side road to walk down to the ferry terminal to gawk at incoming tourists. Or we drove to the park at Kawauchi-toge to enjoy the evening breeze and see the sun set. Knee-high grass covered the slopes and waved slowly with each whisper of wind.

One day when Vicki was at home there came a persistent rapping on the door. At first, she ignored the knocking, thinking it was one or the other or possibly both of the boys. The rapping continued, and she began to think that it might be me and that I had forgotten my key. Finally, she went to the door, and upon opening it found before her an elderly, diminutive, kimono-clad woman whom she did not know but who blithely walked into our apartment. As she did so, she turned to Vicki and said matter-of-factly by way of explanation, "Kenbutsu shimasu," that is to say, "I'm sightseeing." We realized that we were objects of curiosity to many on the island, but we hadn't appreciated that we had achieved the status of a "sight" to see. As she marched about our small apartment, she flung one question after another at Vicki about the apartment itself. We subsequently learned that she was not from Hirado but Ikitsuki. She had come to the hospital to visit a friend, and upon learning that two foreigners were living in the hospital, decided she would see for herself how they lived. We have often wondered what she said to her neighbors when she returned to her home. We suspect that, seeing our tatami, futon, Japanese table and the like, she undoubtedly told them that Americans really did not live very differently from Japanese.

Office space proved much simpler to find. The city government had a large, old, unpainted clapboard building, constructed in the style common at the turn of the century, that was only partially used. It was only a few minutes walk from the hospital through the main section of the town, over a centuries-old gracefully arched stone bridge known to some as the *orandabashi,* the Dutch bridge, although it was actually built by the Chinese during Hirado's golden years. We were given several rooms on the second floor at the top of a set of wide, wooden, frequently oiled stairs to house our professional and clerical staff. Horikawa-san, who had supervised our offices during the Child Health Study, had returned to Japan with us and would again supervise the clerks we would hire to log in and check the completeness of the various census questionnaires which were the basis for the selection of families to be more intensively studied, to code the entries, to send for copies of the koseki of individual families when further verification of the census entries was necessary, and so forth. The actual census itself was to be conducted by midwives and public health nurses, aided by my young medical colleagues in those villages where no midwife resided or which were so remote—some accessible only by boat—that special arrangements had to be made. Meetings were arranged to instruct these groups in how to take a household census, how to probe persistently but sensitively, and how to ensure the participating families of the privacy and confidentiality of the data we were assembling. All individuals then resident on the island or normally there were to be enumerated; the basic unit of enumeration was the household. Japanese law requires a complete listing of the heads of all households within a village, town, or city; this listing, the *jumin tō-roku,* was the point of departure for the household censuses. With 8,229 households to enumerate, the task looming ahead was daunting, and especially so since we proposed, through the good offices of the municipal authorities, to compare our data with the existing records of the Koseki-ka (the office of custody of the koseki), the Hokenshō (the public health office), the Zeimushō (the tax office), and the Nōgyō and Gyōgyō Kumiai (the agricultural and fishing cooperatives).

We had allowed three months to complete this phase of the study, but because we failed to appreciate the remoteness of some of the island's communities, it took us almost five months, exclusive of the training time. Access to some of the remoter communities, such as Kamisaki, Furue, and Ose, required considerable effort. On the trip to Kamisaki, for example, Komatsu, Nakano, Yamamoto, and I drove to Tanoura, a small cold-water spa (*onsen*) near the northeastern tip of the island, parked on the little beach beneath the onsen, making certain that we were above high water, walked for almost an hour in the company of one of the villagers over a narrow footpath that hugged the cliffs to a point where a waiting boatman, who had brought the villager who met us, ferried us around a rock promontory to the hamlet itself. Each way took almost two hours—a heavy investment of time to see possibly twenty families. Scenically, Kamisaki was very rewarding, but scientifically it could have been ignored had we not stubbornly set for ourselves the task of visiting every village and hamlet, however small. Twenty or so households out of over eight thousand could not have distorted our census results significantly.

After interviewing a number of applicants for the clerical jobs, we selected three recent high school graduates. Since they were inexperienced, a period of training was essential, but under Horikawa's tutelage they grasped their tasks quickly and applied themselves conscientiously. Our days of work were adjusted to encourage the maximum possible participation of the island's families in the study, and since Saturday afternoon and Sunday were days of rest for others, we worked these days and took Monday and Tuesday as our holidays. Much of my time initially was spent in the obligatory greetings to explain our program and to solicit the cooperation of the various village heads. The village head at Nakatsura, near Tsuyoshi, had arranged to have one of the teachers at the junior high school present at our meeting. Supposedly she spoke English and could serve as his interpreter and entertain me in conversation. It was soon evident that her skills were not equal to either task; even my simplest remark had to be repeated several times, to our mutual embarrassment. Slowly changing the cadence with which I spoke, and using the simplest of words, we managed an accommodation; and when finally we left, the village head beamed alternately at her and me, as though unaware of the arduousness of our conversation.

Dr. Yamamoto was in charge of our small branch in Tsuyoshi,

which was thirty kilometers away. A telephone was available, but the connection was often so poor or the nature of our business so long that it usually proved easier to drive to Tsuyoshi to consult with him. Isolated as he was, he welcomed these visits; and, as the tempo of work increased, they occurred virtually daily. The roads were so poor that it was over an hour's dusty drive each way. On the way back, Komatsu and I often stopped at one of the little village stores and, with a cold beer, washed down the dust that floated through the open windows of our unairconditioned van. Asahi had only recently put on the market steinies of marvelously thirst-quenching draft lager. Emblazoned on the sides of our van was the name of our study, and when we parked before a store the vehicle served as a quiet advertisement for the survey, or so we told ourselves.

Shishi, a village of thatched-roofed homes distributed haphazardly among paddies of rice, is the largest of Hirado's north-shore communities, and my introduction to it was startling to say the least. We had gone there to examine the household registers. As Drs. Komatsu and Furusho and I entered the branch office of the city hall on a courtesy visit, an arrow darted before our eyes and thumped into a target, followed by immediate and profuse apologies. Seems the head of the branch office was an archery enthusiast and whiled way the many workless hours practicing his skill. A small range had been set up along the corridor that faced the collection of desks of the various secretaries who performed the routine functions of the office.

Once our purpose was stated, book after book of the records were laid before Komatsu and Furusho. As they studied the carefully bound and preserved volumes, they repeatedly commented to me on the uncompromisingly clear calligraphy, art without pretension, perfection in each exquisitely proportioned stroke. They had been fearful that the older records might be scarcely legible, but this was not the case; they would, however, have to contend with an earlier written form of many of the ideograms they knew.

The head of the office himself was a chatty, rough-hewn man, delighted to explain the niceties of Japanese archery but equally helpful in our efforts to involve the community in our proposed study. Like most other officials in the branch offices of the city government, his salary was inadequate and his perquisites were few. He did have on the corner of his desk the branch's sole telephone, a hand-cranked magneto-powered version rarely seen now. It was little wonder that we had such difficulty calling him. To support himself and his family, he regularly fished. The hazards of this occupation were brought home some months later, when our survey team was in Shishi and he and I went fishing together. We sculled away from a primitive dock in a wooden, one-oared boat, possibly twelve to fifteen feet long, whose bottom was drenched with water and whose gunwales were barely inches above the sea. The boat itself rocked in harmony with each pull on the oar; he seemed totally unperturbed by the movements. If he noted how tenaciously I clung to the sides, he was too gentlemanly to comment. As a nonswimmer, I am petrified by water that is more than waist high, and suddenly I found myself on the open sea, albeit only several hundred yards from shore, in water that was fifty or sixty feet deep. He thrust a glass-bottomed wooden pail, a primitive fish locator of a sort, into the water, and schools of fish could be seen swimming a meter or so above the sea's rocky bottom. There were also countless *sazai*, sea snails.

Although not a skillful fisherman under the best of circumstances, I suddenly had ten thumbs, but this did not deter him. He quickly caught several blowfish as bait and carefully showed me how to mount the weights, bait the hook and sense a possible bite. One did not use a rod but a simple lead-weighted line wrapped on a rectangular wooden frame; one draped the line, from which dangled several short extensions, across the index finger. The line was raised and lowered a foot or two rhythmically, by raising and lowering one's hand. I soon became so absorbed in the chase, for I could see the fish nuzzling the bait, that when finally I felt the tug that suggested a bite, I gleefully pulled in my line; only then did I notice that several fish already rested on the bottom of the boat at his feet. We had managed a respectable catch, possibly two dozen or so fish, most of which he caught, when he suggested we return to my colleagues and the inevitable party of sashimi, pickled daikon, and beer.

Conversation was lively, and several laudatory remarks were made about my fishing skills, which I dutifully acknowledged, while intently scanning an inscrutable face for signs of humor. Later, my colleagues would jokingly remark that my most important contribution to the study had been fishing, for this occupied the village head and freed them to do their work uninterrupted.

As it was our practice to call on the village heads of all of the hamlets of fifty or more people, one of these courtesy calls brought me and several of my colleagues to Takagoe, a small hamlet, certainly fewer than two hundred individuals, on the northern aspect of Hirado, at the end of a narrow, tortuous gravel road. We had been accompanied by the head of the branch office of the city hall in Shishi, through which we had to pass to reach Takagoe. He knew the village head, a squat, muscular fisherman to whose house we proceeded. He, like most others in his village, derived his livelihood from a mixture of farming and commercial fishing in the off-shore waters. We explained our program to him, why Hirado had been selected, and what we proposed to do with his help. He listened attentively, asked a few questions about when we expected to be in the village, how long it might take for the interviews we contemplated, whether these could be arranged so as to interfere as little as possible with their work activities, and then nodded his approval. We had brought the traditional gift—two shō of sake—and once the business was conducted, one of the bottles was opened and sashimi appeared to be eaten as we sealed our mutual commitments.

Soon the conversation turned to other matters. The village head looked at me, obviously gauging my age, and asked whether I had been in the war. To this I replied yes, and he asked where. I said that I had been in the Pacific and had seen action on a number of islands, including Bougainville and the Philippines. He nodded knowingly and then told me matter-of-factly that he had been a member of the Japanese Sixth Division, infamous for its rape of Nanking in 1938. He knew, however, that it was not this that would pique my curiosity but rather the fact that the Sixth Division, or at least units of it, had been among our opponents on Bougainville.

Over all of the years I had previously been in Japan, I had wondered what I would do when finally I met "the enemy." Numerous conversations had been rehearsed, none particularly satisfactory. All were predicated on the assumption that a meeting would occur in either a social or work-related context, and that the ensuing conversation would be brief but forthright—no dissembling. But it was difficult to envisage how I might steer remarks through a potentially disturbing situation. I simply could not place myself in the role of the defeated, and worried lest a wholly innocuous statement of mine be seen in an unanticipated context and interpreted as arrogance. As the years passed, and it seemed that I would never encounter anyone who had shared the same specks of land in the South Pacific, these worries faded. I was caught unprepared.

We ate silently for a while, and then he asked what had been my function. I said that I was a surgical technician in a medical battalion supporting an infantry division; he replied that he had been an infantryman. Another pause as we washed down several pieces of sashimi. Some wariness was growing on my part, and because I did not want the past to antagonize the future, I decided to let him direct our conversation. If I was worried about animosity, I should not have been. The village head seemed merely curious about our shared experience. We soon established that we were on Bougainville at the same time, but there was little likelihood that he had shot at me, or I at him. Eventually, he had become an Australian prisoner of war, since Australian troops had replaced us when we moved on in late 1944 to the invasion of the Philippines. Our exchange inevitably drifted to life in the military, something that neither of us had relished. He jokingly recalled that on those occasions when it was possible for him and his fellow soldiers to bathe, access to the hot bath was a function of rank, and he was usually one of the last to bathe. When we parted several hours later, an old soldier's attitude toward one another had developed. Whether this did or did not enhance our program's acceptance is debatable, but it is a fact that no other village had a higher level of participation.

On the drive back to Hirado, I pondered the events in Takagoe as I gazed across the bay and the steady sweep of waves, one following another, on the beach before us.

*   *   *

We were warmly received by the community in Hirado. Undoubtedly the intervention of the Matsuura, the Kakizoes, and the Nakanos—all families of local stature—enhanced our welcome. Drawn from quite different social circles, with little in common between them, nevertheless each sought to make our stay enjoyable and our work profitable, and to share their lives with us. The Matsuura were the socially most illustrious, being the descendants of the family that had ruled the island for seven centuries; the Kakizoe were undoubtedly the wealthiest and privileged as a consequence of profession; the Nakano were unpretentious, a family that had risen to substance not through birthright nor education but by dint of hard work and now had a thriving construction business. Matsuura Tadashi, the younger brother of Motomu, who, as the eldest son, had inherited the family's feudal land and title, was a former member of the Nagasaki Prefectural Assembly.

He lived in one of the family homes near the castle and was a staunch supporter of our survey. His older brother lived near Tokyo and visited the island only once or twice a year, but through Tadashi became familiar with our interests and activities.

Although virtually everyone on the island was aware of its historical contacts with the West, few had dealt with more than the occasional Western tourist, and even these contacts had diminished dramatically after the cessation of the Occupation. Tokyo seemed inordinately far away from western Japan. Our limited contacts came through an English newspaper found in Sasebo or the evening news on television. Only now and then did events in the larger cities intrude on our thoughts, and then primarily they had to do with the Olympics. We were startled, therefore, to learn of the stabbing of Edwin Reischauer, our ambassador. Fortunately, his wound was not life-threatening, but it proved a source of embarrassment not only to Japan but to the United States as well. Most of us had hailed his appointment by John Kennedy. He was a distinguished scholar who spoke Japanese fluently, and his wife, Haruko, came from a socially well-placed family; her grandfather, Matsukata Masayoshi, was the creator of the Meiji financial system. These seemed the ingredients of a period of peace and understanding, but the stabbing suggested that turmoil still lay just beneath the surface of the relationship between the two countries.

At the first opportunity, we sought the reaction of the Nakanos. Predictably, the parents were embarrassed, seeing the stabbing as a poor reflection upon Japan, which indeed it was. Their daughters had a somewhat different perspective. They did not condone the stabbing, but what we had perceived as strengths in our ambassador they saw differently. Reischauer's use of Japanese in occasional television appearances they felt diminished the mystery and position of a foreign ambassador. And the old formalisms he frequently used in conversation were funny to them and their young friends—the speech of their grandparents. We found their attitude perplexing but admired the candor with which they shared their opinions with us. It seemed that skills which were appreciated in an ambassadorial assistant were less well regarded in an ambassador.

As summer waned, and the day drew nearer for us to return to Hiroshima, the tempo of activity accelerated; we wanted no loose ends to interfere with the orderly beginning of the second phase of the study, the examination of parents and their children. Verification of the relationship of some of the parents had proven more difficult than we had envisaged, largely because the parents had been born elsewhere in the prefecture. Dr. Furusho Toshiyuki came to help with these more troublesome cases. Vital events in Japan become official only when they are entered into the koseki. If a marriage is entered in the koseki it is legal; it does not necessarily involve any ceremony, religious or civil. But a single household record does not describe the relationship between the spouses. To determine this one must examine a series of such records. There are potential pitfalls in this search, however, and adoption is one. Japanese families without sons often adopt a male to perpetuate the family name; these adoptees are known as *yōshi*. Two forms of adoption were common. One coupled adoption with marriage to a daughter; the entering husband, who inherited the family's name and fortune, is called a *muko-yōshi*. In the other form, known as *nyūfu*, the entering husband did not take the family's name and did not necessarily inherit. Both practices made it imperative that we be certain that individuals who appeared to be relatives actually were; this entailed the scrutiny of still further household records, and in particular the "struck-off-the-register" books, the *joseki-bō*.

To examine the completeness of what had been done, we convened a small workshop to which other Japanese colleagues were invited. The meeting went exceptionally well, and we were satisfied that our efforts would pass the scrutiny of even the most discriminating geneticists. Simultaneously we began the formal visits to the mayor, the various heads of the branch offices of the city, and the agencies that had supported us, as custom dictates. It was another round of cups of green tea, but this time with a difference. Expressions of goodwill and hope that we had achieved the objectives of this portion of our study and would return punctuated every visit. We had come to like the island, its inhabitants who had patiently endured our quizzing, and our colleagues, and this was sensed. As a tangible gesture of our appreciation for their help, we invited all of the midwives who had served as census-takers to a dinner at one of the local hotels. Some seventeen, actually the bulk, joined us for an afternoon of food, folk dances, and camaraderie. Considering the age of most, over fifty on average, I was surprised at the enthusiasm with which they entered into the dances and their active courting of the laughter that accompanied some of the more humorous ones.

What we had initially seen as a test of our fortitude had become a memorable experience, and the leave-taking was not easy. Shortly before we were to go, the mayor discreetly asked Horikawa-san what would be an appropriate gift to express the city's appreciation of our efforts on its behalf. It was fairly widely known that we were interested in the handcrafts, particularly those that time was displacing. As a result, I received the mayor's own fireman's *happi,* a short sturdy black and red cotton jacket. Such coats have been the conventional dress of Japanese firemen, although now, of course, their fire-fighting garb differs little from that one would expect in any Western city. Conspicuously on the back is a large circle in which the ideograms for Hirado have been dyed, and on the left lapel, if this is a suitable description, is stated that the wearer is the mayor. Few opportunities have arisen for me to don this garment, but periodically it is recovered from the drawer where it normally rests, to be pleasurably fingered and to recall one of our most interesting periods in Japan.

On our last day, when we pulled away from the entrance to the hospital and drove down the little rise on which it sets, our van was as filled as it had been on our arrival six months earlier. Vision to the rear was no less impaired than previously, but this was not troublesome now, for I no longer approached the backing onto the ferry with trepidation. The many trips to and from Kyushu had stilled this apprehension. We drove slowly toward the pier, waving to the various shopkeepers, the Yamamotos and others, with whom we had traded. Some were actually standing before their stores, having learned of our departure. A substantial number of individuals had come to see us off— the Nakanos, the Matsuuras, the Kakizoes, our clerks, many of the midwives and nurses who had worked so diligently on our behalf, and even the mayor. We were touched and proud that we had won their acceptance. Many had brought little rolls of paper tape to throw at our boat as it pulled away from the pier, and others offered the loose ends for us to hold so that the tape would bind us as long as possible. We stood on the ferry ramp bidding each one goodbye; our clerks, who had heretofore been too shy to do more than bid me good morning each day, came to bow tearfully. Finally the ferry backed slowly away from the pier and reversed the thrust of its screws amidst a flurry of paper. With our free hands, we waved until it turned to pass the island in the harbor, and we could no longer see those gathered at the dock. Neither Vicki nor Horikawa nor I had much to say to one another as the ferry labored its way toward Hirado-guchi. It really did not seem as if we were leaving; this was merely another weekend trip. We would return.

RETROSPECT

After more than four decades of acquaintance with Japan and its culture, they remain enigmatic to me, simultaneously enchanting and inscrutably opaque. Superficially, the nation has grown progressively more Western in these years. It has copied, adapted, and integrated numerous aspects of those cultures upon which it has elected to fashion itself. But this malleability is not recent; it characterizes much of Japanese history. China provided its present written language and the cultural bases for many facets of life that we now think of as Japanese, such as ink painting. Buddhism, which dominates the culture, was also imported from China and had virtually displaced the indigenous Shintoism, a religion without dogma, moral code, or historical founder, until 1868 when, in the Meiji era, Shintoism became the state religion, which it remained until 1945.

The pendulum of emulation has swung, and today, the United States and Western Europe are the foci of Japan's imitation. Our technological innovations have been copied, and often their marketability improved, but our sense of worth of the individual, of equality of opportunity, and of confrontation when individual rights are in conflict with those of groups have not. Japanese culture fosters a sublimation of self to the group. Formally, Japan is a democracy, but it is hardly democratic. Japanese newspapers, not with originality, have described the democratic process as a tyranny of the majority, as if to imply that a tyranny of a minority is preferable. Signs of privilege and preference abound, and long have; these are not products solely of the nation's current economic position. Japan has had and continues to have a structured society, and social ostracism or the threat of it perpetuates class distinctions. While consensus may be sought, the perquisites of power are unmistakably clear. Form dominates substance.

Arrogance, or possibly self-pride, colors and distorts many features of Japanese life. Many Japanese are inclined to see outsiders as members of an alien world to which they cannot relate, and they are inclined to ascribe this inability to their presumed homogeneity. But homogeneous, at least physically, they are not. They are tall and short, light and swarthy, fine-featured and coarse, graceful and graceless. The ethnocentricity of the Japanese affects their view of international trade and their perspective on their own history. It is disturbing to hear a major cabinet official, Okuno Seisuke, state that "Japan fought the war in order to secure its safety. The white race had turned Asia into a colony. Japan was by no means the aggressor nation." [1] Clearly, the self-centeredness, even paranoia, that fueled Japanese imperialism persists. But Nazism is not dead in Austria or Germany, nor fascism in Italy, nor anti-Semitism in France, nor racism in our own country. Grievances, real or imagined, often unarticulated, serve to infect, to pollute, to destroy not only Japan's but our own capacity to understand one another. Perhaps the future will be different, but one wonders.

Japan continues to be capable of producing both automobiles that rival the best assembled anywhere and exquisitely delicate cricket cages, fashioned with a care to detail and craftsmanship that defies description, but unessential either to crickets or humans—the material and the evanescent juxtaposed. How should a culture be evaluated that so avidly grasps the present, but seems equally reluctant to relinquish the past? Should it be admired for its firm adherence to traditional values, or faulted for its inability to see beyond its own immediate self-interest? Similarly, how is a nation that prides itself on conciliation to be judged when the language of negotiation overflows with words like "demand" or "smash," and when the opposition parties absent themselves from Diet proceedings instead of debating the merits of their positions? Is consensus a game to be played in a stylized arena, wanting the very ends it purportedly seeks to serve?

There is a streak of violence in the nation's character, as there undoubtedly is in most, but its surfacing is always unexpected in a culture that can be simultaneously as graceful and elegant as the Japanese. Historical incidents of this abound—the virulence of the persecution of the Christian minority, the intemperateness of the conduct of Japanese troops in the course of World War II, or, more recently, the ritual suicide of Mishima Yukio or the Red Army's senseless slaughter of innocent people at the Tel Aviv airport. Possibly no act typifies this more vividly than the assassination in 1960 of Asanuma Inejiro, the portly leader of Japan's Socialist Party, an event graphically captured by a photojournalist present at the political rally at which Asanuma spoke. Before hundreds of on-lookers and a television audience was the stark image of Asanuma, his heavy, horn-rimmed black glasses askew, perched precariously on one ear, his oiled hair awry, slowly falling forward, clutching his abdomen, and nearby a frenzy-haired uniformed college student, a partisan of the extreme right, feet firmly planted, with a drawn short sword grimly held as if to thrust it again into the belly from whence it had just been withdrawn. Coming as it did in times unsettled by the pending renewal of the Japanese-American Defense Treaty, this murder startled the nation and forced it to examine anew the differences with which it was torn. This was a premeditated, consciously planned and implemented act, one totally oblivious of the worth of another human being. One had to ask why? Are behaviors such as these just the pent-up frustrations of individuals in a proscriptive society, alienated and disenfranchised, or do they have other roots? Are they condoned?

Japan's most popularly read literature consists of comic books, often pornographic and so explicit as to bring a blush to a voyeur's cheeks. Yet Japan has an enviable literary heritage, one extending over more than a millenium, from Lady Murasaki's graceful, gossamer-like *Tale of Genji* (the Genji Monogatari) to the fine contemporary prose of Kawabata Yasunari, a Nobel laureate, Mishima Yukio, Endo Shusaku, Tanizaki Junichiro, and many others. Literacy is surely more than just the ability to read; it is a measure of what is read.

Politeness in Japan is proverbial but equally vulnerable to careful inspection. Bowing is its most conspicuous hallmark—one bows at and to almost everything, or so it seems—and this can be carried to ridiculous limits, as in a telephone conversation,where one bows to the unseen speaker at the other end of the wire. However, politeness in Japan is not pursued for its intrinsic merits; it does not go beyond those one knows or those to whom one is indebted (or potentially so), or tourists, where one perceives oneself as a representative of the nation. As a consequence, the commuter in Tokyo or elsewhere soon learns that to step aside to allow an elderly person into a public vehicle is to invite a torrent of others to board as well. Only Westerners hold doors, and then rarely is the courtesy acknowledged. These are obviously not the only signs of politeness or its lack, but they certainly suggest stylized behavior rather than an inherent sense of concern for others.

All is not contradiction; there is a universal regard for education, at all walks of life. A city such as Hiroshima will offer literally hundreds of courses in various aspects of the traditional arts—archery (*kendō*), flower-arranging, calligraphy, sewing, even landscape gardening. Many are taught by repetition rather than by careful analysis and explanation, but they are rewarding nonetheless. Rote learning is not necessarily bad; some skills are gained only by repetition. With the aging of the country, these courses offer a haven of constructive activity for the retired and a means to perpetuate cultural and craft traditions that merit preservation. Japan and the Japanese are to be admired for this; certainly the past, whatever social inequities may have existed, is not to be deplored uncritically.

Affluence has replaced the simple pleasures of the past with material ones. Children jumping rope, older ones tending the younger, the playing of jacks have largely disappeared, and the competition for schools denies many the opportunity to enjoy the brief moments of childhood. Cram schools, the *juku*, once merely a haven for the backward student, have become a way of life for virtually every child; each beavers industriously to gain entrance to the most prestigious possible school, beginning literally with kindergarten, since the first school one enters increases the prospects that the next will be no less important in the hierarchy of schools. The assiduous cultivation of the art of test-taking has replaced the curiosity that should motivate learning.

These remarks may seem overly critical; however, their roots lie not in disenchantment but in admiration of what is and, more importantly, what could be. Japanese society, like most societies, does not do justice to its own values. It temporizes, compromises. Such acts are not inherently wrong, but they do not fulfill the nation's promise.

*   *   *

The places where Vicki and I lived in Japan have fared differently over the years, some better, some worse. Hiroshima has grown substantially since 1950. It is presently Japan's seventh or eighth largest city and almost four-fold larger now than then—over a million inhabitants. It spreads across all of the available flat land, oozes into every nearby valley, however tiny, and resolutely climbs the sides of the surrounding hills and mountains. To gain land, mountains are dismembered and moved into shallows along the coast, but the city's appetite is insatiable. Niho and Oko, rural districts of rice paddies three decades ago, are entirely built over. Toyo Kogyo, the Mazda Corporation, has greatly expanded its industrial site through reclamation, and cars destined for overseas markets can be loaded at the corporation's own pier. Japan Steel and Mitsubishi's shipbuilding and marine complex continue to flourish, although both have recently confronted declining markets and a glut of steel and tankers. They flourish through an emphasis on quality, a rigorous rationalization of unprofitable products, innovation, and a willingness to preserve market share at the expense of immediate gain.[2]

Hiroshima University, originally an undistinguished prefectural institution, has a stature nationally that would not have been thought possible previously. Much of this began during the tenure of former President Morito Tatsuo, a one-time Minister of Education, and has continued under his successors. Unfortunately, for want of land within the city into which to expand, the university is being moved to Saijo to the east, now an administrative part of the metropolitan area, and its influence on local life will diminish. Moreover, the move works a hardship on students who work part-time and will find few such opportunities in the new location.

Traffic grows, and to cross Hiroshima over National Highway 2 at rush hour is as frustrating an experience as any found on a freeway in the United States. Again, however, the old percolates through the new. Oyster culture is said to have originated here some 350 years ago, and oyster beds and shuck houses are still seen between Hiroshima and Kure. Small wooden pails of these bivalves continue to be given as remembrances. And the city prides itself on its three great sake makers, Senpuku, Suishin, and Kamotsuru, all of special class.

High-rise buildings dominate the skyline and have virtually obliterated the view of the Cathedral and the Industrial Exhibition Hall that once stood sentinel-like on the horizon. Yet Hiroshima is much more attractive now than in the early postwar years. Its art museums, some privately endowed and others governmentally supported, house an impressive collection of fine paintings, Japanese and Western. Mori Terumoto's Carp Castle has been rebuilt and the grounds restored to their former park-like quality. Not far way is the stadium in which the city's beloved baseball team, the Carps, annually plays out its destiny. Win or lose, they are supported enthusiastically, and when they have won the championship, sake, donated by the local manufacturers, literally flows in the streets.

In Noboricho is the Memorial Cathedral for World Peace, constructed at the urging of Father Hugo Lassalle, who had himself experienced the atomic bombing, and dedicated on August 6, 1954. Not too much farther away, but in a different direction, is Tange Kenzo's simple but elegant Cenotaph with its perpetual flame memorializing those who died in the bombing, and the Atomic Bomb Museum, with its reminders of the folly of war. They are approached from east and west over a wide, tree-shaded, landscaped boulevard, the Hyaku Metoru Doro, or more properly the Heiwa Dori (Peace Boulevard), originally designed to be a firebreak but now enlivened with stone lanterns and statuary. It is the site of the Spring Festival of Flowers. Once it was merely a naked gash running through the south central part of the city, edged by barbed wire and greened only by weeds.

An attractive, well-kept park now surrounds the Peace Museum, which nestles amidst fountains, the cenotaph, and flowers. The museum itself is arresting, disturbing; one's sympathies are galvanized by the hardships of the survivors. But there is an equally disturbing exploitative aspect. It is a museum decrying the use of nuclear weapons, but it is not a museum of peace. Were it so, there would be some indication of the twenty million Chinese who lost their lives in the war, or the thousands of Filipinos, Guamanians, or Malays, or victims of cruelties perpetrated by the Fascists, Nazis, Russians, or misguided members of our military. The attitude of numerous Hiroshima citizens is, however, well summarized in an rejoinder made in Hiroshima Peace Park to the question, "What does peace mean to you?" The answer: "Peace is when all people shake hands with each other, get on well, and live interesting lives. Men and women must love each other. There is no war in Japan now. I don't care what happens in other countries." [3] To see the first two statements as a praiseworthy aspiration for all humankind is clearly wrong, as the self-centeredness of the final two makes clear. Similarly, the authorities have been slow, indeed reluctant, to allow the Korean survivors to place a memorial to their dead among the others in the park. These attitudes undermine the lesson we should all learn from this catastrophe: the need for a peace founded on mutual respect and understanding.

The Asano Garden or Shukkeien Park, a heritage of the feudal era, offers a delightful, cloistered area in which the world-weary can stroll and muse. Isamu Noguchi's two phallic-symboled bridges contribute to the drama and modernism of Hiroshima, and along the river banks successive mayors have added carefully manicured, attractive grounds through which to amble. This was not always so. Prior to 1967 the river edges were lined with fetid-smelling open sewers and shanties hastily erected by squatters. When the squatters were dispossessed, the community provided alternative housing. City planners still build libraries, broaden roads, and strive to improve the quality of the environment. Moreover, for a community of its size, Hiroshima has impressively few beggars or down-and-outers roaming its streets, and the annual red-feather charity drive is usually oversubscribed.

There is now much evidence, locally and nationally, of caring for the handicapped, once a source of individual and familial embarrassment. All new multistoried buildings have ramps, elevators, and toilets designed for those in wheel chairs. Throughout the busier sections of the city, curbs have been leveled to accommodate invalids or the aged unsteady of gait, but unfortunately bicycles too. The squeak of brakes quickly applied to avoid collision with pedestrians perpetually fills the air. Many sidewalks have strips of inlaid pebbled yellow blocks to guide the visually handicapped, and as stop lights change at intersections, a musical refrain rings out to warn the blind of the direction of flow of the traffic. The heretofore socially disfranchised groups have achieved an acceptance, albeit gradually, that could not have been anticipated forty years ago. However, these advances of the handicapped, physically or socially, did not grow out of a sense of fair play but in response to the achievements of similarly handicapped individuals in countries with which Japan seeks to compare itself. Arguably, the first institution in the country to have a committee charged with the protection of a patient's or participant's legitimate rights and expectations in research was the Atomic Bomb Casualty Commission. Many Japanese institutions treat these issues cavalierly, and few patients with terminal disease are so informed. It is said that they will lose the will to live if aware of a terminal illness. This argument is unpersuasive to one who has had the freedom to make his or her decisions. Fortunately, this notion too changes, and progressively more and more physicians are persuaded of the patient's right to know what he or she confronts.

Public transportation has been and remains good. Numerous buses roll over the streets. Years ago, when the streets were narrower and less straight, drivers of these buses needed all of the assistance they could muster to manipulate sharp, narrow turns and to back their vehicles. Usually they had a whistle-armed conductoress whose tweet-tweet guided the backing. Often the whistle was accompanied by an "All right, all right," which always sounded like "Awry, awry." Today, save for the tourist buses, where the assistant also entertains the passengers, most city buses are conspicuously labeled in kana "unman," that is, one man. Hiroshima is among the few cities in Japan that continues to maintain, although not expand, its streetcar system. Elsewhere, trolleys have been replaced by gasoline-powered buses, with their greater flexibility but added pollution. Even here there have been novel innovations. At many stops, the approach of a streetcar is indicated electronically, and its route identified. It is not necessary to peer anxiously as a tram nears to determine whether it is the appropriate one; the sign specifies its destination.

Four large department stores—Fukuya, Mitsukoshi, Sogo, and Tenmaya—exist where formerly there was only one, Fukuya. All have their gallery sections displaying works of art, old and new. They are closed on different days of the week, and all are open on Sunday, the universal day of rest for most white- and blue-collar workers. The Hondori, the main shopping street, is a covered mall, closed to all but pedestrian traffic through most of the day. Boutiques are everywhere, and stunningly attractive clothes designed by Japan's flourishing community of couturiers are available for those who can afford them. Stores, like Takaki's Andersen, that purvey foreign and exotic foods are crowded. Coffee houses serving cakes worthy of Vienna abound, and fast-food restaurants—McDonald's, Kentucky Fried Chicken, the Sizzler—are on the rise. Even convenience stores—Lawson's and Seven-Eleven—open throughout the day and night, are altering shopping habits. Ice cream parlors such as Baskin-Robbins or Santa Barbara are full of students indulging their cravings for sweets. Multistoried apartment houses, euphemistically called mansions, obscure the horizon but offer little space to the individual occupant.

Hiroshima is known for its bridges, branches, and bars. Almost four dozen bridges span the Ota River or another of the brood it spawns—the Motoyasu, Temma, Kyobashi, or Fukushima—as it flows into Hiroshima Bay. And several hundred of the nation's major industrial enterprises have their regional headquarters, their Chugoku branches, here. Finally, possibly apocryphally, it is stated that over two thousand bars cater to the needs of local citizens, many in a single area, Nagarekawa, immediately to the east of the city's center. Until 1958 portions of this region had been one of Hiroshima's licensed quarters, and the same services are still undoubtedly rendered, though less openly than then.

In 1975, after protracted negotiations between the Japanese and American governments, the Atomic Bomb Casualty Commission was dissolved, to be replaced by an institution known as the Radiation Effects Research Foundation, in which the two nations participate on an equal administrative and financial footing. Administrative authority rests in a Board of Directors, twelve in number, six from Japan and six from the United States. Six of these directors reside permanently in Hiroshima or Nagasaki and constitute the Executive Committee responsible for the day-to-day activities of the Foundation. Although the bulk of the employees continue to be Japanese, there is a sprinkling of American professionals; a Science Council, with equal representation from the two countries, exists to provide scientific advice and research counsel. These administrative changes have not, however, altered the basic research strategy initiated under the Commission. Most of the studies continue to involve the fixed samples defined as an outgrowth of the Francis Committee report.

This new institution is housed in the same quarters on Hijiyama long occupied by the Commission, but for how much longer is not clear. A movement to find an alternative location waxes and wanes. As a result of this uncertainty, buildings and gardens do not always receive the care they warrant; grass is mowed only when an important meeting is to occur at which an unsightly garden would be an embarrassment. Inevitably, the workforce has diminished as the number of survivors has declined, as the technology of research has changed, and as the organization's perception of its mission has evolved. It is now less than half the size it was at its largest, thinned by retirements and death. As medical facilities have improved in the city, its pre-eminence as a diagnostic center has faltered. It still grapples with the problems that have plagued it since its beginning—recruitment of able professional staff, a clear sense of purpose, and achievement of sustained financial support. While these difficulties are easily enumerated, suitable solutions are less readily identified. Restructuring is compromised by employment practices in Japan and by the fact that most potentially outstanding Japanese professionals seek an academic career. The institution is not perceived as a means to this end. American staff, save in exceptional instances, are similarly reluctant to commit their futures to an institution in a foreign land whose tenure is uncertain. And efforts to establish liaisons between the institution and departments at academic centers in the United States capable of providing staff have continued to flounder, however well-intentioned their beginnings.

Sense of purpose looms no less large. While the scientific aim remains unchanged—to describe the long-term effects of exposure to ionizing radiation—how it is to be implemented and the additional functions the organization should serve are not. The survivors and the Japanese governmental agencies would undoubtedly like to see a larger service component, but the American sources of support are interested primarily in the scientific findings and, given the demands for health care in our own nation, are reluctant to fund those in another. Within the recent past American financial support has been exacerbated by the falling value of the dollar and the monetary practices of our government, which provide little latitude to seek the most favorable rate of exchange.

In short, the milieu, scientific and lay, in which the Foundation operates has changed, and its role in the community is necessarily evolving as a consequence. It is, as an organization, an unusual, indeed unique, experiment in binational cooperation, and in a world that grows progressively smaller, it is worthy of support for this reason alone, even were its scientific objectives of lesser importance. Surely, if a binational organization cannot succeed, is there any prospect of a multinational one?

As to the Genetics Program, it too is now different. While the surveillance of mortality among the children conceived by the survivors initiated in 1960 persists, the current research emphasis is upon molecular biology and upon an effort through recent advances in biochemistry to measure mutational damage more directly than was possible at the program's inception. Few of those employees who were involved in the early years of the program remain, and none of the physicians with whom I had so much contact are now with the Foundation. All are in private practice. Their number has been lessened by death, most recently that of Dr. Isoya in 1980. It was autumn when he died. One overburdened heartbeat had ruptured a vessel thinned by time. Life's measure drained in darkness, as it had begun. While we gathered at the temple to pay our respects, the last golden leaves were sauntering earthward, before banners undulating slowly in a breeze tinged with the approaching frosts of winter. As we sat before the altar, bells punctuated the rise and fall of sutra. Tear-stained faces hovered above the plumes of wafted incense, and stifled sobs penetrated the gloom. Isoya was not among the first to join the genetics staff, nor was he ever one of its more dynamic members, a leader. He was a quiet person, but he had shared in the camaraderie we all valued. It was natural, then, that upon learning of his death through Dr. Takeshima, as many of our group as could would attend the funeral services. While the services progressed, in a window of my mind time slowly turned backwards. We were suddenly more than thirty years younger—unlined of face, dark of hair with no traces of gray, quicker and more resolute in step, and imbued with a freshness of spirit uncompromised by the responsibilities of age and position. My thoughts flitted over one event after another. Abruptly, I was nudged back into the present; Koji was pressing his arm into my side. It was time for me to offer incense to the repose of Dr. Isoya's soul. I rose hesitantly, for I always feel ill-at-ease at funerary rites, grabbed one of the thin, elongated sticks of incense, lit it with the coals, thrust it into the ash in the receptacle before the altar amid the numerous others already glowing there, and offered my thoughts briefly so that others might follow. Isoya, you will be missed; some of the youth of all of us has gone with you. Now we are fewer.

> Pain of one who goes,
> Emptiness of one left behind,
> Like the parting of a pair of wild geese,
> Lost in clouds.[4]

*   *   *

Over the last four decades, Nagasaki has fared more poorly than Hiroshima, and as yet, the city is neither served by an express road nor a branch of the Shinkansen. Economically, it has reeled under one blow after another in the past fifteen years. It has grown, of course, as have all of Japan's cities, but its dependence on fishing and heavy industry, notably shipbuilding, and unfavorable geographic position, from a contemporary commercial perspective, have limited its participation in Japan's resurgence. Twenty-five years ago, when the shipbuilding industry flourished, there was a period of growing prosperity, but presently there are few, if any, palpable signs of this era. Efforts to develop its tourist attractions as an alternative to depressed industries have often been poorly conceived, even misdirected, and marginally successful at best. Graceless Disneyesque buildings that would be more appropriate in an amusement park have been constructed; moving sidewalks that serve tourists in the Victorian quarters of the city jangle with their very purpose. Architecturally and historically interesting old buildings such as the English consulate wither and decay through misuse and inattention. Cruise ships call, but not sufficiently often to provide a major source of income to merchants nor the city. Its stores, large or small, and streets, though enchanting to the outsider seeking an earlier era, are poor, almost tawdry versions of those seen in Hiroshima. Its people seem less prosperous.

Streetcars still traverse the broader avenues, and personal automobiles have increased to a point that overtaxes the roads through which access to the center of the city is achieved. The making of sponge-cake, locally known as *castera*, a skill learned from the Spanish, presumably from Castile, continues and is one of the inevitable souvenirs visitors purchase. Its university improves, and its medical facilities grow apace. The bettered position of physicians is conspicuously displayed in the new building that houses the offices of the prefectural medical association, adjacent to an equally new Atomic Bomb Hospital and the Kosei Nenkin Kaikan (literally Health and Welfare in the Golden Years Building), a hotel-like structure available to employees of the Ministry of Health and Welfare. These are governmental, or quasi-governmental, structures, reflecting the investment of funds from the national treasury.

Municipal projects are fewer and seem to move at a slower pace than in Hiroshima—the widening of the old highway that enters the city from the east has been under way for over a decade and is still several years from completion, barred by land-owners here and there who are contesting for the highest possible settlement. Grandiose schemes of large hotels mounted on pylons in the harbor fill the newspapers periodically but are slow to come to fruition. Perhaps Nagasaki will fare better with the national government's current efforts to stimulate its own economy. Since Nagasaki is not a municipality such as Hiroshima, the city's revenues are fewer and must be cautiously used. It maintains, nevertheless, an aura unlike that to be found in any other Japanese city, and the view at night from any of the surrounding peaks is mesmerizingly spectacular.

Nagasaki has made important social, if not economic, strides. Its present mayor, Motoshima, is a Catholic whose election was almost as historic an event as Kennedy's to the presidency of the United States. Its Christian community has expanded, its prosperity has increased, and its self-assurance has grown immeasurably. Nagasaki appears more politically tolerant than most communities in Japan, despite its conservative ways in other regards. It has not aggressively exploited its atomic bombing and projects none of the ironic overtones to be seen in Hiroshima, a community that owed its very importance to Japan's prewar and wartime military establishment and was actively involved in the nation's pan-Asiatic aspirations. Nagasaki's annual tribute to its dead is more a memorial than a political event; there is a decorum about it that fosters reverence, as it should. Moreover, there is and has been a recognition that the Japanese were not the only ones who suffered; American, Dutch, and Korean survivors, prisoners of war, and conscripted labor have been encouraged to participate in the memorial services.

Nagasaki, like Hiroshima, currently has a sufficiently large group of English-reading foreigners to sustain an English monthly magazine, but here the parallel ends. Hiroshima's is slick, glossy and superficial, seemingly uninvolved in Japan or the city that shelters it readers. Nagasaki's now defunct *Harbor Light* had amateur stamped all over it, but this is not a slight, for I use the word in its original meaning—to love or be fond of something. The young people who wrote and edited this monthly, most of whom derived their livelihoods from other activities, had an affection for their city, its history, eccentricities, and current aspirations that made the magazine a delight to read and a source of the unusual. They scrounged local archives for historical notes and relished the sharing of what they found. Their enthusiasm fostered an increased awareness of the past and through it the present, and has served as a nidus of renewed interest in the city's history.

*   *   *

The once charmingly remote and detached island of Hirado has changed in the name of progress, yet one senses a loss. Hirado is now connected with the mainland of Kyushu through an impressive toll bridge; its roads are paved, and some are even national highways, albeit subsidiary ones. Parking lots are proliferating, the previously uncluttered beach at Senrigahama is presided over by a Hawaiian-style resort hotel, and plans exist to connect the island with Ikitsuki to the immediate north with another mammoth bridge bored through the isolation of Shiraishi, one of the few remote villages remaining on the northwestern coast. The castle, illumined of a night, stands stark and white, like a ghost of some former time. All but the most insensitive are deafened by the din of motorcycles; communities such as Miyanoura have more leisure fishermen than population on the weekends. Numerous hotels have been built to accommodate the short-term tourist who is led apathetically, lemming-like to see the sights they are told are of importance. The Japanese have a penchant for traveling, inculcated as children through class outings to scenic and historic spots, and the island authorities obviously see this as a source of economic gain.

Many residents deplore these developments, not that they dislike the better roads nor the bridge, which is a handsome piece of engineering. They are disturbed by the intangible changes, the loss of sense of community, and the making of money through the exploitation of the land with no apparent regard for the island's future. These are certainly not unique problems, but the rapidity with which they have arrived heightens apprehension in what has been a closely knit community. Hirado's population has steadily declined and is now less than three-quarters of that at the time of our study; farming or fishing, despite efforts to ease the hardship and uncertainty, are no longer attractive, and a better life is sought in the larger communities of Kyushu or elsewhere.

Our friends on the island have generally prospered economically, possibly as much because of their professional skills as a percolating down of the increased traffic to and from the island and the economic gains this has brought. Their circumstances have changed in other ways as well. The Kakizoes have expanded their quarters in the hospital, adding space for their sons, who now manage the practice, and with the aid of a California architect have built a stunning new structure on a promontory looking southward from the heights outside the city. It serves both as quarters for guests and as a center for small scientific meetings. The Nakanos still live where they had previously, but within their compound they are building a large, traditional Japanese house to accommodate their growing number of grandchildren when they return to visit. Their construction business is managed by their eldest son, Hiroshi, with occasional counsel from father.

Churches and temples seem better maintained, and many of the commercial buildings impart an air of prosperity not seen before. The decrepit old wooden city structure where we worked has been replaced by a modern one, the seat of the local government, and stoplights have proliferated, as though their number were a measure of growth. Yamamoto's sakeya is still in business, as is our dry cleaner, but more stores now cater to the tourists than to the local population. There is a new ferry terminal building, for ferries remain the only means of connecting Hirado with neighboring islands such as Ikitsuki and Takushima, and small fishing vessels fill the harbor area as they always have. As yet, none of the international fast-food chains have discovered the island, but undoubtedly they will. More distant communities, Himosashi or Tsuyoshi, have changed less, but they too have been altered as a consequence of the improved roads and a greater accessibility.

* * *

What of science—has it too changed? Westerners and the Japanese have been mutually fascinated by one another, their cultures and science, for almost four and a half centuries. Although much has been written about the influence of Western science on Japan, less exists in Western languages about the status of Japanese science and its contribution to Western thinking at the initiation of contact or in the several centuries that followed.[5] A variety of explanations might account for this. Western contacts came later to Japan than most other Asiatic countries, were more limited initially, and were colored by political considerations within Japan and Europe. Most early Western visitors to Japan were either Portuguese merchants, seamen, or priests, and their interest generally centered on matters other than science. Moreover, among the clerics who did come there was apparently none of the scientific bent and intellectual stature of Matteo Ricci, the great sixteenth-century Jesuit scholar in China, but not all were lacking in scientific training. Father Luis d'Almeida, another Jesuit, for example, was practicing and teaching Western medicine in Japan as early as 1556. Although the Portuguese and Spanish carried with them Western skills in cartography, navigation, and medicine, overall their scientific contribution was modest. Possibly this reflected the relatively backward status of Portugal and Spain as scientific nations at that time.

When contacts began, the political environment that obtained was not conducive to a broad exchange of scientific notions. A new political order was evolving in Europe, and Japan was in the throes of its own unification and the development of a system of centralized government. Soon after its emergence, the period of seclusion began, and for two and a half centuries, save for limited economic intercourse, there were no systematic means to further scientific exchange. Matters were made more difficult with the promulgation in 1630 of the edict excluding all books designed to propagate the Christian faith; this order was interpreted so broadly as to include the numerous scientific works written by the Jesuits in Chinese and virtually all books in a Western language. Exclusion of Western books was not a hardship, for only a handful of Japanese read any of the languages in which they were written, but Chinese was read by most of the nation's scholars. Finally, in 1719, under the shogun Yoshimune, amelioration of this draconian step occurred, and the importation of these works of the Jesuits, particularly in mathematics and astronomy, was permitted. Many were promptly translated into Japanese.

Historically, Japan's isolation at this specific time was unfortunate, for in these years modern science in Europe emerged, propelled partially by the demands of industrialization in the West. Among the numerous important advances were Newton's development of the theory of gravity and of infinitesimals, Jenner's description of the utility of vaccination for smallpox, the introduction of microscopy and general anesthesia, as well as the beginnings of modern chemistry and physics. What Western scientists know of Japanese science and technology in the Tokugawa era stems largely from the writings of a very small group of individuals, mostly physicians, who for one reason or another were associated with the Dutch factory at Dejima. Although astute observers and capable naturalists, their knowledge of Japanese science was necessarily fragmentary and their opportunities to travel were limited; moreover, language proved to be an almost insurmountable barrier to an understanding of the status of Japanese medicine and biology, mathematics, and physics.

François Caron, who spent some twenty years in Japan, says little about the biological and physical sciences in sixteenth-century Japan, although he was an able linguist and should have had access to the Japanese science of his day. He did, however, comment on the arithmetic skills of his Japanese contemporaries and noted the rapidity with which sums, multiplication, and division could be performed on the *soroban,* the traditional local abacus. Caspar Schambergen, who served the Dejima factory as its physician in 1650, should, perhaps, be credited with the introduction of Dutch medicine into Japan. He is known to have lectured on surgery in Edo, and his students formed a school of surgery, known as the Caspar-ryu, which persisted until the end of the Tokugawa era. Serious commentary on the status of Japanese science awaited Engelbert Kaempfer, the German, who was the factory's physician in the years 1690–1692. Although he described in some detail acupuncture, which the Japanese had considerably improved since its importation from China, and *moxa,* the burning on the back or feet of a small cone of cottony fibers of the artemisia (a genus of woody plants to which the mugwort and sagebrush belong) to treat a variety of ailments, neither procedure received notice in Europe until quite recently, when a flurry of interest has been awakened in acupuncture, but this interest comes largely from its use in China. Possibly the most important development at this time was the publication in 1774 of the *Kaitai Shinsho;* this book, based on Johan Kulmus' *Tabulae Anatomicae,* underwent many subsequent editions and established the correctness of European, as opposed to Sino-Japanese, anatomical theories.

Carl Peter Thunberg, a friend of Linnaeus and an outstanding eighteenth-century naturalist, identified numerous Japanese plants unknown in the West during his 1775 trip. Like his European medical predecessors, he was often asked to describe current Western therapy, and noted that his Japanese fellow physicians were interested in the treatment of cancer, fractures, hemorrhages of the nose, abscesses, phimosis (a narrowness of the foreskin of the penis), bad teeth, and hemorrhoids, which we may presume were common problems. He was persuaded that the venereal disease he saw had been introduced by Europeans ("as in many countries they have penetrated") and suggested to his Japanese colleagues the use of waters containing mercury in its treatment. Although he was favorably disposed toward the knowledge of architectural principles, astronomy, and several other arts or sciences, and notes with seeming approbation that there were fewer lawyers and even fewer judges, his sketch of the state of Japanese medicine is less than flattering. He observes that physicians generally were ignorant of human anatomy and physiology, essential to the practice of medicine as he saw it, and he opines that "the best instructed physicians, or, speaking more accurately, the least ignorant, are those of the court, and the interpreters who associate with European physicians."[6] He was intrigued with the use of moxa, with which he was not previously familiar, and noted that purveyors of this treatment were prominent in the streets of the cities of an evening. Unfortunately, given the curiosity that most of us have in the perceptions that others have of our culture, few of the accounts of these earliest observers have been translated into Japanese.

Franz von Siebold, one of the most able of all of the physicians who served the Dutch factory, came to Dejima in 1823. Within a year, with the surreptitious endorsement of the local authorities, he had begun to teach Western medicine and the natural sciences on the mainland in Nagasaki. Five years later, however, he was banished for unwisely accepting maps of Japan. Other physicians would follow him; possibly the most important of these was Otto Mohnike, who introduced Japanese physicians to vaccination against smallpox in 1849.

Although the Dutch occasionally chartered American vessels for the trip to Nagasaki, few Americans had visited Japan prior to the arrival of Matthew Perry in 1854; however, the political events he helped to precipitate changed this state of affairs dramatically. Perry and his compatriots, in the narrative on their voyage, commented on many facets of Japanese science. They admired the high quality of astronomy and the accuracy with which eclipses and the like were calculated, and noted the widespread use of telescopes, chronometers, thermometers, and barometers, all modeled on those then available in Europe and the Americas. They were impressed by the extent of knowledge of medical botany but observed that little actual chemistry existed. Indeed, the first book on Western chemistry was not translated into Japanese until 1837, and its impact had undoubtedly not been widespread prior to the mission's arrival. Perry further stated that "the European medical gentlemen, who have contact with their professional brethren of Japan, report favorably of them."[7] Both Western and traditional medicine were practiced, and the medical members of the squadron commented on the use of acupuncture and moxa. Like Thunberg before them, they also remarked on the rarity of postmortem examinations and the limitations this imposed on medical knowledge.

With the Meiji restoration and the concerted effort of the government to achieve status as a modern nation, the tempo of interaction between Japanese and Western scientists accelerated. Johannes Pompe van Meerdevoort, a Dutch naval officer and physician, arrived in Nagasaki in 1857 and, in league with Matsumoto Ryojun, organized the first Western school of medicine in Japan. Soon thereafter the forerunner of the University of Tokyo Medical School, the Vaccination Institute, was founded. In 1859 James Hepburn, the American medical missionary, arrived. Although his method of transliterating Japanese, which continues to be widely used, is probably his greatest contribution to international scholarship, he was also a founder of Meiji Gakuin, its first president, and a teacher of physiology and hygiene at this institution for many years. It was not until 1877 that the results of Japanese medical research were first published in a Western scientific journal. These events suggest that the influence of Western notions on Japanese science was substantially more important than the converse in the years prior to the turn of this century.

After 1900 the contribution of Japanese scientists to world science grew immeasurably. Numerous factors contributed to this flowering. Universities modeled after Western institutions of higher learning were producing a growing number of competently trained, enthusiastic young investigators. Many of these spent several years abroad in research laboratories of prominence in Europe and the New World and returned imbued with new ideas, new techniques, and an earnest desire to contribute to the growth of science not only nationally but internationally. They communicated their sense of purpose to their own students as well as to their colleagues unable to avail themselves of foreign training. Scientific journals were established that provided a medium for the communication of ideas. Industry, particularly the chemical and pharmaceutical, saw the economic advantages to be gained through the furtherance of research and created employment opportunities for promising young investigators. One of the most significant contributions to medicine of Japanese science in these years was the isolation and cultivation of the spirochete causing syphilis by Noguchi Hideyo (Seisaku). However, much of his work occurred in the United States, where he had migrated in 1900 to become a stellar member of the Rockefeller Institute for Medical Research. Other important work can be cited. Nitta Isamu, for example, achieved distinction through his use of x-ray diffraction techniques to determine the lattice structure of organic compounds. This technique proved important in the characterization of the structure of DNA, the complex protein that encodes genetic information.

But these were past events and past perceptions. What are the present ones? Here I can recount only my own impressions, largely of medicine and the biological sciences; my reactions may not mirror those of others. When I first arrived in Japan forty years ago, Japan's scientific community had already begun to repair the ravages of war. Resources and manpower were limited, however, and the isolation of Japanese scientists from the international scientific community occasioned by the war had wrought incalculable damage. The xenophobia of the military oligarchy had led to proscriptions not only on education in Western languages but the importation of scientific journals and equipment. Training in non-war-related branches of science virtually ceased from 1940 until the end of 1945.

Medicine in the early postwar era was poor by international standards. This reflected circumstances beyond the control of the physicians themselves. Many had been hastily trained to meet Japan's war needs, and too much emphasis in this training was placed on didactic, often autocratic instruction and too little on supervised clinical experiences. Moreover, neither antibiotics nor the sulfonamides were widely available for therapeutic use. This is no longer true, and it is a marvelous tribute to the enthusiasm and perseverance of these physicians, many acutely aware of their own training limitations, that the quality of Japanese medicine and medical research now compares favorably with that practiced in Europe and the Americas. Medical training has improved immensely, and the growth in number of private and public medical schools ensures an enviable availability of medical care. Today, Japanese medical equipment ranks with the finest to be found world-wide, and in some instances, such as the use of ultrasound and computerized electrocardiography, knows no superiors.

Academic medicine in Japan, as elsewhere, breeds a competitiveness that is not always wholesome. The indebtedness of the present to the past is treated too cavalierly; all of us stand on the shoulders of our predecessors, and their successes or errors define what we do or fail to do. Frequently this indebtedness is unacknowledged and stultifies scholarship and a rightful recognition of the origin of scientific notions. As a geneticist, I would be remiss if I did not acknowledge Japanese contributions in this area. When I arrived, genetics was already a thriving discipline; the principal figures of the time were Komai Taku, Kihara Hitoshi, Oguma Kan, Furuhata Tanemoto, Tanaka Yoshimaro, Shinoto Yoshito, Kikkawa Hideo, Wada Bungo, and Yasui

Kono, to mention but a few. Many, but not all, of these individuals had received some of their training abroad. Particularly important were their contributions to cytogenetics, to plant genetics, and to the genetics of the silkworm, *drosophila,* and lady beetles. Kihara's work, for example, had elucidated the origin of wheat, and Furuhata had broadened our knowledge of the human red blood cell antigens. In 1948, shortly before my arrival, the Japanese government had established at Mishima the National Institute of Genetics, a more important step than was probably appreciated at the time. It provided a national focus for research in genetics, the most lively of the fields of biological endeavor, and thereby created the intellectual critical mass so essential to sustained research.

Although most of the individuals I have cited are either now dead or no longer active, their positions have been filled by gifted younger investigators. Today the National Institute of Genetics is one of the most celebrated institutions of its kind in the world, and the research of such members as Kimura Motoo, Ohta Tomoko, Maruyama Takeo, Matsunaga Ei, and numerous others is either pace-setting or ranks extremely favorably with work done elsewhere. Particularly noteworthy have been their contributions to molecular evolution and the theory of population genetics. So highly regarded are these achievements that Kimura and Ohta have been repeatedly singled out for national and international recognition. Still other Japanese geneticists have contributed importantly to molecular biology. Honjo Tasuku, for example, has added significantly to our understanding of the genetic control of the immunoglobulins, prominent in the body's response to infection.

Research is not a continuous flow of ideas devoid of the personalities who conceive them; it is the outgrowth of human interactions, and as a result the history of a pregnant notion is frequently impossible to trace. Our age demands scientific heroes and scapegoats, as though each new idea arose unaided from some primordial ooze of ignorance and was not beholden to the times nor what preceded it. Breakthroughs are the catchword, when often all that has been achieved is a perfunctory step, dictated by the spirit of the times, the Zeitgeist, and no novel prescient insight. This pandering to attention belittles not only the history of science but our claim to objective assessment. As our scientific community has grown—and it is now much larger than the sum of all of the intellectuals in Newton's England (or Europe)—publicity is pursued as avidly as truth. Perhaps this was inevitable. Science has moved from the sheltered halls of universities to national laboratories dependent upon the financial whims of a Congress or Diet unable to discipline itself, much less define a vision for the future; or to commercial institutions and consultative agencies, where fame and fortune rest either on our society's compulsion to regulate our lives and the need for a scientific basis for the regulations that are promulgated, or the litigious responses to this compulsion. The sheer enjoyment of lifting a further corner on the mystery of life has too often given way to hucksterism. Japan could avoid this, but the daily newspapers suggest that it too has succumbed. There, as elsewhere, the quest to unravel that which can be unraveled has come perilously near to profit-seeking.

The public's acceptance of the importance of science in society has been colored by these changes. In the United States, although science and scientists are still highly valued and there is general recognition of the importance of research, many people hold the view that the degradation of our environment has come about because of a thoughtless, overenthusiastic endorsement of all applications of research. Nowhere is this view more evident than in the case of nuclear power. Americans are also increasingly aware of the moral and ethical issues that scientific developments presage and are beginning to insist that the public participate in decisions that can profoundly affect collective well-being. Some of these environmental issues have surfaced in Japan as well, driven by unfortunate instances of exposure to methylmercury and organic bromides. As yet, there seems less disenchantment with science and technology in Japan than here. The Japanese see science as the means of gaining and maintaining the economic advantages that have offered to them a new life. Consequently they are, so far, less critical. In any country, in the long run, though, disillusionment and confrontation can only be avoided through education of the lay and the professional segments of society and through the public airing of the benefits and risks that accrue from the introduction of specific technologies. Expectations must be grounded in knowledge and mutual respect; unfortunately, the latter is more difficult to gain than to lose.

Thomas Kuhn, the famed philosopher of science, in *The Structure of Scientific Revolutions,* has maintained that crises provide the necessary tension for the emergence of novel theories, and notes that creative investigators, like artists, must be able to live in a world out of joint. The anomalies that arise between observation and the conceptual framework that guides investigative activities produce the intellectual and emotional challenges that spur revolutions. Scientists respond to the intellectual tensions, the objects of Kuhn's concerns, through modifications or changes in theory, but they cope less well with the more human element, the emotional turmoil. Ultimately, one either rejects science in favor of another occupation or learns to live with its tensions.

Does Japanese culture equip its young scientists with the ability to tolerate the emotional tensions of truly creative scientific research? Are they able to question authority rather than accommodate to the view of the majority? The speed of light, after all, is not determined by consensus, although the resolution of differences in measurement techniques might be. It is questionable whether the prevailing system of education in Japan provides a means for investigators to cope with a scientific world that is forever out of joint. Scientists cannot escape from mystery; it is the character of their universe, and they are driven by the need to unravel it. In this quest gadgets, with which Japanese laboratories are now well-endowed, are important, but science is more than the multiplication of instruments; it involves intellectually creative acts. The separation of creativity from the pull of our own wishful thinking, often culturally inspired, that we misconstrue as scientific can be difficult and lead to absurdities, of which even the brightest are guilty.

*   *   *

Finally, what of the research that has involved me in Japan over these four decades? The findings are not easily nor succinctly summarized, largely because over the years the nature of my participation in the activities of the Atomic Bomb Casualty Commission and its successor, the Radiation Effects Research Foundation, has evolved in unanticipated ways. Time has seen me engaged in research on radiation-related carcinogenesis, the effects of prenatal exposure to ionizing radiation on the developing human brain, the occurrence of radiation-induced cataracts, and of course, the genetic studies. My involvement in many of these occurred, however, after the events and circumstances described here. I focus, therefore, on just two groups of studies, those in which I participated in the years 1949 through 1965: the likelihood of an untoward pregnancy outcome (a pregnancy terminating in a child with a severe birth defect, a stillborn infant, or a child dying before the age of one year) following parental exposure to atomic radiation, and the consequences of consanguineous marriages, that is, marriages between people who are biologically related. Even here, it is difficult to place the findings in perspective.

Some studies—and the follow-up of the children born to parents exposed to ionizing radiation is surely one such—must be done even if scientific wisdom at their inception suggests that they may reveal no measurable effects. Human concerns demand reassurance, and rarely, if ever, is our knowledge so firm and immutable that we can exclude all plausible alternatives to the prevailing scientific notions. The Greek theory of phlogiston, the hypothetical principle of fire, was replaced by the atomic theory, and in physics the principle of parity—that there was a forward and backward symmetry in radiation—was "proved," only to be later disproved. Similarly, scientists once believed that skills acquired by the parents could be inherited by their offspring, and that the genetic potential of the two parents blended together in the child. Today we know that acquired "characters" are not inherited, and that each parent contributes only half of his or her genome to their offspring. Many of the genes a parent contributes will not be expressed in a given child; there is no "blending inheritance." Undoubtedly other contemporary shibboleths of science will prove equally erroneous in hindsight.

Some forty years after the initiation of the first studies of the children of the survivors of the atomic bombing of Hiroshima and Nagasaki—a program that would embrace the physical examination of over 75,000 infants and the surveillance of mortality in an even larger group—the following is clear: We can exclude a doubling of the risk of birth defect or premature mortality under the circumstances of exposure that obtained in Hiroshima and Nagasaki. We can also show that these children are not conspicuously more likely to develop childhood cancers or to die in early adulthood. This should be reassuring to prospective parents who may be exposed to ionizing radiation, wittingly or unwittingly, for clearly there has been no epidemic of birth defects, malignancies, or early mortality among the survivors' children.

We cannot as yet provide the firm estimates we should like of the probability of producing a mutation in a specific gene with a given dose of radiation. We can, however, narrow the window of uncertainty; possibly the biochemical studies currently under way or contemplated will narrow it still further. We believe that some genetic damage has occurred, largely based on experimental investigations, and accept the human data as the best evidence available to us for estimating the mutagenic risk. Although these latter data alone would not satisfy the usual evidential requirements of science, we argue that even if there was no experimental data on humans, every other well-studied animal and plant species exhibits a deleterious effect of exposure to ionizing radiation, and there is no compelling biological reason to believe that the human species would be different. However, our methods of reproduction, the nurturing of preparturient human embryos and fetuses, and postnatal care make hazardous simple extrapolations to human beings from data on animals such as the fruit fly or the mouse.

If one takes at face value the observations in Hiroshima and Nagasaki, then in terms of those mutations that are measurable by epidemiological means, the mutability of human genes is not greater, indeed may be somewhat less, than that seen in experimental animals. The National Academy of Science's Committee on the Biological Effects of Ionizing Radiation and the United Nations' Scientific Committee on the Effects of Atomic Radiation have examined the human and the experimental data and concluded that the genetic risk to a child conceived after parental exposure under most circumstances (save a nuclear war or a nuclear accident more severe than Chernobyl) is less than that to a child whose mother smokes or drinks during her pregnancy.[8] Indeed, the data from Hiroshima and Nagasaki suggest that even at a dose of one gray, the equivalent possibly of 10,000 x-ray films of the chest, the risk of an untoward pregnancy outcome is increased less than 10 percent. Somewhat differently put, since approximately 10 pregnancies in every 1,000 persisting through seven gestational months terminates in an infant conspicuously abnormal at birth, these data suggest that this number would be increased to possibly 11 infants if one or both of the parents of all 1,000 children had received the equivalent of 10,000 x-ray films. Thus, the increased risk following exposure to five rad (one twentieth of a gray)—the dose that one might expect under diagnostic or occupational circumstances—is small, below the level we can presently measure with confidence.

It is important, in a discussion of mutagenic risk, to distinguish between the risk to one pregnancy and the risk to a population where many pregnancies occur. The risk in the first instance is very small, but if this small risk is multiplied by the 2 or 3 million births that occur in the United States each year, then if every prospective parent was exposed to ten rad, for example, there could occur tens, possibly even a few hundred, radiation-affected pregnancies. In the context of public health this number is not large, particularly when one bears in mind that thousands of pregnancies terminate abnormally each year even in the absence of exposure (ignoring the radiation to which we are all exposed from radioactive materials in the earth's crust or from cosmic activity).

The hazards of exposure to ionizing radiation cannot, however, be seen solely in terms of the mutagenic risk. There are other detrimental effects. Of these, the most poignant is the damage done to the developing brain of a fetus if exposure occurs at critical times in gestation. The most vulnerable period is from the beginning of the eighth week following fertilization through the fifteenth week (a lesser risk occurs in the sixteenth through the twenty-fifth week, but apparently not at doses below 0.2 gray or so). This especially critical juncture (the eighth through the fifteenth week) is when the cells that give rise to the neurons of the cerebrum are rapidly dividing and the immature neurons are migrating to their functional sites in the brain. We know now that if the fetus is exposed at this time, the likelihood that the child will be severely mentally retarded is high, either because the immature neurons are killed, or they fail to reach their proper functional sites, or both. Some forty percent of fetuses exposed to one gray or more of radiation during this period of development will be mentally retarded. There is also evidence that those children who are not mentally retarded but were exposed at this same stage perform more poorly in school and on conventional intelligence tests; this finding suggests that they too suffered some impairment in brain development. At lesser doses, below a half gray, we presume the risk is also appreciable, but the data are too sparse to be confident of precisely what it may be. Prior to the eighth week or after the twenty-fifth week following fertilization, we see no evidence of brain damage. This does not necessarily imply that none occurs, merely that we have not been able to see it. We do know that in the first eight weeks following fertilization, a pregnant woman is more likely to lose her pregnancy if she is exposed to an appreciable amount of ionizing radiation, and it may be that the brain-damaged embryos are lost.

Children and adults who were directly exposed to atomic radiation have a greater probability of developing a malignant tumor—especially leukemia, but also cancers of the bladder, breast, colon, lung, ovary, stomach, and thyroid, as well as multiple myeloma—at some subsequent stage in life. This risk increases as the dose increases, and it is higher among individuals exposed in the first two decades of life than those exposed at later ages. At the cellular level, radiation-related cancers are still indistinguishable from cancers attributable to other causes, but perhaps in the future this will not be so. At present, we simply know that we see more cancers among the exposed than non-exposed individuals and that this increase depends upon the dose.

Clearly, if these various burdens of ill-health can be avoided or mitigated, they must be. Complete avoidance is impractical, however. Some occupations necessarily entail exposure to ionizing radiation, as does air travel. Sickness may oblige one to have radiological studies, and some homes contain radon, a radioactive gas, because of their geographical siting. The issue becomes one of restricting exposure to the lowest possible level. Here each of us can contribute, by avoiding unnecessary radiological studies and adhering to occupational procedures established to minimize exposure.

In the case of consanguineous marriages, those involving parental relationships as close as or closer than second cousins (people who share one set of great-grandparents), there is a clearly measurable detrimental effect on the children, one substantially greater than exposure to one, five, or even a hundred rad of ionizing radiation. Children born to such marriages are more likely to exhibit severe birth defects or overt genetic disease, to be retarded somewhat in physical and mental growth, and to succumb prematurely. For example, the average child born to parents who are first cousins will have a probability of dying prematurely (either being stillborn or failing to survive the first year of life) twice as great as that of a child born to parents who are unrelated. This implies that if a liveborn child of unrelated parents has a 3 percent likelihood of not reaching its first birthday, the child of a first-cousin marriage has a 6 percent risk. In terms of intelligence as measured by conventional tests, an average child of first cousins will have an IQ of about 95 rather than 100. A proportional retardation is seen in growth, neuromuscular development, and the like. These are not unexpected findings to a geneticist, for we have long recognized that consanguineous marriages increase the frequency of individuals who are homozygous, that is, who carry two identical copies of a given gene. This happens because the two parents inherited the same gene from a common ancestor. Many genes express themselves only when present in two copies, and if the effects of the gene are detrimental, then, homozygous offspring are more likely to show the detrimental effect.

The studies on Hirado confirmed the earlier findings on Kuroshima: a higher mortality among the children of consanguineous marriages (also seen in Hiroshima and Nagasaki) and a greater fertility in these marriages. However, we have still been unable to document adequately the causes of this increased mortality, largely because at the time of these investigations medical care in the rural areas was not as good as that in the cities, and death often followed brief, undiagnosed illness. Death certificates in these instances are of little help and can be actively misleading.

It would be difficult now to duplicate these studies. The frequency of consanguineous marriages, which in the past represented almost 10 percent of all Japanese marriages, even in a city of the size of Nagasaki, has fallen steadily in the postwar years throughout Japan and is not much greater now than that seen in the United States or most European countries. The causes of this decline are many, including a more mobile population, fewer family-arranged marriages (where cultural and socioeconomic factors loomed large in the choice of a bride or groom), and, presumably, a greater recognition of the biological consequences of marriages between related people.

Studies of the genetic and somatic effects of radiation on the survivors of the atomic bombing of Hiroshima and Nagasaki continue, and surely will for several more decades, for it is still not clear what the full impact of this exposure will be. At present, more of those persons who survived the immediate effects are alive than have died. What their future holds is uncertain. There is much yet to be learned if we are to understand fully the hazards of ionizing radiation and to use productively the wisdom we will have gained.

NOTES

BIBLIOGRAPHY

GLOSSARY

INDEX

MAPS

NOTES

Introduction

1. See Beebe, 1979.

1. Hiroshima, 1949

1. Of the total of 4,535 individuals charged, more were acquitted (985) than received death sentences (805).
2. *Japan Times,* November 14, 1986, p. 4.

2. MacArthur's Japan

1. Green tea is the most common beverage in Japan. It is made from the same kind of tea leaves as western black tea, but it is produced by a process of steaming and drying rather than the fermentation used to make black tea.
2. This arrangement of pine, bamboo, and plum boughs or sprigs is also called a *matsu-take-ume,* a word derived from the alternative readings of the three ideograms. Shochikubai is the *on* or Chinese reading, and matsutakeume the *kun* or Japanese.
3. These New Year's decorations have a variety of names; they may be called a shimekazari or a shimenawa, for example. A house with such an ornament before it is believed to be pure; no devil can enter. Presumably, on a car it wards off evil from the driver, and one would hope from pedestrians encountered.

3. Surveying the Children

1. A gray, usually abbreviated Gy, is a unit of absorbed dose, or energy deposited in tissue, technically equal to one joule per kilogram, or 10,000 ergs per gram. A gray is equal to 100 rad. It is named after the English health physicist L. H. Gray.
2. Bernstein, 1979.
3. Thomas Campbell Cartwright was a prisoner of war from July 28, 1945, until September 1, 1945. Barton has observed, with respect to his crewmen,

the Japanese, who had some records from 1945 or early 1946 on American victims of the atomic bomb, including the six men from Cartwright's plane, remained silent after apparently giving American authorities a list of the names in the months after the Hiroshima bombing."

4. An Honorable Send-off

1. An English version of this rescript can be found in the appendices to Hall's (1949) translation of the *Kokutai No Hongi*, p. v.

5. New Agendas

1. In 1955, a committee, known as the Francis Committee after its chairman, Thomas Francis, a distinguished epidemiologist and member of the National Academy of Sciences, was sent to Japan to evaluate the scientific program and recommend changes if these seemed warranted. Francis, a virologist known for his work with influenza, was accompanied by Felix Moore, the head of the Department of Biostatistics at the University of Michigan, and Seymour Jablon, a mathematical statistician and a member of the Veteran's Follow-up Agency of the National Research Council. Out of their assessment emerged a number of recommendations, but among these the most important was the advocation of a Unified Study Program that was to include a mortality surveillance, a clinical study to assess health and morbidity, and a program of autopsies. The former two were to be centered upon a fixed sample of survivors and suitably age-and-sex-matched comparison persons, whereas the latter was to entail the pathologic study of as many as practicable of those individuals included in the surveillance sample who succumbed.
2. An especially entertaining account of some of these bars, their proprietors, and the events that transpire in Japanese night life will be found in Morley, 1985.

6. Nagasaki, 1959

1. A full account of this survey and its findings appears in Schull and Neel, 1965.
2. See Naruge, 1956, for a complete description of these records.
3. Thunberg, 1796, vol. 4, pp. 17–18.

7. Merchants, Missionaries, and Marriages

1. The Emperor Go-Nara (1497–1557) reigned from 1526 into 1557. In his age, time in Japan was reckoned in eras (*nengō*). Customarily, the accession of an emperor to the throne, or an important event, happy or unhappy, brought a change in the name of the era. Go-Nara's reign spanned three eras—Kyōroku, Tembun, and Kōji. Since the enthronement of the Emperor

Meiji, it has been the practice to apply one era name to the full reign of a given emperor. The recent eras have been named Meiji, Taisho, Showa, and, now, Heishi.

2. The incident of malfeasance involved the daimyō of Arima. See Boxer, 1951, pp. 314–315.

3. Japan's isolation grew out of a series of edicts. The first of these, in 1616, limited foreign trade to the ports of Hirado and Nagasaki and restricted the movement of merchants into the interior of the country. The second, in 1624–1625, severed formal relationships with Spain, Mexico, and the Philippines. In 1633–34, a third edict prohibited Japanese vessels from sailing to foreign ports, and those Japanese who had settled abroad were banned from returning to their country. The fourth, possibly the most famous, was the promulgation in 1636 of the Sakoku, the Closed Country edict. Finally, in 1638, the Portuguese were expelled.

4. Thunberg, 1796, vol. 3, p. 46.

5. Thunberg, 1796, vol. 3, p. 400.

6. Thunberg, 1796, vol. 3, pp. 401–403.

7. Thunberg, 1796, vol. 3, p. 403.

8. Swift, 1726, vol. 2, pt. 3, pp. 150–152.

9. The treaty with France, signed on February 22, 1859, also granted to them freedom of worship, and the rights to construct churches and chapels and lay out cemeteries.

10. In a letter to his superiors, Petitjean wryly noted that "on the next day, workers arrived in triple number; they worked day and night so that the church could be completed on the designated day."

11. Marnas, 1931, vol. 1, p. 645.

12. A copy of this book, *Spiritual Xuguio no Tameni yerabi atcumuru xuguanno Manual,* is in the possession of the Bishop of Nagasaki.

13. Kaempfer, 1726, bk. 4, chap. 9 (vol. 2, p. 250, in Glasgow edition).

14. Ibid. (vol. 2, p. 253, in Glasgow edition).

8. Hirado, 1964

1. A pinnace, a small sailing ship commonly used as a tender for a larger vessel.

2. Cocks, 1883, vol. 2, p. 314.

3. Ibid.

4. Valignano (1954, p. 94, n. 97) described the former as a "crafty, hater of the Christian religion, and a heathen in his bones." Shigenobu fared no better; he was said to be "a pagan, an atheist of a sect that does not believe in any other life, and has always been hateful of Christians, a deceitful man, a son of this century, . . . an old fox."

5. Actually a katabira, a light-weight garment commonly worn by members of the samurai class.

6. Cocks, 1883, vol. 1, p. 169.

7. Ibid.

Retrospect

1. *Japan Times,* 1987.
2. This economic adaptability was adumbrated more than ninety years ago, when one observer, Arthur Knapp (1898, vol. 1, p. 119) noted omnisciently, "Japan's recent and suprising advent in the fields of Occidental commerce, and by her evincing there a spirit of enterprise, such an aptitude for trade, and such an intimate knowledge of the world's modern ways of business, as to make her a most formidable competitor of the leading commercial powers."
3. *Hiroshima Signpost,* Summer 1987, p. 14.
4. Basho, 1968, p. 131.
5. There is no critical study of Japanese science in a Western language comparable to Joseph Needham's (1954–1984) monumental appraisal of Chinese science and technology.
6. Thunberg, 1796, vol. 4, p. 122.
7. Perry, 1856, vol. 1, p. 56.
8. UNSCEAR, 1986, 1988.

BIBLIOGRAPHY

Alcock, Rutherford. 1863. *The Capital of the Tycoon: A Narrative of Three Year's Residence in Japan*. Vols. 1 and 2. New York: Harper and Brothers.

Barthel, Manfred. 1982. *Die Jesuiten*. Federal Republic of Germany: Econ Verlag GmBh. [English translation by Mark Howson, *The Jesuits: History and Legend of the Society of Jesus*. New York: William Morrow and Co., 1984.]

Basho, Matsuo. 1968. *Back Roads to Far Towns (Oku-no-hosomichi)*. Translated and annotated by Cid Corman and Kamaike Susumu. New York: Grossman Publishers, 1968.

Beardsley, Richard K., John W. Hall, and Robert E. Ward. 1959. *Village Japan: A Study Based on Niiike Buraku*. Chicago: University of Chicago Press.

Beebe, Gilbert W. 1979. "Reflections on the work of the Atomic Bomb Casualty Commission in Japan." *Epidemiological Reviews* 1:184–210.

Benedict, Ruth. 1946. *The Chrysanthemum and the Sword: Patterns of Japanese Culture*. Boston: Houghton Mifflin Co.

Bernstein, Barton J. 1979. "Hiroshima's Hidden Victims." *Inquiry*, August 6, pp. 9–11.

Bird, Isabella L. 1893. *Unbeaten Tracks in Japan*. London: John Murray.

Bohours, Dominic. 1688. *The Life of Saint Francis Xavier of the Society of Jesus: Apostle of the Indies and of Japan*. Translated by John Dryden. London: Jacob Tonson.

Bowers, Faubion. 1952. *Japanese Theatre*. New York: Hermitage House.

Bowers, John Z. 1965. *Medical Education in Japan*. New York: Harper and Row.

Boxer, Charles R. 1950. *Jan Compagnie in Japan, 1600–1850: An Essay on the Cultural, Artistic and Scientific Influence Exercised by Hollanders in Japan from the Seventeenth to the Nineteenth Centuries*. 2nd rev. ed. The Hague: Nijhoff.

———— 1951. *The Christian Century in Japan*. Berkeley: University of California Press.

Brinkley, Frank. 1901. *Japan and China*. Oriental Series. Vols. 1–12. Boston: J. B. Millet Co.

Calvino, Italo. 1972. *Invisible Cities*. Translated by William Weaver. New York: Harcourt, Brace, and Jovanovich.

Caron, François. 1935. *A True Description of the Mighty Kingdoms of Japan and Siam*. London: Argonaut Press.

Cocks, Richard. 1883. *Diary of Richard Cocks, Cape-Merchant in the English Factory in Japan, 1615–1622, with Correspondence.* Vols. 1 and 2. Edited by Edward Maude Thompson. London: Printed for the Hakluyt Society.

Cooper, Michael. 1965. *They Came to Japan: An Anthology of European Reports on Japan, 1543–1640.* Berkeley: University of California Press.

Doi, Takeo. 1971. *The Anatomy of Dependence.* Translated by John Bester. Tokyo: Kodansha International.

Elison, George. 1973. *Deus Destroyed: The Image of Christianity in Early Modern Japan.* Cambridge: Harvard University Press.

Funakoshi, Yasutake. 1963. *The 26 Martyrs of Nagasaki.* Tokyo: Bijutsu Shuppan-sha.

Furuno, Kiyoto. 1959. *Kakure Kirishitan.* Tokyo: Sato Publishers.

Griffis, William E. 1876. *The Mikado's Empire.* 5th ed. New York: Harper and Brothers.

Hachiya, Michihiko. 1955. *Hiroshima Diary: The Journal of a Japanese Physician August 6–September 30, 1945.* Translated and edited by Warner Wells. Chapel Hill: The University of North Carolina Press.

Halford, Aubrey S., and M. Giovanna. 1956. *The Kabuki Handbook.* Tokyo: Charles E. Tuttle.

Hall, Robert King, ed. 1949. *Kokutai No Hongi: Cardinal Principles of the National Entity of Japan.* Translated by J. O. Gauntlett. Cambridge: Harvard University Press.

Hamamura Yonezo, Sugawara Takashi, Kinoshita Junji, and Minami Hiroshi. 1956. *Kabuki.* Translated by Fumi Takano. Tokyo: Kenkyusha, Ltd.

Hersey, John. 1946. *Hiroshima.* New York: Alfred A. Knopf.

Hibakusha: Survivors of Hiroshima and Nagasaki. Translated by Gaynor Sekimori. Tokyo: Kosei Publishing Co.

Hildreth, Richard. 1855. *Japan as It Was and Is.* Boston: Phillips, Sampson and Co.

Hironaga, Shuzaburo. 1976. *The Bunraku Handbook.* Tokyo: Maison des Arts, Inc.

Ibuse, Masuji. 1953. *Black Rain.* Translated by John Bester. Tokyo: Kodansha International.

Japanese National Commission for UNESCO. 1958. *Japan: Its Land, People and Culture.* Tokyo: Printing Bureau, Japanese Government.

Kaempfer, Engelbert. 1726. *The History of Japan: Giving an Account of Antient and Present State and Government of that Empire; of Its Temples, Palaces, Castles and other Buildings; of its Metals, Minerals, Trees, Plants, Animals, Birds and Fishes; of Chronology and Succession of the Emperors, Ecclesiastical and Secular; of the Original Descent, Religions, Customs and Manufactures of the Natives, and of their Trade and Commerce with the Dutch and Chinese.* Translated by J. G. Scheuchzer. Vols. 1 and 2. London: Thomas Woodward. Rpt. 1906 in 3 vols. Glasgow: James MacLehose and Sons, publishers to the university.

Kataoka, Yakichi. 1957. *Nagasaki no Junkyosha.* Tokyo: Kadogawa Shoten.

Keene, Donald. 1951. *The Battle of Coxinga.* London: Taylor's Foreign Press.

———— 1952. *The Japanese Discovery of Europe: Honda Toshiaki and Other Discoverers, 1720–1798.* London: Routledge and Kegan Paul, Ltd.

Knapp, Arthur M. 1898. *Feudal and Modern Japan.* Vols. 1 and 2. London: Duckworth and Co.

Laures, Johannes. 1954. *The Catholic Church in Japan: A Short History.* Tokyo: Charles E. Tuttle Co.

Lynn, Leonard. 1986. "Japanese research and technology policy." *Science* 233:296–301.

MacFarlane, Charles. 1856. *Japan: An Account Geographical and Historical.* Hartford: Silas Andrus and Son.

Marnas, Françisque. 1931. *La Religion de Jesus Ressuscitee au Japon: Dans la Seconde Moitie au XIX Siecle.* 2nd ed. Vols. 1 and 2. Clermont-Ferrand: Imprimerie Generale.

Marshall, Eliott. 1986. "School reformers aim at creativity." *Science* 233:261–270.

Maruki, Iri, and Toshi Maruki. 1985. *The Hiroshima Murals: The Art of Iri Maruki and Toshi Maruki.* Edited by John W. Dower and John Junkerman. Tokyo: Kodansha International Ltd.

Morley, John David. 1985. *Pictures from the Water Trade: Adventures of a Westerner in Japan.* New York: Harper and Row.

Murdoch, James. 1926. *A History of Japan.* Revised and edited by Joseph H. Longford. Vols. 1, 2, and 3. London: Kegan Paul, Trench, Trubner and Co., Ltd.

Nagai, Takashi. 1984. *The Bells of Nagasaki.* Translated by William Johnston. Tokyo: Kodansha International.

Naruge, Tetsuji. 1956. *Koseki no Jitsumu to sono Riron.* Tokyo: Nihon Kajo-Shuppan.

Needham, Joseph. 1954–1984. *Science and Civilization in China.* 6 vols. Cambridge: Cambridge University Press.

Okajima, Shunzo. 1975. "Fallout in Nagasaki–Nishiyama District." *Japanese Journal of Radiation Research* 16(suppl):35–41.

Papinot, E. 1948. *Historical and Geographical Dictionary of Japan.* Ann Arbor: Overbeck Company.

Perry, Matthew C. 1856. *Narrative of the Expedition of an American Squadron to the China Seas and Japan. Performed in the Years 1852, 1853, and 1854.* 3 vols. Washington: A. O. P. Nicholson. [These volumes were actually written by Francis L. Hawks, the expedition's historian.]

Pompe van Meerdevoort, Johannes L. C. 1859. "On the study of the natural sciences in Japan." *Journal of the North China Branch, Royal Asiatic Society of Great Britain and Ireland* 1:211–221.

———— 1867–1868. *Viff Jaren in Japan (1857–1862).* Leyden: Henvell and van Sante.

Raucat, Thomas. 1954. *The Honorable Picnic.* Tokyo: Charles E. Tuttle Co.

Redesdale, Lord. 1906. *The Garter Mission to Japan.* London: MacMillan and Co.

Saint Exupery, Antoine de. 1986. *Wartime Writings: 1939–1944.* Translated by Noah Purcell. New York: Harcourt Brace Jovanovich.

Sansom, George B. 1946. *Japan: A Short Cultural History.* Rev. ed. London: The Cresset Press.

———— 1950. *The Western World and Japan.* London: The Cresset Press.

———— 1952. *Japan in World History.* London: George Allen and Unwin, Ltd.

Schull, William J., and James V. Neel. 1965. *The Effect of Inbreeding on Japanese Children.* New York: Harper and Row.

Shohno, Naomi. 1986. *The Legacy of Hiroshima: Its Past, Our Future.* Tokyo: Kosei Publishing Co.

Soka Gakkai. 1978. *Cries for Peace: Experiences of Japanese Victims of World War II.* Tokyo: The Japan Times, Ltd.

Spence, Jonathan D. 1984. *The Memory Palace of Matteo Ricci.* New York: Viking Penguin, Inc.

Swift, Jonathan. 1726. *Travels into Several Remote Nations of the World by Lemuel Gulliver.* Vols. 1 and 2. London: Benjamin Motte.

Tagita, Koya Paul. 1954. *Showa Jidai no Senpuku Kirishitan.* Tokyo: Nihon Gaku Jutsu Shinkokai.

Thunberg, Carl Pieter. 1796. *Voyages de C. P. Thunberg, au Japon, par le Cap de Bonne-Esperance, les Iles de la Sonde, etc.* Paris: Benoit Dandre, Garnery and Obre. 4 vols.

Titsingh, Isaac. 1822. *Illustrations of Japan.* London.

UNSCEAR. 1986. *Genetic and Somatic Effects of Ionizing Radiation.* New York: United Nations.

———— 1988. *Sources, Effects, and Risks of Ionizing Radiation.* New York: United Nations.

Tronson, J. M. 1859. *Personal Narrative of a Voyage to Japan, Kamtschatka, Siberia, Tartary and Various Parts of Coast of China; in H. M. S. Barracouta.* London: Smith, Elder and Co.

Valignano, Alejandro. 1954. *Summario de las Cosas de Japon (1583): Adiciones del Summario de Japon (1592).* Edited by Jose Luis Alvarez-Taladriz. Tokyo: Sophia University Press.

von Siebold, Philipp F. 1841. *Manners and Customs of the Japanese in the Nineteenth Century.* New York: Harper and Brothers. (Reprinted 1973, Rutland, Vermont: Charles E. Tuttle Co., Inc.)

Wells, Warner. 1958. "Dr. Kaoru Shima—his recollections of Hiroshima after the A-bomb." *American Surgeon* 24:668–678.

Yamaguchi, Momoo, and Setsuko Kojima, eds. 1979. *A Cultural Dictionary of Japan.* Tokyo: The Japan Times, Ltd.

GLOSSARY

Many Japanese words with quite different meanings have the same form when romanized; for example, *hana* can mean a nose or a flower. In such instances, I give only the meaning that is pertinent to the use of the word in this book. Similarly, in use many nouns are preceded with the honorific *o*; in the glossary that follows this honorific has generally not been included.

aisatsu: an inaugural or courtesy visit; the courtesy call made at one's retirement from office is called a tainin

aisu kiyande: ice candy

akabō: a luggage porter; literally, a redcap

akamatsu: Japanese red pine

akebono: dawn, or daybreak

amado: the rain shutters on a traditional Japanese home

Amerikajin: an American

Ashikaga: the name of the ruling family during the Muromachi or Ashikaga shogunate (1338–1573)

bachi: a plectrum used with stringed instruments

banzai: hurrah, or long live

bata-bata: a three-wheeled motorcycle used for hauling; its name ostensibly comes from the sound it makes

beijin: an older expression, synonymous with amerikajin

ben: a dialect

benjo: a toilet

bentō: a box lunch, or lunch

bessō: a villa, often now applied to particular inns

bīru: beer

biwa: a Japanese lute

bōnenkai: a year-end party

bonjin: a mediocre person, an ordinary individual

bōzu: a Buddhist priest, a bonze

buchō: the head of an administrative division or department

bugyō: a magistrate or high commissioner

buke: the military class

bunchō: a small paddy bird

bungen: station in life

Bungo: a feudal district in eastern Kyushu, now a part of Oita Prefecture

Bunraku: the puppet theater, or a puppet used in that theater

Bushidō: the way of the warrior; Japanese chivalry

butsudan: a family Buddhist altar

Caspar-ryu: a school of Western medicine in Japan, named after Caspar Schambergen, the Dutch physician under whom the founders studied

chanbara: sword-rattling or samurai films

chanoyu: the tea ceremony; the art of making ceremonial tea

chigaidana: side alcove shelves, usually two, often staggered

chō: a district or street, more or less synonymous with machi

chōnin: a tradesman or merchant

chōzubachi: originally a stone container for water, put in a shrine area for purifying one's hands and mouth before visiting the shrine; the term is not commonly used today; it is synonymous with a mizubachi

chūgakkō: a middle school

dai: great or large

daikon: a Japanese radish

daimyō: literally, the great names; a feudal lord

Dejima: the artificial island created in Nagasaki for the Portuguese merchants but subsequently occupied by the Dutch; the older historical literature often spells this name as either Desima or Deshima

densha: an electric car; a tram-car

deshi: an apprentice

dōri: a street or road

Echizen: a feudal district of Japan on the western side of the country, a part now of Fukui Prefecture

edamame: green soybeans

Edo: the old name of Tokyo

efumi: a copper tablet with a crucifix to be trodden on to prove oneself a non-Christian

eki: a railway station or depot

ekimae: the area in front of a railway station

eta: one of the outcast groups in feudal Japan, including the meat handlers, the buriers of the dead, and the like

fubuki: a snowstorm or blizzard

fude: a writing brush

Fujiwara: the name of an exceptionally influential family during the Heian period in Japan; many of the daughters of this family were consorts or wives of the emperors

fukubuchō: the vice-head of an administrative division or department

fukurotodana: a small cupboard or closed shelf, generally used for storing bags

fundoshi: a loin cloth

furo: a bath or bathtub

furoshiki: a wrapping cloth

fusuma: a sliding door or partition; generally used to describe the interior doors as distinct from the outwardly facing shōji

futon: a quilt or coverlet

gaijin: a foreigner

geisha: a professional beauty and entertainer; a singing girl

Gemmyo: an eighth-century empress of Japan

genkan: a vestibule or foyer

geta: wooden clogs

go: a game played on a ruled board with black and white stones

gohei: a wand of cut paper used in Shinto rituals

go-jū-noto: a five-tiered pagoda

gonin-gumi: a five-family neighborhood unit in the Tokugawa era

Go-Tairo: the five members of the council of state selected by Hideyoshi who were to assist his son during his minority; one of the members was Tokugawa Ieyasu

gyūnyū: cow's milk

hachimaki: a headband

hakama: a divided skirt for formal men's wear

han: a name stamp

hana: a nose; a flower

hanami: flower viewing

hanamichi: an elevated passageway running from the stage to the rear of the theater through the audience

hanare: literally, the separatists, the hidden Christians who have not rejoined the Catholic Church

hanga: a color print

haori: a coat

harakiri: a ritualized form of suicide through disembowelment; also called seppuku

haramaki: a belly band

hashi: chopsticks; a bridge; the edge

hashiwade: the ritual claps before a shrine or altar

hatsu shu sho: a police box

hibachi: a charcoal brazier

hidari: left

higanbana: an autumnal flower, *Lycoris radiate*

hikidemono: a present or keepsake, traditionally one made to an equal or an inferior

hinin: one of Japan's feudal outcast groups

hinoki: the Japanese cypress

hiragana: the cursive Japanese syllabary

hitotsu: the number one

hōanden: the small shrine-like structure before schools housing the Imperial Rescript on Education

Hojo: the name of the ruling family during the Kamakura shogunate (1200–1333), descendants of Taira Sadamori

hotaru: a firefly

hyōsatsu: a name or door-plate

hyōshigi: wooden clappers

ihai: a Buddhist mortuary memorial tablet

Imari: a town in western Kyushu, now in Saga Prefecture, associated with pottery and porcelain making

inumaki: a bush or small tree with an elongated thin leaf, a podocarp, *Podocarpus macrophylla* ise ebi: a lobster or prawn

janken: a game generally played by children, very similar to the game played by Western children known as paper-scissors-stone

jichinsai: a Shinto ritual to placate the earth prior to the initiation of new construction

jidai: a period, epoch, era, or age

jikaku: self-consciousness, or self-awakening

jimbei: a light-weight cotton undergarment worn in the summer

jinja (or jinsha): a Shinto shrine

Jizō: a guardian deity of children and pregnant women

jōruri: the chanter or narrator in a bunraku play; the style of chanting is sometimes called Gidayu, after Takemoto Gidayu, possibly the most famous of all chanters

jūmin tōroku: resident registration

ka: a mosquito

Kabuki: a kabuki performance

kachō: a section chief

kaki: a persimmon

kakure kirishitan: the so-called hidden Christians of Japan and their descendants; properly, this term includes both the separatists, the hanare, as well as those Catholics descended from the hidden families who have rejoined the Catholic Church

kama: a kiln

kamaboko: a boiled fish paste; a cake made from this paste

kamabokojō: fishcake castle

kamabokotei: fishcake palace

kami: paper

kamishibai: a picture-card show

kana: the Japanese alphabet

kanji: a Chinese ideogram

kanjō: a reckoning; a bill

kanna: a carpenter's plane

kappōgi: a sleeved apron

kasuri: a splash patterned kimono or kimono material, previously of cotton but now commonly a synthetic fiber katakana: the square form of the Japanese syllabary

katsuo: a bonito; a skipjack tuna

katsura: the Japanese Judas-tree

kawa (or gawa): a river

kawata: a tanner or leather-dresser; akin in meaning to eta

kekkonshiki: a wedding or the marriage ceremony

kempei tai: the Japanese wartime thought control police

kimono: a wrap-around Japanese garment

kiseru: the traditional Japanese pipe, about a foot long with a metal mouth-piece, a slender stem, and a tiny metal bowl into which a pill-sized pellet of tobacco can be placed

kōden: an obituary gift

kokoro: heart; sincerity; spirit

koku: a measure of capacity, commonly of rice, equal to 180 liters, or about 5 bushels; feudal fiefdoms were measured in terms of their rice yield, e.g., 100,000 koku

konnyaku: a kind of edible root

konro: a portable cooking furnace

kōri: ice

kōro: an incense burner

kotatsu: a foot warmer

kotōshōgakkō: a higher elementary school

ku: a ward in a city

kuchi: mouth

kuge: the aristocracy surrounding the imperial court

kun: a Japanese rendering of a Chinese ideogram

kura: a storehouse

kuroi: black

kusunoki (or kusu): a camphor tree

kutsubako: a shoe box

kyōgen: a comic or farcical interlude in a Noh play, or a subsidiary character in a kabuki performance necessary for the development of the plot

kyōiku-chokugo: the Rescript on Education promulgated by the Emperor Meiji on October 30, 1890

machi: a town, or often a district in a city; synonymous with chō, another reading of the ideogram machi bugyō: the feudal administrator of a machi

maiko: a dancing girl, a dancer, usually used in connection with student geisha

mamushi: a viper or adder found commonly in western Japan

matsuri: a festival or fete

matsutake: a mushroom, *Armillaria matsudake*

meisaku: a fine piece of literature or art; a masterpiece

meishi: a calling card

mejiro: a small white-eyed bird, the Japanese white-eye

mie: to assume a posture

migi: right

mikan: a mandarin orange

mingei: folk craft

miokuri: to see off or send off

miokurinin: a person who sees or sends off another

misaki-dōrō: a particular kind of stone lantern

mirin: a sweet rice wine, often used in cooking

miso: a bean paste, usually of soya beans

mitsugi-mono: a tribute, or a tributary payment

mizubachi: a water container commonly seen in Japanese gardens

mizushōbai: literally, the water trade, but an expression applied to the entertaining or gay professions

mochi: a glutinous rice cake

mokugyo: a wooden gong

mompe (or monpe): women's work pants gathered at the ankles, a kind of pantaloon

mon: a crest or family coat-of-arms

moxa: the burning on the back or feet of a cone of soft combustible material derived from artemisia as a medicinal remedy

mukoyōshi: an adoption of a son coupled with his marriage to a daughter in the family

nakōdo: a go-between at a wedding

nama: fresh or raw, implying draft in connection with beer

ninja: near-supernatural intruders of the dark, some good, some evil

nishinbako: a box or container for herring

Noh: a kind of Japanese drama

nōgyo: an agricultural cooperative

nyūfu: an entering husband, that is, one whose name is removed from his family's household census and inscribed in his wife's

oiran: prostitutes of a high class

on: a phonetic reading of Chinese characters

onnagata: men who play women's roles in kabuki

o-shibori: a small hand towel

O-shōgatsu: the New Year

owaiya: men, usually farmers, who collected nightsoil

pachinko: a pinball game

pan: bread

ramune: a carbonated nonalcoholic drink somewhat reminiscent of lemon soda

rōka: a passage or corridor

rokujizō: the six jizō, the god of pregnant women and small children

rōnin: a masterless samurai

ryokan: a Japanese-style inn or hotel

sake: rice wine

samurai: a warrior, a member of the warrior class

sanmon: the two storied gate to a Buddhist temple

sashimi: raw fish

senbei: Japanese rice crackers

Sengoku jidai: the period of countless warring factions in Japan; it terminated with the rise to power of Oda Nobunaga, Toyotomi Hideyoshi, and Tokugawa Ieyasu

sensu: a folding fan

shaku: a linear measurement, somewhat less than one English foot

shi: an administrative city

shibui: astringent, but commonly used to imply tastefulness or in good taste

shichigosan: a festival celebrating children who are seven (shichi), five (go), or three years (san) old

shichirin: a portable cooking furnace, synonymous with konro

shida: a fern

shidare zakura: a drooping or weeping cherry tree

shi-ju-shichi gishi: the tale of the 47 rōnin

shima (or jima): an island

shimenawa: a rope-like ornament, generally of straw, found at Shinto shrines or wrapped around objects, such as trees, thought to have mystical properties

shingeki: modern drama

shishochō: head of a branch office of city government

shitsuchō: literally, the head of a room, but also an expression applied to the head of several sections

shō: a liquid measure, slightly more than 1.5 quarts

shōchikubai: the ornamental arrangement of plants used at the New Year; also called matsutakeume

shōgi: a game superficially similar to chess

shōji: a paper sliding door, usually facing a rōka or exterior corridor

shōyu: soy sauce

shūmon aratame: the ritualized proof that one was not a Christian

sievert (abbreviated Sv): a unit of radiation dose equivalent; a means of expressing all kinds of radiation on a common scale for calculating the effective absorbed dose; it is the product of the absorbed dose in rad or gray and the modifying factors; the dose equivalent unit is the rem, one hundredth of a sievert

sokutatsu: quick-delivery postal service

soroban: an abacus

suika: watermelon

suimono: a clear brothlike soup

sukiyaki: a Japanese dish of meat, tofu, noodles, and the like cooked over a brazier

sumi: ink

tabako ire: a pouch for tobacco, about the size of a hand, closed with a drawstring to which an ivory or stone ornament is attached; it can be inserted under a waist sash

tai: the sea bream

takenoko: a bamboo shoot, or sprout

takotsubo: an octopus trap

tankō bushi: the traditional coal-miner's dance

tanzen: a quilted kimono

tatami: the Japanese floor mat

tenbinbō: a shoulder pole for balancing buckets or baskets

tobinouo: a flying fish

tōfu: soybean curd

tokkuri: a sake flagon

tokobashira: the wooden pillar at the end of the tokonoma

tokonoma: an alcove, the recess in a Japanese room in which a hanging scroll can be placed

tonarigumi: a neighborhood association

torii: literally, the bird's seat, the gate to a Shinto shrine

tozama: the so-called outside daimyō who were not hereditary feudatories of the Tokugawa

tsubaki: a camellia, *Camellia japonica*

tsuitate: a small screen-like divider often used in older Japanese restaurants to separate one table from another

tsuzuri: an inkwell

wakizashi: the short sword worn by a samurai

washi: handmade paper, customarily made from fibrous plants

washoku: Japanese-style food

yae-zakura: a double-blossomed cherry tree

yamadōrō: a particular type of stone lantern, literally, a mountain lantern

yamazakura: a wild cherry tree

yashiki: a mansion, commonly applied to the quarters of a daimyō

yōshi: an adopted child (son)

yōshoku: Western-style food

yukata: an unlined cotton kimono worn in the summertime or as a bathrobe

yuki: snow

yukimi: a particular type of stone lantern, literally "snow viewing"

zabuton: a kneeling or floor cushion

INDEX

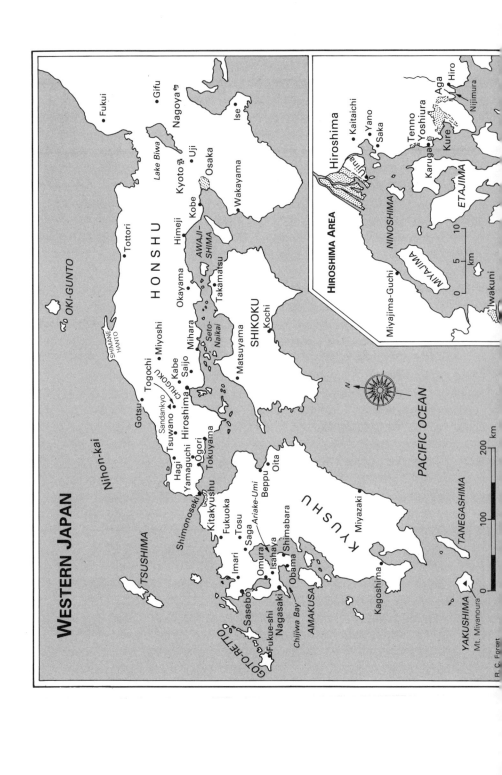

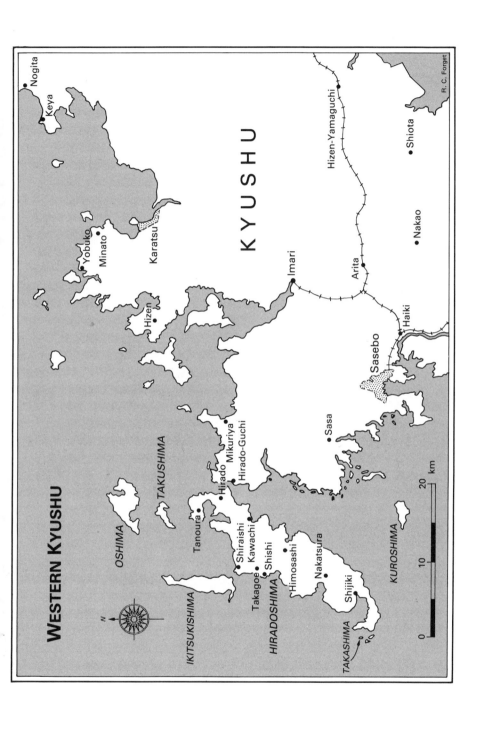

pecific information about how to ingest
rticular medication.

the boxes were rigged to record, time-
and transmit via phone lines all lid open-
which the researchers equated with the
of an actual medication. The research-
ted that the pillbox comes with such a
-monitoring system, for patients and their
vers to use as desired.

THE RESULTS

researchers found that electronic pill-
boosted drug adherence. With the boxes,
ts prescribed more than a single dose per
any particular drug took one pill more
iy on average, the authors found. In ad-
, the number of days when patients acci-
ly skipped their drug regimen altogether
ed to just 6% when using an electronic
x—from 12% without the box.

, the proportion of doses taken at, or near,
ie they should be taken went up with the
onic pillbox.

EXPERT REACTION

id Flockhart, MD, PhD, director of the di-
of clinical pharmacology at Indiana Uni-
's School of Medicine in Indianapolis, said
tion of an electronic pillbox draws critical
on to a major public health concern.

mpliance with medications is a huge
m in general, and in particular among the
," he observed. "It is even more problem-
nong those who take a lot of medication,
is a lot of people, given that the majority
iors who take medications take more than
escriptions a day. So the value of some-
ike this is potentially very large."

wever," Dr. Flockhart added, "the ques-
lways comes up as to whether these
of benefits seen in a clinical trial would
translate to the real world. Because the
ts in a study like this know that they're
monitored, they might be remembering
something when the box beeps that they
not actually remember in real life. So I
encourage the investigators to follow up
iding with a strictly observational study,
than a clinical trial, to see how this will
n a natural setting."

KEEPING TRACK OF MEDICATIONS

To take control of their prescription medication regimen, the US Food and Drug Administration recommends that senior citizens use
a calendar or a pillbox to help adhere to drug
routines. They point out that pillboxes with
multiple compartments are particularly helpful
for older patients dealing with complex multi-
pill regimens, as well as for those who have difficulty opening safety sealed drug containers.

The FDA also encourages seniors to undergo
a yearly "Medicine Check-Up," as an opportunity to both toss out expired medicines and to
discuss possible drug side effects and interactions with a pharmacist and/or doctor.

info For more information on safe prescription medication use for senior citizens,
visit the Web site of the US Food and Drug
Administration at *www.fda.gov*, and type "older adults" into the search box.

Saving Money on Medications

Marjory Abrams, publisher, *Bottom Line* newsletters,
Boardroom Inc., 281 Tresser Blvd., Stamford, Connecti-
cut 06901.

My eyebrows went up when I read recently about a North Carolina insurance program that saved $6.6 million
over a three-year period through prescription
drug "cost-control interventions." Individuals
covered by the plan saved, too. Their average
monthly drug expenditures dropped by about
11%, from $11.52 to $10.23 per prescription.

The study, published in *The American Journal of Managed Care*, was led by David P.
Miller, Jr., MD, assistant professor at Wake Forest University School of Medicine in Winston-
Salem, North Carolina. The study focused on
the university medical center's own health insurance plan, which covers more than 22,000
people.

The bulk of the savings came from prescribing generic instead of branded drugs—an option that is becoming more widely available.

your doctor. He may be able to recommend a dietary supplement or to allay your fears concerning the medication.

•**Feeling better.** Once the immediate health problem improves, many people cut back or stop taking the prescribed medication.

Solution: Realize that doctors prescribe medication according to specific dosing schedules so that the medication builds up in the bloodstream. If you stop taking an antibiotic, for example, you may not eliminate all the bacteria causing an infection. "Drug holidays" should be avoided, because skipping days may cause fluctuations in the blood levels of medication, which can make the drug less effective.

•**Forgetfulness.** Memory problems, as well as having other priorities, cause many people to miss taking medication.

Solutions: Count doses in advance and store them in a compartmentalized pill storage box. One- or two-week containers are available at most supermarkets and drugstores.

It also helps to put medication in a place where you are most likely to see it. For example, if you take it in the morning, store the medication near your coffee mug or in the utensil drawer. If the medication must be refrigerated, place a reminder note near your mug or in the utensil drawer.

People who own cell phones, personal digital assistants or computers with an alarm feature can set these devices to ring at the same time every day. Pocket-sized alarms used solely as a reminder to take medication are available in drugstores for about $6 to $10.

•**Lifestyle.** Travel, inconsistent work or home hours, or a generally busy schedule can interfere with medication compliance.

Solution: Keep your medication stored in a pill box and leave it where you will be sure to see it—such as on a bedside table—whether you're at home or traveling.

•**Side effects.** Nausea, headache, drowsiness and upset stomach can occur with many medications.

Solution: Your doctor usually can change the prescription, give suggestions about other ways to take the medication—for example, with food—or prescribe an ad teract the side effects.

Electronic Pil Helps Seniors Meds Properl

David Flockhart, MD, PhD cal pharmacology, Indiana U cine, Indianapolis.
American Geriatric Society

Older adults follow men are less like reminded by an both beeps at the appoi and announces the numl how to take them, new r

The study, which was Institute on Aging, was pre Geriatric Society meeting by co-authors Vesta Brue of Lifetechniques Inc., of cilla Ryder, of the Depar cal Health Services Rese of Maryland School of Ph is the manufacturer of th pillbox that was used in

THE ST

The interactive pillbox of patients between the were each following a p at least four medications self-sufficient with respec their own medications.

Researchers tracked p ing patterns for three we more weeks using "Me cially available electroni

The pillbox holds up of medications, with s for up to four drugs. As beeps at pill-taking time priate compartment and of pills to take on a scre ment lid is lifted, a progr announces the number

with the

Al stam ings, takir ers r phor careg

Th boxe patie day per condition dent drop pillbo

La the ti elect

Da vision versit the n atten

"C prob elder atic a whic of se five thing

"H tion kinds really patie being to do migh woul this f rathe work

Popular medications that have recently become available in generic versions include the antihistamine Allegra (*fexofenadine*)…the sleep aid Ambien (*zolpidem*)…the heart drug Coreg (*carvedilol*)…the herpes/shingles drug Famvir (*famciclovir*)…Lamisil (*terbinafine*), which treats nail fungus…the cholesterol-lowering drug Zocor (*simvastatin*)…and the antidepressant Zoloft (*sertraline*).

Pharmacists are your best source of information about new generics. They also can tell you which ones are not yet available legally. Some Internet suppliers sell what are purportedly generic forms of drugs that are still under patent protection.

Examples: Drugs such as Lipitor (*atorvastatin*), a cholesterol-lowering drug, and Viagra (*sildenafil*) do not have generic equivalations.

People with prescription drug coverage save with generic drugs because co-payments are lower (typically $10 versus $25 or more). Savings can be much more substantial for people who lack such coverage and for Medicare recipients whose annual drug expenses are approaching the infamous "doughnut hole," when no drug costs are covered.

Good for everyone: Some large chain stores, such as Target and Walmart, now charge only $4 for a 30-day supply and $10 for a 90-day supply of any of hundreds of generic drugs.

More ways to save…

•**Buy OTC.** In the study, medications were moved out of the insurance formulary (the list of drugs that an insurance plan covers) if there were comparable over-the-counter (OTC) drugs available. Dr. Miller says that consumers, especially those without prescription drug coverage, may indeed save in two particular categories—the nonsedating antihistamines, such as Claritin (*loratadine*), and the acid-reducing proton pump inhibitors, such as Prilosec (*omeprazole*). In some cases, even if a person has insurance, the cost of the OTC drug still may be lower than the co-payment. Ask your physician which medications may help you the most, and then compare costs.

•**Split pills.** Because many medications are priced about the same regardless of dose, buying a double-dose pill and halving it gives you two pills for the price of one. Check with your pharmacist to make sure that a pill can be split safely. Pill-splitting devices can be purchased in drugstores and on-line for less than $10.

•**Avoid drugs altogether.** In this study, the plan limited the quantity of sleep aids supplied to members. Members were provided with only enough tablets for every-other-day use, instead of the usual 30-day supply. Because sleep medications can have serious side effects, including daytime drowsiness and slower reaction time, Dr. Miller urges people with insomnia to try to restore a healthy sleep pattern without medication, whenever possible.

Bottom line: If you eat right, exercise regularly and guard your health in other ways, you are likely to need fewer drugs altogether.

■ ■ ■ ■

Prescription Drug Moneysaver

Get help paying for prescription drugs, health care, utilities and more. At BenefitsCheckUp (*http://benefitscheckup.org*), sponsored by the National Council on Aging, older Americans can get information about more than 1,500 programs to help with health care, taxes, energy costs and more. Once you find a program, you can print out application forms or apply on-line for benefits.

Do You Really Need Statin Drug Megadoses?

Mark A. Stengler, ND, naturopathic physician in private practice, La Jolla, California…adjunct associate clinical professor at the National College of Natural Medicine, Portland, Oregon…author of many books, including *The Natural Physician's Healing Therapies* and coauthor of *Prescription for Natural Cures* (both from Bottom Line Books)…and author of the *Bottom Line/Natural Healing* newsletter.

Financially speaking, cholesterol-lowering statin drugs are hugely successful for pharmaceutical companies. Pfizer Pharmaceuticals had worldwide sales of $12.7 billion in 2007 of *atorvastatin* (Lipitor). Now Pfizer has launched an aggressive marketing campaign to convince doctors and patients that an 80-mg dose of Lipitor offers better protection against cardiovascular disease (CVD) than the usual 10-mg dose.

Yet to me, the big question is not whether people need this higher dose—it's whether people need statins at all.

Pfizer's push is based on a study that the company sponsored, published in *The Journal of the American Medical Association*, to compare dosages and gauge the safety of reducing LDL "bad" cholesterol below the currently recommended maximum of 130 milligrams per deciliter (mg/dl). Participants included 10,001 CVD patients whose LDL was less than 130 mg/dl. For eight weeks, one group took 80 mg daily of Lipitor and the other took 10 mg daily. *Findings…*

•**Average LDL fell to 77 mg/dl** in the high-dose group…versus 101 mg/dl in the low-dose group.

•**Over the five-year study period,** there were 126 CVD-related deaths in the high-dose group…versus 155 in the low-dose group.

•**The high-dose group had 2.2% fewer nonfatal cardiovascular events** than the low-dose group.

The study results are interesting—but are they reason enough to put millions of people on statin megadoses? The answer must take into account statins' possible side effects—muscle pain, muscle damage, fatigue, weakness, and potentially fatal liver and kidney damage. *In the Pfizer study…*

•**Among high-dose patients,** 8.1% had side effects…versus 5.8% in the low-dose group.

•**In the high-dose group,** 7.2% of participants quit taking Lipitor because of side effects…versus 5.3% of low-dose patients.

•**Among high-dose users,** 1.2% had elevated enzymes indicative of liver inflammation and/or injury…versus 0.2% of low-dose users.

Given the risks, I think statins are hugely overprescribed. Most people can reduce cholesterol with diet, exercise and supplements of niacin (vitamin B-3)…Sytrinol (a brand-name

vitamin E and citrus extract formula)…guggul (a plant resin)…red yeast rice…and/or fish oil.

Furthermore, as a marker for CVD risk, cholesterol may be less significant than elevated levels of *C-reactive protein* and *Interleukin 6*, blood markers of plaque-promoting inflammation…*homocysteine*, an amino acid that is toxic in excess…*lipoprotein* a, which transports artery-clogging LDL…and *apolipoprotein b*, a major component of LDL. Another important marker is a reduced level of *apolipoprotein a*, the major constituent of HDL "good" cholesterol.

My opinion: Long-term statin use usually is appropriate only for people with LDL above 200 mg/dl and total cholesterol above 350 mg/dl who do not respond to natural therapies. Short-term statin use (about one year) may be beneficial after a heart attack to reduce inflammation. In such cases, the question of dosage remains. Best: If it is absolutely necessary to take a statin, use a "start low, go slow" approach. The 10-mg dose may be effective for you—and if not, your doctor can increase it gradually to determine your optimal dose.

Antidepressants May Not Work as Well as Folks Think

Adil Shamoo, PhD, founder, Citizens for Responsible Care and Research, and professor, biochemistry and molecular biology, University of Maryland School of Medicine, Baltimore.

Julio Licinio, MD, professor and chairman, psychiatry and behavioral sciences, University of Miami Miller School of Medicine, and editor, *Molecular Psychiatry*.

A. Mark Fendrick, MD, professor, internal medicine and health management and policy, University of Michigan, Ann Arbor.

Erick Turner, MD, assistant professor, psychiatry and pharmacology, Oregon Health Sciences University, and medical director, mood disorders, Portland, Virginia.

New England Journal of Medicine.

A systematic review of studies on antidepressants concludes that the positive effects of these drugs are probably overstated in the medical literature.

It's not clear, however, if the bias comes from a reluctance to submit negative manuscripts or decisions by journals not to publish them, or a combination of both, according to Oregon Health and Science University researchers.

THE STUDY

The researchers compared drug efficacy inferred from published studies with drug efficacy from studies reported to a mandatory US government registry of clinical trials, in which all results, including raw data, must be included.

Only 51% of studies in the US Food and Drug Administration registry were considered by the agency to have positive results.

In the published medical literature, however, 94% of studies appeared positive.

The increase in the effectiveness of the drug ranged from 11% to 69% for individual drugs, and was 32% overall. Published studies reflect that the antidepressant *bupropion* (Wellbutrin) appeared to show a high level of bias.

The authors also noted that studies that were not positive were often published with a slant that made them seem positive.

EXPERT REACTION

One expert hailed the finding.

"This is, in my opinion, an excellent paper and what it shows is really consistent with the past five years," said Adil Shamoo, PhD, founder of Citizens for Responsible Care and Research and a professor of biochemistry and molecular biology at the University of Maryland School of Medicine in Baltimore.

"Publications have indicated that [the] industry basically has a tendency to bias publications towards positive results and to not publish negative results," he said.

"Research is not regulated by anyone, so therefore they don't have to submit to anyone, and that's really the key," Dr. Shamoo added. "It's a systemic failure."

Other experts presented slightly differing points of view.

"It's not that the drugs are ineffective, but that the public's perception is that they are more effective than they are [something the authors also point out]," said Julio Licinio, MD, chairman of psychiatry and behavioral sciences at the University of Miami Miller School of Medicine. "The data was good enough for the drugs to get approval."

"The concern that selective publication of clinical studies routinely overestimates the benefit of medical services is not new," added A. Mark Fendrick, MD, a professor of internal medicine and health management and policy at the University of Michigan.

"While the creation of clinical trial registries will enhance access to unpublished data, the availability of information alone will likely not be very useful to patients and clinicians unless the results can be analyzed in a careful manner, as performed in this important study," Dr. Fendrick pointed out.

Dr. Licinio believes the bias stems from a combination of issues. He says people are not submitting papers and editors/publishers are not printing them.

As the editor of the journal *Molecular Psychiatry*, Dr. Licinio receives more than 700 submissions a year and publishes an average of 144. "It becomes very difficult to select," he noted.

"There should be a national repository for these kinds of reports [that don't get published in journals]," Dr. Licinio added. "Those papers don't see the light of day."

STUDY AUTHOR GIVES ADVICE FOR THE FUTURE

"For the average patient, yes, these drugs are effective, just less effective than they appear," said study author Erick Turner, MD, an assistant professor of psychiatry and pharmacology at Oregon Health Sciences University.

"Don't be overly disappointed if you don't have an excellent response to the first antidepressant you try," he advised. "You might have to try two or three, and you might need to add psychotherapy."

info For more information on safety issues related to the use of antidepressants, visit the Web site of the Food and Drug Administration at *www.fda.gov* and type "antidepressants" into the search box.

Fish Oil—You Get What You Pay For

Joseph C. Maroon, MD, clinical professor of neurological surgery at the University of Pittsburgh School of Medicine and team neurosurgeon to the Pittsburgh Steelers. He is coauthor of *Fish Oil: The Natural Anti-Inflammatory* (Basic Health) and serves as medical adviser to Nordic Naturals, a manufacturer of fish oil supplements.

Stroll through any drugstore, and you'll see an array of fish oil supplements. You've heard that fish oil—or more specifically, the omega-3 fatty acids it contains—may combat many health problems, including heart disease, chronic pain and inflammation, asthma and depression.

But which one should you buy? The fancy-looking 100-capsule bottle for $25...the plain-looking store brand with twice as many capsules for $8.99...or something in between?

In general, with fish oil supplements, you get what you pay for. Less expensive products may contain as much "fish oil" as more expensive ones, but they typically contain less *eicosapentaenoic acid* (EPA) and *docosahexaenoic acid* (DHA), the most biologically active and beneficial components.

Why it matters: The American Heart Association recommends supplementation with about 1,000 mg of combined EPA/DHA per day for people with known heart disease. It recommends 2,000 mg to 4,000 mg for people who need to lower triglycerides (a type of blood fat). I recommend a starting dose of 2,000 mg to 2,500 mg to all of my patients for general good health. Fish oil supplements are not a good choice for those taking *warfarin* (Coumadin) or other blood-thinning medications—consult your doctor. Because the typical low-priced "1,000-mg" fish oil capsule contains only 180 mg of EPA and 120 mg of DHA, you would have to take six to eight capsules each day to reach the recommended dose of EPA/DHA. With each capsule, you're getting 700 mg of oil that is not EPA or DHA. Day after day, that's a lot of oil you don't really need—and it can cause loose stools and other unpleasant gastric effects.

Best: Buy a fish oil product that will require you to take the fewest capsules per day to reach your desired dose of EPA/DHA. This is better for your health than a less concentrated product and may not be as expensive as it seems when you consider that you'll consume fewer capsules.

■ ■ ■ ■

Beware Foods with Added Omega-3s

Foods with added omega-3s often contain little or none of the omega-3s that are best for the heart. The omega-3s *docosahexaenoic acid* (DHA) and *eicosapentaenoic acid* (EPA) are far more prevalent in fish than in other sources. Six ounces of farmed Atlantic salmon provides about 3,500 milligrams (mg) of DHA and EPA...six ounces of sardines, about 3,000 mg. Foods supplemented with omega-3s provide much less—one Land O'Lakes Omega-3 egg, 350 mg...one cup of Silk Plus Omega-3 DHA Soy Milk, 32 mg.

Best: Get omega-3s from fish, such as salmon, sardines, mackerel and herring.

Katherine Talmadge, registered dietitian, Washington, DC, and spokesperson, American Dietetic Association, Washington, DC.

■ ■ ■ ■

High Octane Soda

Citrus sodas may contain more caffeine than popular colas. Caffeine content can vary widely from brand to brand and even among products within a brand. The Food and Drug Administration (FDA) requires labels to disclose if caffeine is an ingredient but does not require the amount to be listed. The FDA limits the caffeine content in carbonated beverages to about 72 milligrams (mg) per 12-ounce glass. In a recent study, Coca-Cola had 33.9 mg....Diet Coke 46.3 mg...Pepsi 38.9 mg...and Diet Pepsi 36.7 mg per 12 ounces. But Mountain Dew had 54.8 mg...and Diet Mountain Dew had 55.2 mg. Major manufacturers, including Coca-Cola and PepsiCo, have started to phase in new labels that include caffeine amounts.

Leonard Bell, PhD, food researcher, Auburn University, Auburn, Alabama, and coauthor of a laboratory study of caffeine in 131 brands of sodas, published in *Journal of Food Science*.

Don't Let Medical Mistakes Cost You Your Life

Charles B. Inlander, a consumer advocate and health care consultant based in Fogelsville, Pennsylvania. He was the founding president of the nonprofit People's Medical Society, a consumer advocacy organization credited with key improvement in the quality of US health care in the 1980s and 1990s. Mr. Inlander is author of 20 books, including *Take This Book to the Hospital With You.* St. Martin's.

Medicare has announced that it will no longer pay hospitals for the extra costs of treating patients with complications due to medical mistakes—infections, hospital-acquired bedsores, etc.

Medical mistakes are the eighth leading cause of hospital deaths among Americans.

Examples: Mistakes in drug use, such as giving the wrong dose or even the wrong drug, account for tens of thousands of deaths annually in hospitals and outpatient facilities. Hospital-acquired infections result in 80,000 to 100,000 deaths—the vast majority of which are preventable.

Most hospital mistakes are relatively minor. A patient might be given an unnecessary test, or a drug that's not optimal for his/her condition. But some mistakes are life threatening—and, in most cases, completely avoidable.

MEDICATION MISTAKES

It's estimated that hospitals in the US make more than 1.5 million medication errors annually, causing the deaths of 40,000 to 80,000 patients. Name confusion is the most common cause of drug-related errors.

Example: A patient who's supposed to get Celebrex for arthritis might be given Cerebyx (for seizures) or Celexa (an antidepressant) instead.

Handwritten prescriptions should not be allowed in hospitals. It's common for pharmacists to accidentally substitute drugs with similar-sounding names…give the wrong dose (by confusing the numbers "3" and "8," for example)…or miss a decimal point (confusing 1.5 mg with 15 mg). *To prevent a medication mistake in the hospital…*

•**If possible, choose a hospital with a bar-coding system** that automatically matches drug prescriptions/labels with coding on patient bracelets. Only 7% to 8% of hospitals now use this technology. Prescriptions are entered directly into computers—and the computers are programmed to recognize potential errors.

•**Question any drug that looks different** from what you've been getting. If you've been taking blue pills for two days, and then are given a green pill, ask what it is…what it's for…why you're being given something different from before…and who ordered it.

•**Notice changes in side effects** and report them to your doctor or nurse. You may have been given a wrong drug and/or dose.

INFECTIONS

The Centers for Disease Control and Prevention estimates that an average of 4% to 6% of hospital patients develop hospital-acquired infections annually. In some hospitals, the percentage is much higher. With the emergence of antibiotic-resistant organisms, such as the potentially deadly methicillin-resistant *Staphylococcus aureus*, some of these infections are life threatening. *Ways to prevent unnecessary infections…*

•**Watch health care workers wash their hands.** Not washing hands accounts for half of all hospital-acquired infections. Don't let health care workers touch you—or equipment that's attached to you, such as intravenous (IV) lines—unless they wash their hands. They should wash even if they're putting on gloves—the outside of the gloves could get contaminated from germs on the hands.

• **Check those catheters.** Bladder or blood-vessel catheters are open doors to infection. Every time a nurse comes into the room to monitor your catheters, ask him to check them for cleanliness, and to make sure that the entry site is clean and sterilized each time. It's also important to sterilize the connections every time a catheter bag is changed.

•**Food-service workers who come into your room** should touch *only* the trays they bring and retrieve and the food on the trays. If they touch anything else—for instance, they prop up a patient in the bed next to you—ask them to change their gloves before giving you your tray.

Key protection: If you have time, get a pneumonia vaccination from your doctor as far in advance as possible before entering the hospital. Hospital patients often have weakened immune systems—and pneumonia is among the leading causes of hospital deaths. In patients who develop pneumonia while hospitalized for another condition, the mortality rate can be as high as 70%. The pneumonia vaccine protects against 23 strains of the most common bacterial pneumonia.

BEDSORES

Also known as pressure ulcers, bedsores occur when circulation to the skin and underlying tissues is blocked in immobile patients. This usually occurs on the elbows, hips, buttocks and/or heels. Older patients, and those with diabetes or other diseases that cause circulatory problems, have an especially high risk for bedsores.

Warning: Bedsores are often painless until damage is already extensive. Patients and health care workers have to watch for them. *Self-defense...*

- **Move your arms and legs as often as you comfortably can.**

- **Change position** or ask friends/family members to help you change position about every hour or two.

Also important: Ask them (or nurses) to inspect areas that you can't see to make sure sores aren't developing.

FALLS

Falls are the second leading cause of accidental deaths in hospitals, and the main cause of broken hips. They often occur when patients need to use the bathroom but don't get a response after pressing the call button—and attempt to climb out of bed without waiting for assistance.

Other causes: Unsteadiness due to sedation from medications...tripping over IV lines and other obstacles...and slipping on wet floors. *To ward off falls...*

- **If a nurse doesn't respond when you press the bedside call button,** use the phone (assuming you are able) to call the main switchboard. Ask for the nursing station on your floor. Someone will always answer.

- **Have a friend/family member bring you a pair of slippers with nonskid soles.**

Important: For safety's sake, don't get up without assistance, even if you think you're strong enough to do so. Some hospitals have bed alarms that alert the nurses if you try to get up on your own—but many don't.

SURGICAL SAFETY

Under the new rules, Medicare won't pay hospitals for treating complications from serious preventable errors, such as recovering a sponge or other object that was left inside a patient after surgery.

Patients have little control over surgical outcomes, but they can pick a surgeon who has a lot of experience with a particular procedure. Statistically, the error rate is significantly lower when surgeons do a procedure at least 100 times a year.

Also important: Before surgery, use a marker to write an "X" on the body part—knee, hip, elbow, etc.—that's supposed to be operated on. Some hospitals already do this, but some don't.

There aren't reliable statistics on the frequency of "wrong-site" surgeries, but it happens frequently enough that the Joint Commission on Accreditation of Health Care Organizations, a nonprofit organization that evaluates and accredits health care organizations, issued an emergency alert in 2001 about the rise in frequency. It now advises marking the surgical site ahead of time.

Should You Opt Out of Medicare?

Steven Podnos, MD, CFP, principal, Wealth Care, LLC, Merritt Island, Florida. Dr. Podnos, a practicing specialist in respiratory medicine and a fee-only financial planner, is author of *Building and Preserving Your Wealth, A Practical Guide for Affluent Investors*. Oak Hill.

About 44 million people are enrolled in Medicare, the federal health insurance program for seniors. Generally, it's available to Americans age 65 and older.

Most participants choose "original" Medicare (the term used by the government). They can select any doctor or hospital that accepts Medicare and Medicare picks up most of the cost.

Alternative: Almost nine million Medicare beneficiaries—nearly 20%—have opted out of the original Medicare program.

These seniors have so-called *private* Medicare coverage, run by companies, not by the federal government. Several forms of private Medicare are available, collectively known by the government as Medicare Advantage programs. *Why people choose private Medicare…*

•**Medicare Advantage programs may be less costly to participants than original Medicare.**

•**More comprehensive services are offered to patients.**

•**Participants in original Medicare sometimes have to pay considerable out-of-pocket expenses.** Private Medicare can help you avoid this.

Example: Under original Medicare, you typically pay 20% of doctors' bills while Medicare pays 80%. Your share might be thousands of dollars a year. Many original Medicare beneficiaries buy a Medicare supplement ("Medigap") insurance policy to cover these potential expenses. However, the premiums for such policies can be substantial.

•**With private Medicare Advantage plans,** there is no need to buy a Medigap policy. You can get comprehensive care, often at a lower out-of-pocket cost.

But there are trade-offs. Before making any decisions, it's vital to understand the fine print.

TYPES OF PRIVATE MEDICARE

Most Medicare Advantage programs are health maintenance organizations (HMOs) or preferred provider organizations (PPOs).

Many Americans become familiar with HMOs and PPOs during their working years. A network of doctors and hospitals is offered to covered individuals, who bear little or no cost beyond their premiums if they stay within that network. Generally, PPOs are more expensive than HMOs but offer more coverage for out-of-network care than HMOs.

Downside: You have less freedom to choose providers than with original Medicare.

In recent years, yet another private option—Medicare private fee-for-service (PFFS) plans—have stolen the spotlight from Medicare HMOs and PPOs. According to the Congressional Budget Office, enrollment in Medicare PFFS plans grew from 200,000 in late 2005 to 1.35 million in early 2007.

Why they grew: These plans claim to offer the advantages of Medicare HMOs and Medicare PPOs (comprehensive care with, potentially, lower costs than original Medicare) without the restrictions of staying in a provider network.

How they work: If you're in a PFFS plan, you may see any doctor or go to any hospital that accepts the plan terms. You show a card to prove that you are in the plan. But it is up to the doctor or hospital to decide whether to treat you—they decide on a case-by-case basis.

Trap: Some seniors have signed up for PFFS plans, only to discover that many physicians won't treat them.

My advice: Do not use a PFFS plan. They are poorly designed. Just the fact that you may be denied care by any doctor at any visit makes them a poor choice.

HOW TO CHOOSE

When deciding among Medicare options, the key is comparing what you'll pay and get with original Medicare with what you'll pay and get with a selection of private Medicare plans.

Reality: You can never know with absolute certainty whether original or private Medicare will turn out to be the best deal for you, because you cannot perfectly predict your own future health needs. But, in general, the more health care that people need, the more they will benefit financially from using a private network, because their out-of-pocket costs (copayments, etc.) will probably be lower than with Medicare.

Typical costs: Although each Medicare HMO and Medicare PPO has its own rules and costs, participants in both types must be enrolled in Medicare Part A (hospital coverage) and Part B (coverage for doctors' visits). For the vast majority of people, those who have worked and paid Medicare taxes for at least 40 quarters in their lifetimes, there is no monthly premium for Part A. In 2007, the monthly Part B premium ranges from $93.50 to $161.40, depending on the participant's income.

Some private plans do not charge anything more, but many others do…

•**Medicare HMOs** charge an average of $20 a month, over and above the Part B premium. For HMOs that offer a prescription drug benefit, the average extra charge is nearly $40 a month.

•**Medicare PPOs** charge about $20 to $40 above the Part B premium without prescription drug coverage, and about $40 to $50 more with the drug benefits. Prices vary widely depending on competitive market forces in each region.

In return for paying those extra fees up front, private Medicare participants may realize cost savings overall. *Examples…*

•**You won't need a Medigap policy.** In fact, private Medicare enrollees are *prohibited* from buying this insurance.

•**Private Medicare plans also may offer checkups,** dental care and vision coverage— so you might be able to avoid having to pay for these services.

•**Copayments may be lower with private Medicare** than with original Medicare.

Bottom line: Some private Medicare plans can be good deals. Others will wind up costing you more than you would pay with a Medicare-Medigap combination.

If you have used an HMO or a PPO during your working years, and you know how these systems work, a Medicare version may be a practical choice because you'll know how to "work the system" to get adequate coverage while enjoying the cost savings. If you would rather not be confined to a network, opt for original Medicare with a Medigap policy.

Best: Check with your current physicians before signing up for any kind of private Medicare. Some of these programs are very limited in terms of participating doctors and hospitals.

Read the plan documents closely. Generally, seniors may move in and out of private Medicare plans during the annual election period from November 15 to December 31. (You can do the same from January 1 to March 31, but you cannot add or drop Medicare prescription drug coverage during this time.)

After you make such a choice, you're usually locked in until the next annual election period. However, seniors who were misled into signing up for a private Medicare plan can change coverage right away.

Potential cutbacks: *The Children's Health and Medicare Protection Act,* the details of which are currently being worked out in a House-Senate conference committee, is expected to sharply decrease funding for Medicare Advantage plans. If the measure is enacted as expected, such cuts likely would reduce the number of providers participating in private Medicare.

Safety net: You can choose to drop private coverage and enroll in original Medicare within specific periods, as described above.

WHAT TO DO NEXT

To start comparing and shopping for private Medicare plans, visit the Medicare Web site (*www.medicare.gov*). Click on "Learn more about plans in your area." Enter your state and county to find details on what's available, with contact data for each plan.

Also: Some Medicare Advantage plans are evaluated by the National Committee for Quality Assurance, with results published periodically in US News & World Report. To see how several plans rank, go to *http://health.usnews. com,* type "best health insurance" into the search box, then click on "Medicare Plans" under "Rankings."

■ **More from Dr. Podnos…**

If You Were Misled…

People in private Medicare plans can disenroll if they signed up after receiving misleading information, according to the US Department of Health & Human Services' Centers on Medicare & Medicaid Services. To find out if you can change your coverage, call the Centers' hotline at 800-633-4227. Provide as much detail as possible, including the name of the person who provided the wrong information, along with the date and time.

Claims of misleading information can be made whenever the plan member discovers that a promised benefit isn't available.

Examples: If the member was told that all health care providers participate in the plan when in fact they do not…incorrect assurance that the member can transfer to traditional Medicare at any time.

Beware These Medical Scams, Cons And Rip-Offs!

Chuck Whitlock, a journalist whose work exposing scams has been featured on many television programs, including *Inside Edition* and *Extra*. He is author of several books about scams, including *MediScams: Dangerous Medical Practices and Health Care Frauds*. St. Martin's Griffin. *www.chuckwhitlock.com.*

Bogus "miracle cures" and quack physicians probably have been around as long as the health care profession. Con artists prey upon the unhealthy because sick people may be so desperate to find a cure that they will try any possible treatment, however expensive and farfetched.

These snake oil salesmen have been on the rise in recent years, as our aging population has more medical problems…millions of underinsured and uninsured Americans search for health care options that they can afford…and the surging popularity of "alternative medicine" makes unscientific treatments seem more mainstream.

Sometimes it's obvious when a claim is fraudulent—but certain scams and unethical practices are difficult to spot…

SCAMS

•**Cheap health insurance.** Many of the more than 45 million Americans who are without health insurance are desperate to find affordable coverage. Disreputable insurance companies offer these people exactly what they want—health insurance at a low price, sometimes as little as $50 per month.

Most buyers are so thrilled to find insurance they can afford that they don't pay attention to the fine print. Many of these policies have huge deductibles, scant benefits and other restrictions that make them virtually worthless.

Most policyholders do not discover the problems until they have a serious health condition and receive the bill.

What to do: Look for a policy that covers doctor visits and protects against a major medical expense. A way to spot-check the quality of a policy is to look at the benefits for hospital stays. A good policy should pay $500 or more per day for the hospital room, with additional coverage for other hospital costs. An inadequate policy might provide only $100 a day.

Wise: Contact your state's insurance commissioner to find out if complaints have been filed against the company. Blue Cross/Blue Shield offers many reputable plans, and Kaiser Permanente has a good reputation for a health maintenance organization (HMO).

•**Natural appetite suppressants.** Unscrupulous marketers claim that there are natural supplements that help you lose weight—but these supplements don't work or have dangerous side effects. The most heavily promoted "appetite suppressants" include ephedra, garcinia cambogia and hoodia.

Ephedra is an herbal stimulant, and like other stimulants, it does suppress appetite but at the price of increased heart rate, nervousness and agitation. In large doses, ephedra has killed people, and the FDA has removed it from general sale though it's still available through herbalists and on the Web.

Garcinia cambogia (a fruit from Asia) and hoodia (from a succulent African plant) are not stimulants and seem to be safe—but no studies have shown them to be effective.

What to do: The only reliable cures for excess weight are consuming fewer calories and getting more exercise.

•**Organ transplants in developing countries.** America's organ donation program can't keep pace with the demand for transplants. For patients languishing on waiting lists, flying to a developing country where organs can be purchased on the black market might seem viable.

Before you board a plane to India, the Philippines, Hungary or Russia and agree to pay $1,000 to $100,000 for an organ, consider that you're putting yourself in the hands of people who are more interested in taking your money than saving your life.

These people might find an organ for you—but you might not receive it until you've been milked for many times the agreed-upon price. Some patients are told at the last minute that someone else will pay more for the organ, which might or might not be true. To get the organ, you have to top this other offer. Even when patients do get their organs, their transplants often are performed according to medical standards that are not as strict as they are in the US.

What to do: Talk to your doctor about the best options.

UNETHICAL PRACTICES

•**Bonuses from HMOs and PPOs to doctors who skip useful tests.** HMOs and preferred provider organizations (PPOs) often give cash bonuses to doctors who don't perform pricey tests—even when those tests are in the patient's best interest. They are essentially bribing doctors to scam their patients.

Example: Your HMO or PPO doctor tells you that he/she is going to spare you the invasive thallium stress test (where radioactive dye is injected into your bloodstream) and perform a routine treadmill test without thallium, though the thallium test is warranted.

What to do: When your HMO or PPO doctor tells you that you have a particular health condition, research that condition on a reliable Internet Web site, such as WebMD.com or MayoClinic.com. If the site mentions a test that your doctor has not performed, ask him why it was skipped. If the doctor's response seems evasive, consider getting a second opinion.

•**Unqualified plastic surgeon.** Many doctors have switched to plastic surgery in recent years, drawn by the lucrative nature of the specialty and its lower reliance on insurance payments. (Most plastic surgery procedures are elective and not covered by insurance.) No law or regulation prevents doctors from changing their specialty to plastic surgery—even if they have no background or training in this field.

Patients likely have no idea that they are trusting their lives and appearances to what are essentially unqualified, untrained novices.

What to do: If you're considering plastic surgery, ask the surgeon…

Are you board-certified in plastic surgery? He/she should be.

At what hospital do you have physician's privileges? A general hospital is fine, but a university hospital is even better—university hospitals tend to have very high standards.

Who will be handling my anesthesia? Don't trust a plastic surgeon who says he'll handle it himself. He may be trying to cut corners and putting your health at risk in the process.

Medical "Discount" Plans That Cost You Plenty

Charles B. Inlander, consumer advocate and health care consultant based in Fogelsville, Pennsylvania. He was the founding president of the nonprofit People's Medical Society, a consumer advocacy organization credited with key improvements in the quality of US health care in the 1980s and 1990s, and is the author of 20 books, including *Take This Book to the Hospital with You*. St. Martin's.

Forty-seven million Americans have no health insurance, but there are also 50 million Americans—including 30% of adults over age 65—who are said to be "underinsured." This means that their insurance does not sufficiently cover some key aspect of medical care, such as physical therapy, psychological counseling or prescription medication. If you don't have health insurance—or it's inadequate—it can be tempting to respond to those ubiquitous advertisements for "discount" medical cards. With come-ons like "Discounts of up to 60%" and "No deductibles or co-pays," these services sound like the answer to your prayers. But are they? *What you should know…*

•**Medical discount cards are not health insurance.** Medical discount providers make arrangements with doctors, dentists, pharmacies, hospitals, chiropractors and other providers to offer a discount off their normal retail prices for services. But you still must pay the bill out of your own pocket. In addition, most medical discount cards charge a monthly fee ($9.99 to more than $50) for the use of their services.

•**Discounts may not be worth it.** Over the years, several states, including New York and Florida, have investigated some of the companies providing medical discount cards and found that many offer only small—or no—savings.

Example: If the doctor you see offers a 10% discount, and his/her fee is $100 per visit, you save $10 per visit. But if the card fee is $10 per month, you won't save anything unless you see the doctor more than 12 times per year! The discounts also can vary widely from provider to provider, making it difficult for you to get the best use out of the service unless you invest considerable time in finding the best discounts. Eyeglasses, medical equipment and foot care are among the few instances in which a valid discount card may be worthwhile.

•**Provider lists are often out-of-date.** Medical discount plans give clients a list of participating providers. But quite often, those lists are outdated, or the providers accept only a limited number of customers from one card sponsor. If you are considering signing up for a card, be sure that the providers or pharmacies that you want to use participate in the discount plan. Call each provider and ask if he accepts the card, what the discounts will be and the retail prices for the services that you are most likely to need.

•**You can negotiate discounts yourself.** Most people don't realize this, but many doctors, dentists and other providers will offer a discount if you ask—particularly if you are a longtime patient. For example, a doctor may charge a discounted fee for a physical exam, while a dentist may discount fees for fillings and crowns. AARP and other organizations, including drug companies, have discount programs for medications that are worth checking out. Many pharmacies also have their own discount plans, with discounts ranging from 10% to 70%, depending on the drug and whether it's a brand-name or generic. By negotiating your own discounts with your health care providers, you cut out the middleman fee, which means you'll receive the greatest possible savings.

14

Stroke Prevention

Even Moderate Aerobic Exercise Lowers Stroke Risk by 40%

 A moderate level of aerobic fitness can significantly reduce stroke risk in men and women, according to a large, long-running study.

The study, presented at the American Stroke Association's International Stroke Conference in New Orleans, showed 30 minutes or more of brisk walking, or an equivalent aerobic activity, five days a week could lower stroke risk by about 40%.

Fitness has a protective effect regardless of the presence or absence of other stroke risk factors, including family history of cardiovascular disease, diabetes, high blood pressure, elevated cholesterol levels and high body mass index, said study author Steven Hooker, PhD.

BACKGROUND

This study is the first to suggest there may be a significant independent association between cardiorespiratory fitness (CRF) and fatal and nonfatal stroke in men and nonfatal stroke in women, said Dr. Hooker, director of the Prevention Research Center at the University of South Carolina Arnold School of Public Health.

About 780,000 US adults suffer a stroke each year, and stroke is a leading cause of serious, long-term disability in the United States, according to the American Stroke Association. About 150,000 people die from strokes annually, making it the nation's third leading cause of death.

THE STUDY

Researchers analyzed data on more than 60,000 people—46,405 men and 15,282 women—who participated in a long-term study at the Cooper Aerobics Center in Dallas between 1970 and 2001.

American Heart Association/American Stroke Association, news release.

The participants, ages 18 to 100 and free of known cardiovascular disease when they entered the study, were followed for an average of 18 years. During that time, 863 people—692 men and 171 women—had strokes.

Upon entering the study, each participant took a test to measure cardiorespiratory fitness. The fitness test required participants to walk on a treadmill at an increasing grade and/or speed until they reached their maximal aerobic capacity.

Men in the highest quartile (25%) of CRF had a 40% lower relative risk of stroke compared with men in the lowest quartile.

That difference stayed constant even after adjusting for other factors such as smoking, alcohol intake, family history of cardiovascular disease, body-mass index (a measure of body fat based on height and weight), high blood pressure, diabetes and high cholesterol levels, Dr. Hooker said.

Among women, those with a higher cardiorespiratory fitness level had a 43% lower relative risk for stroke than those female participants in the lowest fitness level.

The overall stroke risk dropped substantially at the moderate CRF level, with the protective effect persisting almost unchanged through higher fitness levels.

IMPLICATION

According to Dr. Hooker, "We found that a low-to-moderate amount of aerobic fitness for men and women across the whole adult age spectrum would be enough to substantially reduce stroke risk."

Physical activity is a major modifiable cardiovascular disease risk factor. Increasing the nation's fitness level through a regular regimen of physical activity could be a vital weapon to lower the incidence of stroke in men and women, he said.

info To learn more about improving your cardiovascular fitness level, visit the Web site of the American Heart Association, *www.americanheart.org*, and search "exercise and fitness."

Mammograms Might Spot Stroke Risk

Paul S. Dale, MD, chief, surgical oncology, University of Missouri Ellis Fischel Cancer Center, Columbia.

Suzanne Steinbaum, DO, cardiologist, with a subspecialty in women and heart disease, Heart and Vascular Institute, Lenox Hill Hospital, New York City.

International Stroke Conference, New Orleans.

In addition to detecting breast cancer in its early stages, new research suggests that mammograms may also help predict which women are at risk for strokes.

Calcifications found in the blood vessels of the breasts—what doctors call benign arterial calcifications—were more commonly found on the mammograms of women who had suffered a stroke, said Paul S. Dale, MD, chief of surgical oncology at the University of Missouri's Ellis Fischel Cancer Center. He is the lead author of the research, which was presented at the American Stroke Association's International Stroke Conference in New Orleans.

BACKGROUND

Previous studies, including some by the University of Missouri team, have found a link between these calcifications, which are not cancerous, and the risk of diabetes, heart disease and stroke. But Dr. Dale said he believes this latest study has found an even stronger association.

Stroke is the third leading cause of death for women over 40, Dr. Dale noted.

THE STUDY

In all, they looked at 793 mammograms of women ages 40 to 90. On the screening mammograms, 86 of the 793 women, or about 10%, had the calcifications. But 115 of the 204 women in the group (56%) who had had stroke had the calcifications.

"Of those who had a stroke, 56% of them had these calcifications on their mammogram, compared to about 10% of women in the general population," Dr. Dale said. "The important thing here is, we adjusted for age, because age increases your risk of stroke and also of having calcifications on your mammogram."

EXPERT REACTION

"There have been other studies to show this," said Suzanne Steinbaum, DO, a cardiologist with a subspecialty in women and heart disease at the Heart and Vascular Institute at Lenox Hill Hospital in New York City.

If a patient turns out to have calcifications, she said, that could potentially help inform doctors about a woman's risk for cardiovascular disease. "As a cardiologist, I would love this information," said Dr. Steinbaum.

If she got a mammogram from a patient with that information, she said, "It would lead me to believe I need to screen this woman for cardiovascular disease."

FUTURE RESEARCH

Dr. Dale's team is continuing to study the link between the calcifications and cardiovascular diseases and diabetes. It's probably too soon, he said, to advise women to ask their doctor if they have calcifications on their mammogram. After more research is done, that might be a consumer-savvy step.

info To learn more about stroke, visit the Web site of the American Stroke Association, *www.strokeassociation.org*.

Magnesium May Lower Risk for Some Strokes

JAMA/Archives journals, news release.

Increased consumption of magnesium-rich foods such as whole grains may reduce male smokers' risk of cerebral infarction. This type of stroke, also known as an ischemic stroke, occurs when blood flow to the brain is blocked, a new Swedish study suggests.

THE STUDY

Researchers at the Karolinska Institute in Stockholm analyzed the diets and other health/lifestyle habits and characteristics of 26,556 Finnish men, ages 50 to 69, who smoked but had never had a stroke. During an average of 13.6 years of follow-up, 2,702 of the men had cerebral infarctions, 383 had intracerebral hemorrhages (bleeding into the brain tissue), 196 had subarachnoid hemorrhages (bleeding between the brain and the tissues that cover it), and 84 had unspecified types of strokes.

After they adjusted for age and cardiovascular risk factors (such as diabetes and cholesterol levels), the researchers concluded that men who consumed the most magnesium (an average of 589 milligrams per day) had a 15% lower risk for cerebral infarction than those who consumed the least amount of magnesium (an average of 373 milligrams per day). The association was stronger in men younger than 60.

There was no association between magnesium consumption and risk for intracerebral or subarachnoid hemorrhage, said the researchers, nor was calcium, potassium and sodium intake associated with risk for any type of stroke.

The findings were published in the *Archives of Internal Medicine*.

POSSIBLE EXPLANATION

"An inverse association between magnesium intake and cerebral infarction is biologically plausible," according to the study authors. Magnesium lowers blood pressure and may also affect cholesterol concentrations and the body's use of insulin to turn glucose into energy, both of which would affect the risk for cerebral infarction, but not hemorrhage.

LOWERING STROKE RISK

Recent studies have suggested that changes in diet may help reduce stroke risk, according to background information in the study. High blood pressure is a risk factor for stroke, which means that dietary changes that lower blood pressure may reduce stroke risk.

While encouraged by the results of the study, the study authors added, "Whether magnesium supplementation lowers the risk of cerebral infarction needs to be assessed in large, long-term randomized trials."

info For more information on stroke prevention, visit the Web site of the National Institute of Neurological Disorders and Stroke, *www.ninds.nih.gov*. Then type "stroke prevention" into the search box.

New Ways to Prevent a Stroke

Larry B. Goldstein, MD, professor of medicine and director of the Duke Center for Cerebrovascular Disease and the Duke Stroke Center at Duke University Medical Center, Durham, North Carolina. He was the chairman of the committee that prepared the American Stroke Association's 2006 Primary Stroke Prevention Guideline and is a former chair of the American Stroke Association and the Stroke Council of the American Heart Association.

Most people worry more about having a heart attack than a stroke. Heart attacks are more common, but even what may seem like a mild stroke can have serious consequences, leaving you with functional deficits that affect everyday life. Stroke is the third-leading cause of death in the US—and 15% to 30% of patients who have a stroke are permanently disabled.

New strategies: The American Stroke Association's guidelines include both well-known preventive measures (such as regular exercise) and measures that have only recently come to light.

Example: Premenopausal women who suffer from migraines have about twice the risk of stroke as those who don't have migraines (although the absolute risk of stroke for women with migraines is small). It's not known if treating migraines reduces stroke risk—but these patients need to be especially careful about controlling other risk factors for stroke.

Important steps everyone can take...

MANAGE BLOOD PRESSURE

Hypertension (high blood pressure) is the single most important treatable stroke risk factor. It increases the risk of both types of stroke—*ischemic*, in which a clot blocks circulation to the brain, and *hemorrhagic*, when a blood vessel in the brain leaks or ruptures.

Because hypertension doesn't cause symptoms in most patients, millions of Americans who have it don't know it. Among those who have been diagnosed, less than half manage hypertension successfully.

Good news: For every 10-point reduction in systolic pressure (the upper number) and five-point reduction in diastolic pressure (lower number), the risk of stroke drops by about 30%.

What to do: Get your blood pressure checked. A reading below 120/80 is optimal for stroke prevention. Patients who have had a prior stroke or a *transient ischemic attack* (TIA, sometimes referred to as a "ministroke," which typically causes symptoms for an hour or less) should almost always take steps to lower their blood pressure. Available data suggest that those who have had a stroke and have blood pressure below 120/80 benefit from lowering pressure further.

Most patients can significantly lower blood pressure with lifestyle changes, such as losing weight, exercising more and consuming less salt. If the numbers stay elevated—140/90 is the threshold for hypertension—patients should be treated with antihypertensive drugs.

CHECK YOUR HEART'S RHYTHM

Atrial fibrillation is an irregular rhythm in the upper chambers (atria) of the heart. It prevents blood from circulating efficiently, which makes clots more likely. This, in turn, increases the risk of ischemic stroke.

Some patients with atrial fibrillation have no symptoms. Those who have symptoms may experience palpitations, confusion, light-headedness or shortness of breath.

What to do: Check your wrist pulse. See a doctor if the rhythm seems irregular.

Atrial fibrillation often can be treated with daily aspirin to reduce clotting. Patients with additional stroke risk factors, including those who have had a prior stroke or a TIA, have a higher risk. They may require *warfarin* (Coumadin), a more potent anticlotting drug. There is no evidence that treating the rhythm problem itself decreases stroke risk.

CONTROL CHOLESTEROL

Patients with high cholesterol—above 240 milligrams per deciliter (mg/dL)—have an increased risk of ischemic stroke. The Asia Pacific Cohort Studies Collaboration, which looked at more than 352,000 people, found a 25% increase in ischemic stroke rates for every 38.7 mg/dL increase in cholesterol.

What to do: Most people can significantly improve cholesterol with regular exercise and a healthy diet—less saturated fat, more fiber, etc.

Patients with existing heart disease, or those who have had a prior stroke or a TIA, should talk to their doctors about cholesterol-lowering statin drugs. The use of statins can decrease the risk of stroke in some patients who have seemingly "normal" cholesterol levels.

DON'T IGNORE SNORING

Loud snoring may be a sign of sleep apnea, in which breathing intermittently stops and starts during sleep. Moderate-to-severe sleep apnea raises blood pressure, can worsen atrial fibrillation and increases stroke risk.

What to do: A person with apnea symptoms—loud snores/snorts, gasping for breath, daytime fatigue and morning headaches—needs to be evaluated by a sleep specialist. Talk to your primary care provider about sleep specialists in your area.

AVOID SECONDHAND SMOKE

People who smoke have approximately twice the risk for ischemic stroke as nonsmokers, and two to four times the risk for hemorrhagic stroke. Smoking is believed to contribute to at least 12% of all stroke deaths.

Secondhand danger: Exposure to environmental (secondhand) cigarette smoke is nearly as dangerous as active smoking. People who are exposed to secondhand smoke regularly are 50% more likely to have a stroke than unexposed nonsmokers.

What to do: If you smoke, talk to your doctor about behavioral and pharmacological treatments for smoking cessation...avoid secondhand smoke whenever possible.

EAT MORE FRUITS AND VEGETABLES

The average American diet is high in sodium and low in potassium—two factors that can increase blood pressure and stroke risk. A diet that's high in fruits and vegetables (and relatively low in saturated fat) naturally increases potassium and lowers sodium.

Data from the Nurses' Health Study and the Health Professionals' Follow-up Study indicate that each daily serving of fruits and vegetables can reduce the risk of ischemic stroke by 6%.

What to do: Eat a minimum of three servings of fruits and vegetables daily—but the more produce you eat, the lower your risk of stroke.

ASK ABOUT ASPIRIN FOR WOMEN

Low-dose aspirin therapy is not typically recommended for stroke prevention in men, but it does make a difference in women—particularly those with an elevated stroke risk, for whom the benefits of aspirin can outweigh the potential side effects (such as intestinal bleeding).

What to do: Women who are at high risk for stroke should discuss aspirin therapy with their doctors. The recommended dose is 81 mg daily.

■ **More from Dr. Goldstein...**

Five Warning Signs of Stroke

One treatment for ischemic stroke is a drug called *tissue plasminogen activato*r (tPA)—but it has to be given within three hours of the onset of symptoms to be effective.

Unfortunately, about half of American adults can't name a single stroke symptom. It's common for patients to ignore mild symptoms and fail to get to an emergency room in time.

What to do: You should call 911 immediately if you or anyone you are with has any of these symptoms...

•**Sudden unexplained weakness,** tingling or numbness in the face, arm or leg—usually on just one side.

•**Sudden difficulty speaking** or understanding words.

•**Sudden changes in vision,** such as blurred or decreased vision.

•**Sudden and severe headache,** sometimes with nausea/vomiting.

•**Sudden dizziness** or difficulty with walking/coordination.

■ ■ ■ ■

Easy Ways to Spot a Stroke

Detect warning signs of stroke with the acronym HELP NOW. *Headache*—a sudden, severe headache, particularly in someone with no history of headaches. *Eyesight*—a sudden change in vision in either or both eyes. *Language*—sudden difficulty talking or understanding speech. *Paralysis*—sudden onset of complete inability to use a part of the body, especially an arm and a leg on the same side. *Numbness*—sudden onset of numbness in the face, an arm and/or a leg, especially on the same side. **Orientation**—sudden onset of disorientation or confusion. **Weakness**—sudden onset of weakness in an arm and/or leg, especially on the same side.

Tedd Mitchell, MD, president and medical director, Cooper Clinic, Dallas, Texas.

How to Prevent a Major Stroke After A Minor Event

Keith Siller, MD, medical director, Comprehensive Stroke Care Center, New York University Medical Center, and assistant professor, NYU School of Medicine, both in New York City.

The Lancet Neurology, news release.

Patients who've suffered a transient ischemic attack, or TIA—a temporary stroke-like condition with full recovery of neurological function within 24 hours—have a substantial risk of suffering a major stroke within seven days.

But that risk of a second stroke is lowest among TIA patients treated as emergency cases in specialist stroke units, says a new British review of previous studies that included a total of 10,126 patients.

STUDY RESULTS

The researchers concluded that there's a 5.2% risk of a major stroke within seven days of a TIA, meaning that about 1 in 20 TIA patients will have a major stroke within a week. But the risk among TIA patients who received emergency care in specialist stroke units ranged from 0% to 9%, while the highest risk—1%—was for patients who didn't receive emergency care.

The findings were published in *The Lancet Neurology*.

Studies of major stroke risk after TIA have produced conflicting findings, with seven-day stroke risk ranging from 0% to 12.8%. The inconsistent findings among the previous studies that were included in the new review could be explained by differences in study method, setting and treatment, the review authors said.

"Our study almost fully explains why the results of previous studies have been conflicting and illustrates the importance of methods used by a medical study when interpreting its results," they concluded.

IMPLICATIONS

"The risk of stroke reported amongst patients treated urgently in specialist units was substantially lower than risks reported among other patients treated in alternative settings. These results support the argument that a TIA is a medical emergency and that urgent treatment in specialist units may reduce the risk of subsequent stroke," said the study authors, from Oxford University's Stroke Prevention Research Unit.

The researchers said reliable estimates of major stroke risk following a TIA could maximize the benefits of early treatment, allow effective planning of service for patients, assist in the design of clinical trials, and justify investment in public education.

EXPERT RECOMMENDATIONS

Another expert, Keith Siller, MD, medical director of New York University Medical Center's Comprehensive Stroke Care Center, called the study important for several reasons.

The likelihood that a major stroke will follow a minor stroke or TIA shows that these patients are not out of the woods, even though they may appear to be back to normal, said Dr. Siller. "This is analogous to a patient with chest pain who may be having angina as a warning sign for impending heart attack and is admitted for additional testing to avoid sending them home and having them suffer a fatal heart attack outside the hospital, Dr. Siller said.

Dr. Siller, who's also an assistant professor at the NYU School of Medicine, added that the study also emphasizes that "patients with TIA or minor stroke should be treated in stroke centers with specialized units since the care provided is specifically targeted for stroke and does result in better outcomes compared to a general medical ward."

What's more, Dr. Siller said, "these results need to be understood by insurance companies and HMOs that have unofficially discouraged doctors from admitting these same patients and prefer them to be worked up electively as outpatients. The reality is that completing all of the necessary testing as an outpatient within one week is often not possible, during which time the patient may have the second more devastating stroke that might have been prevented had they been in the hospital setting."

info To learn more about TIA, visit the Web site of the American Heart Association, *www.americanheart.org* and search "transient ischemic attack."

■ ■ ■ ■

Better Stroke Recovery

Patients treated within 90 minutes of the onset of an ischemic stroke (caused by a blood clot) are almost three times more likely to have a favorable outcome than those who wait longer for care. The longer the wait, the worse the outcome. Compared with those who received a placebo, stroke patients treated in 91 to 180 minutes had only a 50% better chance of a favorable outcome...and a 40% better chance if treated in 181 to 270 minutes. (Consent was obtained from all patients.) If you experience paralysis in the arm, face or leg, slurred speech, vision loss, dizziness and/or headache, call 911 immediately.

Thomas G. Brott, MD, professor of neurology, Mayo Clinic, Jacksonville, Florida.

■ ■ ■ ■

Hot/Cold Therapy Brings Better Recovery for Stroke Patients

When added to standard physical therapy treatment, thermal stimulation—in which hot and cold packs are alternately applied to the patient's impaired hand and wrist over 20 to 30 minutes—resulted in significantly better recovery rates in wrist extension and sensation and grasping strength than results in patients who were given only standard physical therapy. Thermal stimulation was given five times a week for six weeks.

Fu-Zen Shaw, PhD, associate professor, Institute of Cognitive Science, National Cheng Kung University, Taiwan, and leader of a study published in *Stroke*.

Is Your Blood Pressure Really Under Control?

Stevo Julius, MD, ScD, professor emeritus of medicine and physiology in the division of cardiovascular medicine at the University of Michigan in Ann Arbor. Dr. Julius is a recipient of the American Heart Association's Irvine Page-Alva Bradley Lifetime Achievement Award, presented annually to an individual with at least 25 years of outstanding service, research and teaching in the field of hypertension. In 1998, the executive committee of the International Society of Hypertension established the Stevo Julius Award in Hypertension to honor Dr. Julius's contributions to hypertension education.

It's been widely reported that an estimated 72 million Americans have high blood pressure (hypertension)—but the fact that only 35% of the people who are being treated for the condition actually have it under control comes as a surprise to most people.

There are a variety of effective treatments, ranging from diet, exercise and supplements to dozens of medications. So why don't more people keep their hypertension in check?

250,000 AMERICAN LIVES ARE AT STAKE

Even though one in three adults has hypertension, about one-third of them don't know it. That's largely because hypertension—defined as a systolic (top) number of 140 millimeters of mercury (mm Hg) or higher and/or a diastolic (bottom) number of 90 mm Hg or higher—usually produces no outward symptoms until organ damage has occurred. (For example, blood vessels in the kidneys can be harmed by untreated hypertension, leading to kidney failure.) That's why it's important to have your blood pressure checked at least once a year.

Danger: If you have hypertension that is not being adequately treated, your risk for stroke, heart attack, heart failure and kidney failure is greatly increased. This year alone, uncontrolled hypertension will cause or contribute to the deaths of more than a quarter-million Americans, according to the American Heart Association (AHA). *How to protect yourself...*

TREAT IT EARLY

Many doctors still "start low and go slow" when prescribing antihypertensive medications

—a holdover from the days when blood pressure drugs invariably produced unwanted side effects, such as cough, dizziness and headache. But many of today's antihypertensives, particularly the angiotensin II receptor blockers (ARBs)—a class of hypertension drugs including *valsartan* (Diovan) and *losartan* (Cozaar)—are generally well-tolerated.*

Recent research shows that aggressive *early* treatment of hypertension can be lifesaving for patients who are at high risk for cardiovascular disease.

Latest scientific evidence: At the University of Michigan, we recently completed a large-scale international clinical trial comparing the effectiveness of an ARB with that of a calcium channel blocker (CCB)—a class of hypertension medication including *amlodipine* (Norvasc) and *felodipine* (Plendil)—for reducing cardiac deaths in people with hypertension and at least one additional cardiovascular risk factor (such as diabetes, high cholesterol, previous heart attack or stroke). Over an average of four years, we didn't see a significant difference between the two drugs' abilities to control blood pressure and reduce related deaths. By the end of the study, patients in both treatment groups had achieved good blood pressure control.

What did appear to make a crucial difference was the speed with which blood pressure was lowered. Those patients who responded quickly to treatment—that is, their systolic blood pressure dropped by 10 mm Hg or more within the first month—were significantly less likely to suffer sudden cardiac death (an abrupt cessation of the heartbeat), heart failure or stroke.

Self-defense: If you have hypertension and at least one other cardiovascular risk factor, work with your doctor to quickly bring your blood pressure under control. If you have hypertension but no other risk factors, aim to lower your blood pressure within six months.

Be aware that a diuretic (a water-excreting drug often prescribed for high blood pressure), such as *furosemide* (Lasix) or *chlorothiazide* (Diuril), when used alone probably won't bring your blood pressure under control. Most people require two or more medications to achieve a

*Women of childbearing age should not use these drugs—they can harm a fetus.

normal blood pressure of less than 120/80 mm Hg. Two or more low-dose drugs can achieve significantly greater reductions in blood pressure, with fewer side effects, than a single high-dose drug, according to research.

If you find it difficult to take more than one pill a day, ask your doctor about combination blood pressure medication (two antihypertensive drugs in a single pill), such as *enalapril* and *hydrochlorothiazide* (Vaseretic). In an analysis of studies, Columbia University researchers recently concluded that such medication reduces the risk for noncompliance (not taking medication as prescribed) by 24%.

NEVER TOO OLD FOR TREATMENT

Two-thirds of Americans age 75 and older have high blood pressure—and 90% of people in that age group will develop it by the end of their lives. As a result, many doctors accept high blood pressure as a "normal" part of aging and do not aggressively treat it in their oldest hypertensive patients.

Latest scientific evidence: A recent British trial comparing active treatment (with an ACE inhibitor plus a diuretic) with a placebo unequivocally showed that lowering blood pressure reduces both stroke and death in adults in their 80s. The advantages of treatment were so striking that the researchers prematurely halted the study to allow all patients to receive the drugs.

Self-defense: If you're over age 80 and have hypertension, lowering systolic blood pressure by 12 mm Hg to 15 mm Hg can reduce your stroke and heart failure risk by as much as 35%, and your heart attack risk by as much as 20%.

Warning: A normal or low diastolic pressure does *not* offset an elevated systolic pressure. In fact, isolated systolic hypertension (high systolic pressure of 140 mm Hg or higher accompanied by normal diastolic pressure of less than 90 mm Hg) is the most common—and most dangerous—type of hypertension in older adults. This condition indicates that the arteries are stiff, which makes the heart work harder. If you have isolated systolic hypertension and your doctor is not treating it aggressively, find one who will.

TOO SOON TO TREAT?

In May 2003, the AHA created a designation for "prehypertension"—systolic pressure of 120 mm Hg to 139 mm Hg and/or diastolic pressure of 80 mm Hg to 89 mm Hg.

Latest scientific evidence: In our recent Trial of Preventing Hypertension (TROPHY) study, researchers found that people with a systolic pressure of 130 mm Hg to 139 mm Hg have a 63% chance of developing full-blown hypertension within four years. By giving pre-hypertensive patients an ARB drug for two years, and then stopping it, our study found that the progression to hypertension can be effectively postponed for up to two years, without further side effects.

Self-defense: The TROPHY study results hold promise for the estimated 37% of American adults on the verge of high blood pressure, but more research is needed before we can recommend antihypertensive medication for people with prehypertension. If you are prehypertensive, speak to your doctor about blood pressure monitoring (it should be checked every six months) and preventive steps you can take, such as improving your diet and getting more exercise. Aim to exercise for 30 minutes every other day, with an activity, such as brisk walking, that raises your heart rate to 110 beats per minute.

I also recommend following the Dietary Approaches to Stop Hypertension (DASH) eating plan, which calls for eight to 10 servings of fruits and vegetables and two to three servings of low-fat dairy foods daily…sodium intake of no more than 2,400 mg daily…and moderate alcohol consumption (up to two drinks daily for men and up to one drink daily for women).

BLOOD PRESSURE GUIDELINES

•**Normal blood pressure**—less than 120/80 mm Hg

•**Prehypertension**—120–139 mm Hg (top number) and/or 80–89 mm Hg (bottom number)

•**Hypertension**—140/90 mm Hg or higher.

Stroke Risk Factors Mean Memory Loss

George Howard, DrPH, professor and chairman, biostatistics, University of Alabama in Birmingham.

Argye Hillis, MD, associate professor, neurology, Johns Hopkins University, Baltimore.

American Stroke Association's International Stroke Conference, New Orleans.

Older people whose health conditions put them at high risk for stroke are more likely to suffer from memory loss, even if they never actually have a stroke, new research shows.

The cause could be mini-strokes that people don't notice but that nonetheless contribute to the brain's deterioration. "Stroke risk factors really matter, and they matter even if you don't have a [major] stroke," said study author George Howard, DrPH, chairman of biostatistics at the University of Alabama in Birmingham.

Three risk factors in particular were linked to memory loss—high systolic blood pressure (the top number in a blood-pressure reading), diabetes and left ventricular hypertrophy (thickening of the muscle of the main pumping chamber of the heart, often caused by high blood pressure).

THE STUDY

Researchers interviewed 17,626 people with an average age of 66. Nearly 40% of participants were black, and none had suffered a stroke. The average systolic blood pressure in the group was 127.9, 56% had hypertension, 19.3% had diabetes, 21.9% had heart disease and 6.5% had left ventricular hypertrophy.

To test participants' mental skills, researchers gave them three common words during a phone call. Later they asked the participants to repeat the words.

Then the researchers tried to find any connections between scores on the mental test and risk factors for stroke.

THE RESULTS

The findings were reported at the American Stroke Association's International Stroke Conference in New Orleans.

The rate of cognitive decline in one group of participants, those with a 22% chance of having a stroke in the next 10 years, was double that of those who had a 2% chance of stroke in the next 10 years.

POSSIBLE EXPLANATION

"What I think is happening, and what a lot of other people think is happening, is that people develop small strokes that are not significantly big enough to disable you, but cumulatively knock your [mental] condition down," Dr. Howard said.

Mini-strokes are not that uncommon, he said. Starting at age 55, about 10% of people who report never having had a stroke show evidence of one when they undergo brain scans, he said. The number grows to about half by age 80. Those mini-strokes damage areas of the brain where blood flow has been restricted.

EXPERT REACTION

Argye Hillis, MD, an associate professor of neurology at Johns Hopkins University, said the study "provides the strongest evidence of the association between the risk factors and cognitive decline."

It's not clear, though, if the risk factors directly cause brain deterioration or if they're part of a larger complex of conditions, according to Dr. Hillis. For example, people with diabetes and high blood pressure may be prone to have inflammation of the blood vessels, leading to mini-strokes.

Other factors, such as a sedentary lifestyle, could also play a part, she said.

info For more information about stroke, visit the Web site of the National Institutes of Health at *www.nlm.nih.gov/medlineplus/stroke.html*.

■ ■ ■ ■

Blood Pressure Drugs That Stop Memory Loss

In a six-year study of 1,074 patients who took blood pressure drugs, those using ACE inhibitors that cross the blood-brain barrier (a membrane that normally protects the brain) had 50% less cognitive decline, on average, than those using other blood pressure drugs, including ACE inhibitors that don't cross the blood-brain barrier.

Theory: ACE inhibitors that cross the blood-brain barrier, such as *captopril* (Capoten) or *fosinopril* (Monopril), may reduce cell damage in the brain.

If you take blood pressure medication: Ask your doctor if these potentially brain-protecting drugs are right for you.

Kaycee Sink, MD, assistant professor of medicine, Wake Forest University School of Medicine, Winston-Salem, North Carolina.

Damaged Blood Vessels—The Fearsome Silent Killer

Bruce A. Perler, MD, the Julius H. Jacobson, II professor of surgery at Johns Hopkins School of Medicine, and chief of the division of vascular surgery at Johns Hopkins Hospital, both in Baltimore. He has authored or coauthored more than 100 papers on cardiovascular disease.

Few people realize that the same condition that sets the stage for a heart attack (blockage of blood vessels leading to and from the heart) can also damage or weaken blood vessels throughout the body. These so-called vascular (blood vessel) diseases can lead to leg amputation, stroke—and even death.

Good news: Medicare recently began paying for a onetime screening of qualified individuals for abdominal aortic aneurysm (AAA), a type of vascular disease characterized by a weakened area in the main artery of the abdomen.* The condition typically causes no symptoms, but is fatal in 80% of cases if a rupture occurs. Each year, about 200,000 Americans are diagnosed with AAA.

OTHER SILENT KILLERS

Besides AAA, the two main categories of vascular diseases are…

•**Carotid artery disease (CAD),** due to blockages in one or both of the carotid (neck)

*The screening is offered within the first six months of Medicare enrollment to any man who smoked 100 or more cigarettes during his lifetime and men and women with a family history (parent or sibling) of AAA.

arteries that carry blood to the brain. Each year, more than 750,000 Americans suffer a stroke as a result of CAD.

•**Peripheral artery disease (PAD),** resulting from blockages in blood vessels in the kidneys, stomach, arms and/or legs. More than eight million Americans are believed to have PAD, and there are about 80,000 leg amputations annually in the US due to the disease.

COMMON CAUSES

Vascular diseases are caused in part by *atherosclerosis*, buildup of fatty deposits (plaque) in the arteries. As the blood vessels get thicker and stiffer (typically due to aging) and/or become more blocked, blood circulation is reduced, increasing the risk for clots.

Many of the same factors that increase the risk for heart attack also increase the risk for other vascular diseases. These include high blood pressure, elevated cholesterol, smoking, diabetes, obesity, family history and a sedentary lifestyle.

SCREENING FOR VASCULAR DISEASE

If you are age 55 or older and have any of the risk factors listed above, you should undergo noninvasive screenings—ultrasound for abdominal aortic aneurysm and carotid artery disease…and blood pressure measurements taken at the ankles and the arms for peripheral artery disease.

All three tests, which can be ordered by your primary care physician, can be completed in about 30 minutes and are often covered by insurance. If your insurer does not pay for screening, the total cost (approximately $110) is a small investment to protect your health. If the results are negative, the tests generally should be repeated in three years. If the results are positive, treatment may be required.

ABDOMINAL AORTIC ANEURYSM (AAA)

The aorta, which is the largest blood vessel in the body, extends from the heart down through the chest and into the abdomen. An aortic aneurysm may occur in areas where the walls of the artery have weakened. The damaged area stretches and bulges outward, like a bubble on a car tire.

If an AAA ruptures, patients typically experience massive internal bleeding. About half don't survive long enough to get to a hospital. AAAs rarely cause symptoms until they rupture. They're usually discovered during tests for unrelated conditions, such as kidney disease. In rare cases, patients may notice a pulsing sensation in the abdomen with each heartbeat and/or severe pains in the abdomen or lower back. Anyone with these symptoms should have an ultrasound test.

•**Treatments.** Most aneurysms are small, slow-growing and unlikely to rupture—and don't require treatment. However, every patient with an AAA should get an ultrasound every six months to determine if it's getting larger.

In addition, it's critical to stop smoking…and manage blood pressure (this helps prevent AAAs from getting worse). Beta-blocking drugs, which are used to lower blood pressure and treat some heart conditions, are particularly effective at reducing pulse pressure on the aortic wall.

•**Surgical options.** An AAA that is more than 5 cm (about two inches) in diameter is likely to rupture and should be surgically removed.* The traditional approach is open surgery—the surgeon makes an incision in the abdomen or chest, cuts out the damaged section of artery and replaces it with a synthetic tube (graft).

Newer approach: A stent graft—instead of cutting out any of the aorta, the surgeon threads a catheter into an artery in the leg and up into the aorta, where a wire mesh tube is attached to the weakened section to strengthen the artery wall. About half of AAAs are now being repaired with stent grafts because the procedure has a lower risk for complications and a shorter recovery period than open surgery.

CAROTID ARTERY DISEASE (CAD)

More than half of ischemic strokes (when a clot blocks circulation to the brain) are due to blockages not in the brain itself, but in one of the carotid arteries.

•**Treatments.** Some blockages can be treated with medical and lifestyle approaches, such as managing blood pressure (120/80 or lower is ideal)…taking a statin drug, such as *simvastatin* (Zocor)—it can lower the risk for stroke

---

*To find a physician who specializes in vascular disease, consult the Society for Vascular Surgery, 800-258-7188, *www.vascularweb.org.*

by about 20%—and/or taking aspirin daily to reduce clots. Patients who can't tolerate aspirin may be given another anticlotting drug, such as *clopidogrel* (Plavix).

•**Surgical options.** If CAD causes symptoms, such as a stroke or a milder "ministroke," surgery is typically required. Surgery also is considered for patients with a severe (70% to 80%) blockage, even if they don't experience any symptoms.

The main surgical approach is carotid endarterectomy—the surgeon makes an incision in the neck to expose the carotid artery...cuts open the damaged portion...and peels away the blockage. It can reduce the risk for stroke by an average of about 50%.

With another option, known as angioplasty/stenting, a balloon catheter is inserted into the artery to flatten deposits against the artery wall. A mesh stent is then pressed into place to keep the artery open. This procedure is mainly used in high-risk patients who aren't healthy enough for open surgery.

PERIPHERAL ARTERY DISEASE (PAD)

PAD occurs when circulation to the peripheral arteries, such as those in the legs, is diminished. Most people who have PAD, particularly those who are sedentary, have no symptoms initially. Some may experience leg pain when they walk, a condition called intermittent claudication.

In more-severe cases, PAD can cause leg numbness and/or weakness, cold legs or feet, or foot infections. It also may lead to a need for amputation. Because PAD can be a marker for atherosclerosis elsewhere in the body, sufferers are three times more likely to die from a heart attack or stroke than those without it.

•**Treatments.** Patients with PAD should quit smoking...maintain healthy cholesterol levels (total cholesterol below 200 mg/dL)...achieve good blood pressure and diabetes control (keeping blood sugar levels within limits set by their doctors)...and sometimes take daily aspirin or other anticlotting drugs.

Daily walking is also key. Up to 60% of PAD sufferers who walk about half an hour, five days a week, notice a significant improvement.

Important: The legs *should* hurt when patients are walking regularly. It means they're

reaching a level of exertion that enables the leg muscles to use oxygen more efficiently.

•**Surgical options.** Angioplasty is considered the first-line procedure for PAD—the surgeon inflates a small balloon inside an artery to clear a blockage and also may insert a wire mesh stent to keep the artery open.

Severe blockages (that extend over a long portion of the artery) typically require open bypass—the surgeon cuts out the damaged section of artery and replaces it with either a synthetic tube or a blood vessel taken from another part of the body.

■ ■ ■ ■

New Help for Folks with PAD

Arm exercises as well as walking may help, in patients with peripheral artery disease (PAD). Blood flow to arteries, particularly in the legs, is blocked by fatty deposits, causing pain and cramping in the legs while walking. People with PAD are at high risk for heart attack or stroke.

New research: Thirty-five PAD patients were assigned to either a no-exercise control group or to groups that performed aerobic arm exercises (using a device with bicycle-like pedals cranked by hand), treadmill walking or a combination of the two activities. All exercisers worked out for one hour, three times a week for 12 weeks in a supervised setting.

Result: After three months, people in all three exercise groups, but not in the control group, could walk about one-and-a-half blocks farther without pain and up to three blocks farther after they had rested.

Theory: Aerobic arm exercises, which are an especially good option for PAD patients who have trouble walking, are believed to have a systemic effect that improves overall cardiovascular function and fitness.

If you have PAD: Ask your doctor whether walking and/or doing aerobic arm exercises for one hour at least three times a week could improve your overall fitness.

Diane J. Treat-Jacobson, PhD, RN, assistant professor of nursing, University of Minnesota School of Nursing, Minneapolis.

Four Natural Ways to Clean Out Your Arteries!

Mark A. Stengler, ND, naturopathic physician in private practice, La Jolla, California…adjunct associate clinical professor at the National College of Natural Medicine, Portland, Oregon…author of many books, including *The Natural Physician's Healing Therapies* and coauthor of *Prescription for Natural Cures* (both from Bottom Line Books)…and author of the *Bottom Line/Natural Healing* newsletter.

Amid the clamor created by pharmaceutical companies hyping statins, the drugs that lower cholesterol, important information is going unheard.

You need to know: Four natural therapies can protect, and even restore, the health of your arteries—without the dangerous side effects of statin drugs.

These natural substances help to *reverse* the process of plaque buildup—to dissolve plaque, eliminating the danger it presents. They are available over-the-counter from health food stores, on-line and/or from holistic physicians. Unless otherwise noted, they have no side effects, can be taken indefinitely and are safe for everyone, including people who take statin drugs. *Important…*

•**If you have atherosclerosis** or any other cardiovascular condition, you need to be under a cardiologist's care.

•**If you take blood-thinning medication,** such as aspirin or *warfarin* (Coumadin)—typically given to improve blood flow and prevent blood clots—talk to your doctor before using the natural therapies below, because they could affect your medication dosage.

•**If you are pregnant,** speak to your doctor before taking these or any other supplements or drugs.

•**Discontinue use of these substances 10 to 14 days prior to scheduled surgery** to reduce the risk of excess bleeding. Resume use according to your doctor's instructions (typically 10 days after the procedure).

•**To determine which substance or combination of substances to use,** see "The Right Regimen for You" on page 307.

1. TOCOTRIENOLS

There are two principal categories of vitamin E, called *tocopherols* and *tocotrienols*. Each of these has four subcategories—alpha, beta, gamma and delta.

Some vitamin E supplements contain synthetic alpha-tocopherol. If that's what is in your cupboard—indicated on the label as "dl-alpha-tocopherol"—throw it out! Synthetic alpha-tocopherol leaves less room for the more healthful tocopherols you ingest from food or natural supplements (labeled "d-alpha-tocopherol"). Natural tocopherols reduce free radicals, helping to prevent new plaque, but they have not been shown to reduce plaque that is already present. For that, you need tocotrienols.

Tocotrienols are found in rice bran, coconut, barley and wheat germ—but only in small amounts. Supplements with mixed tocopherols/tocotrienols are sold in health food stores—but they are not the best for reducing plaque.

Better: A tocotrienol-only supplement. I like Allergy Research Group Delta-Fraction Tocotrienols (800-545-9960, *www.allergyresearchgroup.com*), available from holistic doctors. The dosage is 300 mg per day.

2. VITAMIN K

Vitamin K protects against harmful arterial calcification. Forms include *phylloquinone* (K1) and *menaquinone* (K2). Vitamin K1 is abundant in dark green leafy vegetables, such as lettuce, spinach and broccoli. However, vitamin K2 is better absorbed and remains active in the body longer than vitamin K1. Good food sources of vitamin K2 include natto (fermented soybeans) and, to a lesser degree, fermented cheeses (the type with holes, such as Swiss and Jarlsberg), beef liver, chicken and egg yolks. Vitamin K also promotes beneficial bone calcification.

People taking warfarin are at higher risk for atherosclerosis and osteoporosis (brittle bones) due to the drug's effects on calcification. A study in *Pharmacotherapy* demonstrated the safety and benefit of low-dose vitamin K supplementation in patients taking warfarin. Vitamin K also promotes beneficial bone calcification.

Note: It is vital that a person who takes blood thinners use vitamin K only under the close supervision of a doctor, because the medication dosage may need to be adjusted.

Research suggests that most people get too little vitamin K. I think daily supplementation with 150 mcg to 200 mcg of vitamin K2 is appropriate for all adults—and especially important for those with atherosclerosis. If you take warfarin, the daily dose may be modified, depending on blood tests that indicate how your clotting mechanism interacts with vitamin K. A brand I like is Jarrow Formula's MK-7 (800-726-0886, *www.jarrow.com*).

3. GARLIC EXTRACT

Legend tells us that garlic protects against vampires—but its true power lies in its ability to protect arteries. Most beneficial is aged garlic extract (AGE), available in capsule or liquid form. A study from the University of California, Los Angeles, involved 19 cardiac patients who were taking statin drugs and aspirin daily. Participants took either a placebo or 4 milliliters (ml) of a liquid brand called Kyolic AGE for one year.

Findings: Participants who took AGE had about a 66% reduction in new plaque formation compared with those who took a placebo.

Research demonstrates that AGE also can…

•**Reduce LDL cholesterol by up to 12%…** total cholesterol by up to 31%…and triglycerides by up to 19%.

•**Protect against the LDL oxidation** that can trigger arterial plaque formation.

•**Thin the blood.**

•**Lower blood pressure.**

•**Reduce blood levels of *homocysteine*—** the amino acid that, when elevated, raises cardiovascular disease risk.

•**Reduce *C-reactive protein*,** a blood marker of inflammation and a risk factor for atherosclerosis.

•**Combat carotid plaque.**

I recommend supplementing daily with AGE.

Dosage: 4 ml to 6 ml of liquid AGE or 400 mg to 600 mg in tablet or capsule form.

4. POMEGRANATE JUICE

Once considered exotic, pomegranate juice now is sold in supermarkets and is a proven boon to arterial health. Israeli researchers verified this in a three-year study of 19 men and women, ages 65 to 75, with severe carotid artery blockage. Ten participants drank 50 ml (about two ounces) per day of 100% pomegranate juice, and nine participants drank a placebo.

Results: Among juice drinkers, plaque thickness decreased by an average of 13% after three months and 35% after one year. This is phenomenal—no drug can come close to reducing plaque like this! Placebo drinkers had a *9% increase* in plaque after one year.

Pomegranate juice is loaded with antioxidants called *polyphenols*, which prevent cholesterol oxidation and improve blood flow.

Note: Choose 100% juice with no added sugar. Pomegranate juice contains a lot of naturally occurring sugar, so dilute two to four ounces of juice with an equal amount of water. Drink it with meals to slow the absorption of sugar into the bloodstream and help to maintain stable blood glucose levels. Use twice daily.

THE RIGHT REGIMEN FOR YOU

Arterial plaque buildup can be detected and its severity gauged using imaging tests, such as computed tomography (CT), magnetic resonance imaging (MRI) scan and/or ultrasound. Results help to determine the appropriate therapy for you. Tests may be repeated periodically to see if your regimen needs to be modified.

The four natural substances described in this article, used alone or in combination (following the dosage guidelines in the main article), can improve the health of your arteries.

Note: If you take blood-thinning medication, talk to your doctor before trying any of these substances.

•**If you do not have atherosclerosis, take one or more of these:**

  •Vitamin K

  •Aged garlic extract (AGE)

  •Pomegranate juice

•**For mild atherosclerosis, take all of these:**

  •Vitamin K

  •AGE

  •Pomegranate juice

•**For moderate or severe atherosclerosis, take *all* of these:**

  •Tocotrienols

- Vitamin K
- AGE
- Pomegranate juice
- Statin medication—but only if you and your doctor are not satisfied with your level of improvement after 12 months of daily use of all four natural substances.

Deadly Dangers of Sleep Apnea

Ralph Downey III, PhD, chief of sleep medicine at Loma Linda University Medical Center and associate professor of medicine, pediatrics and neurology at Loma Linda University School of Medicine, both in Loma Linda, California. He is the lead author of several medical journal articles on obstructive sleep apnea and related sleep disorders.

D octors have long known that obstructive sleep apnea (repeated interruptions in breathing during sleep) can harm the overall health of men and women who suffer from the condition.

WHAT IS SLEEP APNEA?

This common disorder occurs when breathing is temporarily interrupted during sleep due to narrowing of the airway. The condition affects an estimated 12 million Americans. Obesity, especially in older adults, is a leading cause of obstructive sleep apnea.

Now: New research shows that sleep apnea is even more dangerous than experts had previously realized, increasing the sufferer's risk for heart attack, stroke, diabetes and fatal car crashes.

What you need to know...

NO ROOM TO BREATHE

Sleep apnea occurs about twice as often in men as in women, but it is overlooked more often in women. An estimated 70% of people with sleep apnea are overweight. Fat deposited around the neck (men with sleep apnea often wear a size 17 or larger collar, while women with the disorder often have a neck circumference of 16 inches or more) compresses the upper airway, reducing air flow and causing the passage to narrow or close. Your brain senses this inability to breathe and briefly awakens you so that you can reopen the airway.

The exact cause of obstructive sleep apnea in people of normal weight is unknown, but it may involve various anatomical characteristics, such as having a narrow throat and upper airway.

Red flag 1: About half of all people who snore loudly have sleep apnea. One telling sign is a gasping, choking kind of snore, during which the sleeper seems to stop breathing. (If you live alone and don't know whether you snore, ask your doctor about recording yourself while you are sleeping to check for snoring and other signs of sleep apnea.)

Red flag 2: Daytime sleepiness is the other most common symptom. Less common symptoms include headache, sore throat and/or dry mouth in the morning, sexual dysfunction and memory problems.

DANGERS OF SLEEP APNEA

New scientific evidence shows that sleep apnea increases risk for...

- **Cardiovascular disease.** Sleep apnea's repeated episodes of interrupted breathing—and the accompanying drop in oxygen levels—takes a toll on the heart and arteries.

New finding: Heart attack risk in sleep apnea sufferers is 30% higher than normal over a four- to five-year period, and stroke risk is twice as high in people with sleep apnea.

- **Diabetes.** Sleep apnea (regardless of one's weight) is linked to increased insulin resistance—a potentially dangerous condition in which the body is resistant to the effects of insulin.

New finding: A 2007 Yale study of 593 patients found that over a six-year period, people diagnosed with sleep apnea were more than two-and-a-half times more likely to develop diabetes than those without the sleep disorder.

- **Accidents.** Sleep apnea dramatically increases the risk for a deadly mishap due to sleepiness and impaired alertness.

New finding: A study of 1,600 people, presented in 2007 at an American Thoracic Society meeting, found that the 800 sleep apnea sufferers were twice as likely to have a car crash over a three-year period. Surprisingly, those who were unaware of being sleepy were just as likely to crash as those who were aware of being sleepy.

DO YOU HAVE SLEEP APNEA?

If you think you may have sleep apnea, see a specialist at an accredited sleep center, where a thorough medical history will be taken and you may be asked to undergo a sleep study.* This involves spending the night in a sleep laboratory where your breathing, oxygen level, movements and brain wave activity are measured while you sleep.

BEST TREATMENT OPTIONS

The treatment typically prescribed first for sleep apnea is *continuous positive airway pressure* (CPAP). A stream of air is pumped onto the back of the throat during sleep to keep the airway open. The air is supplied through a mask, most often worn over the nose, which is connected by tubing to a small box that contains a fan.

In recent years, a larger variety of masks have become available, and fan units have become smaller and nearly silent. A number of adjustments may be needed, which may require trying several different devices and more than one visit to a sleep lab.

Other treatments for sleep apnea are usually prescribed to make CPAP more effective, or for people with milder degrees of the disorder who have tried CPAP but were unable or unwilling to use it.

These treatments include…

•**Mouthpieces.** Generally fitted by a dentist and worn at night, these oral appliances adjust the lower jaw and tongue to help keep the airway open.

•**Surgery.** This may be recommended for people who have an anatomical abnormality that narrows the airway and for whom CPAP doesn't work. The most common operation for sleep apnea is *uvulopalatopharyngoplasty* (UPPP), in which excess tissue is removed from the back of the throat. It works about 50% of the time.

HELPING YOURSELF

Several measures can make sleep apnea treatment more effective and, in some cases, eliminate the condition altogether. *What to do…*

*To find a sleep center in your area, visit the on-line directory of the American Academy of Sleep Medicine, *www.sleepcenters.org.*

•**Lose weight,** if you are overweight. For every 10% of body weight lost, the number of apnea episodes drops by 25%.

•**Change your sleep position.** Sleeping on your side—rather than on your back—typically means fewer apnea episodes. Sleeping on your stomach is even better. Some obese people who have sleep apnea do best if they sleep while sitting up.

•**Avoid alcohol.** It relaxes the muscles around the airway, aggravating sleep apnea.

•**Use medication carefully.** Sleep medications can worsen sleep apnea by making it harder for your body to rouse itself when breathing stops. If you have sleep apnea, make sure a doctor oversees your use of sleep medications (including over-the-counter drugs).

Watch Out! Women Are Raising Their Stroke Risk

Amytis Towfighi, MD, assistant professor, neurology, University of Southern California, Los Angeles.

Daniel T. Lackland, PhD, professor, epidemiology, University of South Carolina, Charleston and spokesperson, American Stroke Association.

American Stroke Association's International Stroke Conference, New Orleans.

Middle-aged American women are gaining weight, especially around the waist, while their risk of stroke has increased significantly, a new study finds.

"In this study, we can't determine exact cause and effect, but it suggests there might be a relationship," said Amytis Towfighi, MD, an assistant professor of neurology at the University of Southern California, who reported the findings at the American Stroke Association's International Stroke Conference in New Orleans.

BACKGROUND

Most stroke studies focus on older people, but the incidence of stroke in women ages 35 to 54 is twice as high as in men of the same age, Dr. Towfighi said. Using data from the National Health and Nutrition Studies done in 1988–1994 and 1999–2004, she and her colleagues looked at whether the risk of stroke in middle-aged women has increased and what the causes of such an increase might be.

STUDY RESULTS

The increase is real, the study found. In the earlier study, 0.6% of women in the age group reported strokes, but that rose to 1.8% in the later study. Stroke incidence among men of the same age remained stable, with an incidence of about 1%.

"In women, waist circumference increased significantly, as did the prevalence of obesity," Dr. Towfighi noted. "There was no difference in the percentage of women who had diabetes, were smokers or who had hypertension."

Women in the later study had an average waist circumference that was 4 centimeters wider than women in the earlier study. Average body-mass index (or BMI, a measure of body fat based on height and weight) rose from 27.11 in the earlier study to 28.67 in the later study. (A BMI of 25 to 29.9 is considered overweight, 30 or above is obese.) And 14.8% of the women in the later study reported using medications to lower blood pressure, up from 8.9% in the earlier study. Almost 4% of women in the later study said they were taking medication to lower cholesterol, compared to 1.4% in the earlier study.

POSSIBLE EXPLANATION

"Abdominal obesity is a known predictor of stroke in women and may be a key factor in the midlife stroke surge in women," said Dr. Towfighi.

The relationship makes sense, said Daniel T. Lackland, PhD, a professor of epidemiology at the University of South Carolina, and a spokesperson for the American Stroke Association. "People have shown that obesity does make a big difference in increasing the risk of stroke and cardiovascular disease," he said.

OTHER RISK FACTORS

But it's difficult to disentangle the various risk factors that circle around obesity, Dr. Lackland said. "What many have shown is that if you increase obesity, you increase the risk of high blood pressure, diabetes and also lipid abnormalities," he said, a reference to unhealthy cholesterol and triglyceride levels. "Is obesity an independent risk factor for stroke? The study was not designed to show that."

REDUCING STROKE RISK

The study does reinforce the standing advice to avoid obesity, as well as other stroke factors, Dr. Lackland noted. "By losing weight, you lose abdominal circumference, you reduce the risk of diabetes and lipid abnormalities, all of which are known risk factors for cardiovascular disease," he said.

info For more information on risk factors for stroke, visit the Web site of the National Stoke Association at *www.stroke.org*, and click on "Risk Factor."

■ ■ ■ ■

Migraines Raise Stroke Risk

In a study of 1,000 women, those who had migraine headaches at least twice a year preceded by visual auras (bright flashing dots, blind spots or distorted vision) had a 50% greater risk for ischemic stroke (caused by a blood clot) than women who did not have migraines.

If you suffer migraines with aura: Ask your doctor to monitor your stroke risk.

Leah R. MacClellan, PhD, former researcher, University of Maryland School of Medicine, Baltimore.

■ ■ ■ ■

Minor Leg Injuries Increase Blood Clot Risk

Injuries that are not serious enough to be treated by a doctor, such as ankle sprains and pulled muscles, cause about 8% of all *thromboses*—serious clots that form inside the veins of the legs. The risk is low, but be aware of the potential danger. If you have a minor leg injury and pain and swelling increase over time or if you experience pain or difficulty breathing (possibly a sign of a pulmonary embolism, which is a blockage of the blood vessels to the lung caused by one or more blood clots, often originating in the leg), seek medical attention.

Frits Rosendaal, MD, researcher, department of clinical epidemiology, Leiden University Medical Center, Leiden, The Netherlands, and leader of a study of more than 6,000 people, published in *Archives of Internal Medicine*.

15

Women's Health

Lack of Sleep Hurts Women's Hearts!

**W**omen can and do suffer more damage to their cardiovascular health because of lack of sleep than men do, and researchers at Duke University Medical Center believe they've determined why.

They found that poor sleep is associated with greater psychological distress and higher levels of biomarkers associated with increased risk for type 2 diabetes and heart disease. They also found that these associations are stronger in women than in men.

"This is the first empirical evidence that supports what we have observed about the role of gender and its effects upon sleep and health," said study author Edward Suarez, MD, an associate professor in the department of psychiatry and behavioral sciences at Duke.

"The study suggests that poor sleep—measured by the total amount of sleep, the degree of awakening during the night, and most importantly, how long it takes to get to sleep—may have more serious health consequences for women than for men," he said.

The study was published online in the journal *Brain, Behavior and Immunity*.

BACKGROUND

Even though women are twice as likely as men to report sleep problems, most sleep studies in the past have focused on men, said Dr. Suarez, who added that this pattern has been slowly changing in recent years.

NEW STUDY

The study included 210 healthy middle-aged women and men without any history of diagnosed sleep disorders. None of them smoked or took any medications on a daily basis. The participants filled out a standard sleep quality questionnaire and were assessed for levels of depression, anger, hostility and perceived social

---

Duke University, news release.

311

support. Blood samples from the participants were analyzed for levels of biomarkers associated with increased risk for diabetes and heart disease.

RESULTS

About 40% of the participants were classified as poor sleepers, meaning they had frequent problems falling asleep or awoke frequently during the night. While the men and women in the study had similar sleep quality ratings, their risk profiles were dramatically different, the researchers said.

"We found that for women, poor sleep is strongly associated with high levels of psychological distress, and greater feelings of hostility, depression and anger. In contrast, these feelings were not associated with the same degree of sleep disruption in men," Dr. Suarez said.

Women who were poor sleepers also had higher levels of C-reactive protein and *interleukin-6*—inflammation biomarkers associated with increased risk for heart disease and higher levels of insulin, a hormone that regulates blood glucose.

"Interestingly, it appears that it's not so much the overall poor sleep quality that was associated with greater risk, but rather the length of time it takes a person to fall asleep that takes the highest toll. Women who reported taking a half hour or more to fall asleep showed the worst risk profile," Dr. Suarez said.

THEORY

He suggested the sleep/health risk differences between men and women may be partly due to variations in the activity of the number of naturally occurring substances in the body, such as the amino acid *tryptophan*, the neurotransmitter *serotonin*, and the neurohormone *melatonin*. "All of these substances are known to affect mood, sleep, onset of sleep, inflammation and insulin resistance," said Dr. Suarez, who plans further research into the link between poor sleep and health risk in women and men.

info For more information about sleep, visit the Web site of the US National Institute of Neurological Disorders and Stroke, *www.ninds.nih.gov* and search "understanding sleep."

Diseases Women's Doctors Miss Most

Marianne J. Legato, MD, professor of clinical medicine at Columbia University College of Physicians and Surgeons, and founder and director of the Foundation for Gender-Specific Medicine, both in New York City. She is the author of *The Female Heart: The Truth About Women and Heart Disease*. Perennial Currents.

She died of a heart attack in the parking lot outside the emergency room—after the ER physician sent her away with a prescription for tranquilizers and a referral for gastrointestinal tests. The doctor had seen that she was sweaty and short of breath, and she had told him she felt tired, weak, achy and nauseous, so he considered it an obvious case of nerves. None of her complaints had sounded like classic heart attack symptoms—a feeling of crushing pressure on the chest and/or pain radiating down the left arm.

Fact: Fatigue, nausea, upper-body aches, perspiring and shortness of breath are indeed symptoms of a heart attack—in about 20% of women who have heart attacks.

The "classic" signs are typical primarily among men. I hear about case after case like the one above. The medical community still struggles, if not with actual gender bias, at least with persistent ignorance about women's unique health concerns.

Here are specific diseases that doctors most often overlook in women...the warning signs to watch for...and the steps you can take to protect yourself.

HEART DISEASE

Heart disease is the number-one killer of women in the US. The incidence of the disease is on the rise even among women in their 20s and 30s, though this increase may be due to our improved diagnostic abilities. Still, test results are too often misinterpreted.

New finding: A common test for heart disease is *coronary angiography*, in which a doctor injects dye into the arteries and then takes an angiogram (X-ray) that reveals blockages of fatty plaque inside blood vessels.

A recent study shows that for many women, this test is not accurate—because instead of forming large blockages (as plaque does in men), in women the plaque is spread more evenly throughout the artery walls.

Consequence: As many as 3 million American women who are at high risk for heart attack may be incorrectly told that they're at low risk—so they go untreated.

Self-defense: Other warning signs of heart attack include insomnia, irregular heartbeat, jaw pain, or pain in the upper abdomen or back during physical activity or intense emotion.

If you experience any such symptoms—especially if you smoke, have diabetes or have a family history of heart attack before age 55—it is vital to be tested for heart disease.

Initial tests may include an electrocardiogram (a recording of the heart's electrical activity)… echocardiogram (ultrasound of the heart)…exercise (treadmill) stress test…and/or noninvasive nuclear scan (in which an injection of radioactive liquid produces three-dimensional images of the heart).

Important: Memorize the warning signs of heart attack in women mentioned above. If you experience several of the severe symptoms, call 911 immediately. If you suffer from a less worrisome symptom like insomnia, contact your doctor instead.

OVARIAN CANCER

Ovarian cancer strikes about 22,000 women in the US yearly, most of them over age 50. It is the second most common gynecologic cancer—and the deadliest.

If the cancer is caught before it spreads beyond the ovary, a woman has a 90% chance of living at least five years. Unfortunately, only 20% of ovarian cancers are discovered at this early stage because the disease exhibits only vague symptoms.

Recent study: Up to a year before being diagnosed, women with ovarian cancer were more likely than cancer-free women to experience abdominal pain and swelling, pelvic pain, gas, constipation or diarrhea and/or frequent urination.

Self-defense: If any of the above symptoms persist for more than a month, see a gastroen-terologist. If no digestive problem is found, ask your gynecologist about ovarian cancer screening tests—ultrasound, magnetic resonance imaging (MRI), computed tomography (CT) and a blood test for CA-125, a protein produced by ovarian cancer cells.

THYROIDITIS

One of the most commonly missed autoimmune diseases is *Hashimoto's thyroiditis.* This disease impairs the thyroid gland's ability to produce proper amounts of the thyroid hormones *T3* and *T4*, which affect the body's metabolic rate (speed at which calories are burned) and other biological processes.

Symptoms of Hashimoto's thyroiditis vary according to the progression of the disease. They may include fast heart rate, anxiety, swollen neck, brittle nails, thinning hair, heavy or frequent periods, muscle cramps or weakness, weight gain, constipation, chills, fatigue and/or depression.

Some people experience no symptoms at all. Untreated, Hashimoto's can lead to high cholesterol, heart disease, birth defects and other problems.

Self-defense: If you have several of the symptoms above, ask your doctor about blood tests that measure levels of hormones and antibodies associated with the disease. Low levels of T3 and T4 and/or high levels of *thyroid stimulating hormone* (which the pituitary gland produces more of when the thyroid gland is underactive) indicate Hashimoto's.

Treatment: Thyroid hormone therapy—generally, for life.

SLEEP APNEA

Many doctors describe a typical sleep apnea patient as "a fat man who snores and gasps in his sleep." Yet this sleep disorder—in which a sleeping person repeatedly stops breathing because the tongue blocks the airway or the brain fails to signal muscles that control respiration—is just as common in postmenopausal women not on hormone replacement therapy as it is in men.

Symptoms include morning headaches, weight gain, lethargy and memory loss. Without treatment, sleep apnea can contribute to high blood pressure, heart disease and accidents brought on by sleep deprivation.

Self-defense: Before bed, set a tape recorder to "voice activation." If the playback is a cacophony of snorts and gasps, ask your doctor for a referral to a sleep disorder center. Sleep apnea may be relieved by weight loss...an oral appliance that positions the jaw and tongue...or a continuous positive airway pressure (CPAP) machine, which gently blows air through a face mask to keep your airway open.

■ **More from Marianne J. Legato, MD...**

Get the Care You Deserve

To make sure your doctor takes your health concerns to heart...

•**Be assertive.** Arrive at appointments with a written list of questions, and don't leave until you've gotten clear, complete answers.

•**Keep track of all medical tests performed,** and phone the doctor if you are not notified about results within the expected time frame.

•**Go elsewhere for your health care if your physician rushes through exams** or brushes off your questions.

■ ■ ■ ■

Women Less Likely to Control Cholesterol

In an analysis of 194 health care plans, researchers found that women were significantly less likely than men to maintain their LDL "bad" cholesterol at the recommended levels (below 100 mg/dL for healthy women and below 70 mg/dL for women with heart disease or risk factors).

Theory: Women and their health care providers underestimate risk for high cholesterol in women. If you are a woman, ask your doctor to check your cholesterol. If it's high, ask how you can lower LDL levels through diet, exercise and/or medication.

Ileana Piña, MD, professor of medicine, Case Western Reserve University, Cleveland, Ohio.

Exercise During Pregnancy Benefits Your Baby, Too

Experimental Biology 2008, news release.

Exercise does a body good—two bodies, in fact, when the one exercising is a pregnant woman. A new study shows that when a mom-to-be works out, her fetus reaps cardiac benefits.

The findings were presented at the Experimental Biology 2008 annual meeting in San Diego, California.

THE STUDY

Ten women participated in the study. Five women were exercisers and the other five were not. Fetal movements, such as breathing and body and mouth movements, were monitored and recorded from 24 weeks into pregnancy to term.

THE RESULTS

The researchers found significantly lower heart rates among fetuses that had been exposed to maternal exercise throughout the study period. The heart rates among the fetuses not exposed to exercise were higher, regardless of the fetal activity or the gestational age.

IMPLICATION

"This study suggests that a mother who exercises may not only be imparting health benefits to her own heart, but to her developing baby's heart as well. As a result of this pilot study, we plan to continue the study to include more pregnant women," said study co-author Linda E. May, of the Department of Anatomy at Kansas City University of Medicine and Biosciences.

info For more information on prenatal care, visit the Web site of the National Women's Health Information Center at the US Department of Health and Human Services, *www. womenshealth.gov* and type "prenatal care" into the search box.

■ ■ ■ ■

Exercise May Help Prevent Fibroids

Benign uterine tumors, fibroids may cause infertility, bleeding and pain, are the leading cause of hysterectomies in the US.

Recent finding: Women who do recreational exercise or walk seven or more hours a week are less likely to develop fibroids than sedentary women.

Donna Day Baird, PhD, epidemiologist, division of intramural research, National Institute of Environmental Health Sciences, Research Triangle Park, North Carolina, and lead author of a study of 1,189 women, published in *American Journal of Epidemiology.*

Nonsurgical Way to Treat Fibroids

John Lipman, MD, medical director, Atlanta Interventional Institute, and director, interventional radiology, Emory-Adventist Hospital, Atlanta.
William Romano, MD, interventional radiologist, William Beaumont Hospital, Royal Oak, Michigan.
Society of Interventional Radiology annual meeting, Washington, DC.

Uterine artery embolization is an effective, nonsurgical therapy for fibroids—common, benign growths of the muscular wall of the uterus that can result in pain and heavy bleeding—a new study says.

But many American women are unaware of embolization as a treatment option, according to another team of researchers.

The two studies were presented at the Society of Interventional Radiology's annual meeting, in Washington, D.C.

"If you're suffering with symptomatic fibroids, you don't have to have surgery. Uterine artery embolization is an excellent option," said the author of one of the studies, John Lipman, MD, director of interventional radiology at Emory-Adventist Hospital in Atlanta. "There are some women who are silent sufferers. They basically hemorrhage each month because they know they don't want a hysterectomy and they feel it's

the only option. So, they just sit on the sidelines because they don't want the surgery."

FIBROID TREATMENT OPTIONS

Uterine artery embolization (UAE), which is sometimes called uterine fibroid embolization, cuts off the blood supply to fibroids by blocking the uterine arteries. The procedure is minimally invasive, doesn't require general anesthesia, and often allows women to return to normal daily activities within a week, instead of the six weeks or so needed for recovery from a hysterectomy.

The downside to this procedure is that it's not effective for everyone, and in rare cases, the fibroids may recur, according to the Society of Interventional Radiology (SIR).

Other treatment options—such as hysterectomy (surgical removal of the uterus) and myomectomy (surgical removal of individual fibroids)—are available, as is magnetic resonance-guided focused ultrasound. This technique is also known as focused ultrasound surgery (FUS) or ablation, and involves using high-intensity ultrasound waves to destroy (ablate) fibroid tissue, according to SIR.

STUDY #1

If ablation fails to relieve fibroid symptoms, uterine artery embolization is still an effective treatment option, according to one of the studies presented. The study evaluated the efficacy of uterine artery embolization after failed ablation, and found that the technique could still provide a reduction in fibroid size and in symptoms.

"FUS is a newer procedure for uterine fibroids and whether it provides sustained symptom relief is still being evaluated. My research shows that when FUS fails, these women could benefit from uterine fibroid embolization," said the study's lead author, Dr. Alisa Suzuki, an interventional radiologist at Brigham and Women's Hospital.

STUDY #2

But Dr. Lipman suspected that—despite the procedure's efficacy—many women aren't being made aware of UAE. In his study, the Emory researcher interviewed 105 women being seen at an Atlanta-area fibroid practice to assess whether or not their gynecologists had told them that uterine artery embolization was an option. Lipman pointed out that gynecologists don't perform UAE, interventional radiologists do. So, if women were not informed about the procedure by their gynecologist, Dr. Lipman asked how they learned about UAE.

He found that only 33% of the women were given uterine artery embolization as an option, and most of these patients were referred from a single health care organization—one that requires doctors to present all options. The most common way that the remaining women found out about uterine artery embolization was either from radio advertisements or the Internet.

RECOMMENDATIONS

"While I encourage patients to use the Internet—it's important for patients to empower themselves—it's up to doctors to give their patients all the options. Gynecologists need to take the lead and tell their patients about this option. Patients shouldn't have to direct their own therapy," Dr. Lipman said.

William Romano, MD, an interventional radiologist at William Beaumont Hospital in Royal Oak, Michigan, added, "Embolization is a good treatment option and it should at least be considered in all patients with fibroids. It would be nice if the knowledge of it was more widespread."

info For more on uterine artery embolization and other fibroid treatment options, visit the Web site of the Society of Interventional Radiology, *www.sirweb.org* and type "fibroid treatment" into the search box.

New Fixes for Fibroids

Elizabeth A. Stewart, MD, professor of obstetrics and gynecology at Mayo Medical School and senior associate consultant at Mayo Clinic, both in Rochester, Minnesota. She is the author of *Uterine Fibroids: The Complete Guide*. Johns Hopkins.

Imagine having a softball-size growth in your uterus. That's a real threat for the one in four American women in their reproductive years who have fibroids—benign muscular masses on the inner or outer uterine walls.

Fibroids can number from one to hundreds and range in size from a speck to a melon.

Though not cancerous, fibroid tumors can cause intense pelvic pain, constipation, incontinence and prolonged menstrual periods. The tumors typically shrink after menopause, but some symptoms may continue. Fibroid sufferers—even those who are symptom-free—are at increased risk for infertility, miscarriage and premature labor. Some fibroid patients bleed so heavily that they develop anemia and/or require a blood transfusion. In some cases, fibroids can be life-threatening if untreated.

Diagnosis can be made during a pelvic exam and confirmed with ultrasound. Hysterectomy (surgical removal of the uterus) is the only way to guarantee that no new fibroids will form. Fortunately, less drastic treatments—some quite new—can reduce the symptoms of and risks for fibroids.

Note: All fibroid treatments carry a risk for impaired fertility.

•**Focused ultrasound (FUS).** High-intensity ultrasound waves are directed at individual fibroids to destroy them. FUS involves no incisions or hospital stay and requires one to four days of recovery. Though fibroids may return, symptom relief usually lasts up to 24 months. FUS is FDA-approved, but some insurers label it "experimental" and so do not cover it.

•**Myolysis and cryomyolysis.** A probe is inserted into the fibroid and heat (myolysis) or cold (cryomyolysis) destroys the mass. Both outpatient procedures are usually done laparoscopically, so incisions are small—but they work only for fibroids on the outer surface of the uterus. Recovery takes about two weeks.

•**Uterine artery embolization (UAE).** Tiny beadlike particles are injected into the uterine artery to block the blood supply to the fibroids. Tumors soften over time, easing pain and reducing menstrual bleeding.

•**Myomectomy.** This technique removes individual fibroids with a laser, electrical current and/or scalpel. Depending on the location of the fibroids and the surgical procedure used, recovery may take from several days to several weeks. Though fibroids may grow back within 10 years, half of patients get sustained relief.

Important: Among fibroid treatment options, myomectomy is most likely to preserve fertility.

Vitamin D Protects Against Female Cancers— Are You Getting Enough?

JoAnn E. Manson, MD, DrPH, a professor of medicine and women's health at Harvard Medical School and chief of the division of preventive medicine at Brigham and Women's Hospital, both in Boston. A lead investigator for two important studies on women's health, Dr. Manson is coauthor of *Hot Flashes, Hormones & Your Health.* McGraw-Hill.

Up to half of American women don't get enough vitamin D, even though this nutrient has significant health benefits. Some experts now recommend a particular form of vitamin D, in amounts significantly higher than current governmental guidelines.

You may have a deficiency of vitamin D if you don't get much sunshine, because ultraviolet B (UVB) rays trigger vitamin D synthesis in the skin...if you have dark skin, which is less efficient at converting UVB to vitamin D...or if you have liver or kidney disease or a digestive disorder that impairs vitamin D absorption. *Among its benefits, vitamin D may...*

•**Lower death rates.** Researchers analyzed 18 studies in which 57,000 adults took vitamin D pills or a placebo for about six years.

Result: Vitamin D supplementation cut risk for death by about 7%.

•**Protect against breast, ovarian, colon and other cancers.** In a trial involving 1,179 postmenopausal women, participants who took daily vitamin D at 1,100 international units (IU) plus calcium were 60% less likely to develop cancer than those who took a placebo or calcium alone.

Reasons: Vitamin D strengthens the immune system and may inhibit out-of-control cancerous cell growth by reducing blood supply to tumors.

•**Strengthen bones** by promoting absorption of such nutrients as calcium and phosphorous from the intestines.

•**Help prevent diabetes** by improving sugar metabolism.

•**Protect against autoimmune disorders,** such as multiple sclerosis and rheumatoid arthritis, by fighting inflammation.

•**Lower risk for heart disease and stroke** by inhibiting accumulation of artery-clogging plaque.

Current guidelines call for intakes of 200 IU daily until age 50...400 IU between ages 51 and 70...and 600 IU after age 70.

New thinking: Adults should get at least 800 IU to 1,000 IU of vitamin D daily from foods and/or supplements in addition to whatever they get from sunlight. *How...*

• **Food sources of vitamin D include cod liver oil** (1,360 IU per tablespoon)...fatty fish, such as salmon, mackerel and tuna (200 IU to 300 IU per three ounces)...fortified milk (100 IU per cup)...and some cereals (about 40 IU per cup).

•**Choose supplements that provide the more active form—vitamin D-3** (cholecalciferol), the type produced by our bodies—rather than the less active D-2 (ergocalciferol) from plants.

•**Don't go overboard.** Excessive vitamin D can cause high blood calcium levels, which in turn can contribute to gastrointestinal problems, cognitive impairment, heart rhythm abnormalities and kidney stones. Multivitamins, calcium pills and osteoporosis drugs often include vitamin D—so check all labels to tally your daily total, making sure not to exceed 2,000 IU per day.

info The Office of Dietary Supplements, part of the National Institutes of Health, has a number of downloadable fact sheets on nutrients. Visit their Web site, *http://dietary-supplements.info.nih.gov/.*

The Secret to Weight Loss After 40

Harriette R. Mogul, MD, MPH, associate professor and director of research in the division of endocrinology and metabolism at New York Medical College in Valhalla, New York. Dr. Mogul is author of *Syndrome W: A Woman's Guide to Reversing Midlife Weight Gain*. M. Evans. *www.SyndromeW.com.*

What is it with women's waistlines? Even if you work out and watch what you eat—and even if you have always been reasonably trim—at around age 40, your midsection starts expanding and the pounds start creeping upward. This frustrating experience often is due to a metabolic disorder that I call *Syndrome W*. It is responsible for midlife weight gain in most women who eat sensibly and exercise regularly.

A woman's risk for this condition is influenced by genetics. Having this syndrome greatly increases the chances of developing diabetes.

Fortunately, with the right treatment, you can get back into the pants that fit you at age 40—and improve your overall health, too.

WHAT IS SYNDROME W?

Based on research involving more than 800 patients, my colleagues and I conclude that Syndrome W affects at least one in five women. It typically develops during *perimenopause*, the years leading up to menopause. The defining characteristics that give Syndrome W its name include *weight gain* and *waist gain* (an accumulation of fat at the waistline). The disorder can also involves high blood pressure.

Women with Syndrome W have elevated levels of *insulin* (the hormone produced in the pancreas that is responsible for regulating blood sugar levels), even though their blood sugar is normal. Evidence suggests that this can produce an increase in appetite that makes it hard to control eating.

If a woman has gained more than 20 pounds or gone up more than two clothing sizes since her 20s, Syndrome W may be to blame. Untreated, she may progress to *metabolic syndrome*, a combination of obesity and abnormalities in blood sugar, blood pressure and blood fats.

As women approach menopause, declining levels of estrogen, progesterone and other hormones appear to trigger a buildup of fatty tissue. Fat cells in and around the abdomen secrete hormones that adversely affect metabolic rate and cause *insulin resistance*—the inability of the body's cells to use insulin. Genetics and other unknown factors can cause a disproportionate amount of fat to accumulate around the midsection—so women with insulin resistance start to resemble eggs rather than hourglasses.

As these women gain additional weight, insulin resistance worsens—so the pancreas produces more insulin to compensate. This sets off a cycle of progressive hunger and weight gain as the women experience a dramatic increase in appetite, especially for sweets and carbohydrates. Insulin resistance also is believed to contribute to the blood pressure elevations seen in women with Syndrome W.

Even when Syndrome W women are disciplined about diet and exercise, they suddenly find that the same health strategies they used in their 20s and 30s no longer keep them slim. Each heroic effort to lose weight ends in failure.

THE W DIAGNOSIS

The best way to confirm a diagnosis of Syndrome W is with simultaneous glucose tolerance and insulin level measurements. Blood tests measure both glucose and insulin *before* and again 30, 60 and 120 minutes *after* you swallow a glucose solution. Elevated insulin levels combined with normal blood sugar levels indicate Syndrome W.

Catching and treating the disorder before weight gain gets too out of hand enhances the chances of avoiding diabetes. Among my patients, the vast majority of women who have been treated with a unique combination of medication and diet have lost 10% to 15% of their body weight. Only a very small percentage of these women have progressed to diabetes.

Added benefit: Successful treatment of Syndrome W also may lower the risk for heart disease and cancer.

THE TREATMENT THAT WORKS

My colleagues and I have developed an effective protocol for treating Syndrome W by normalizing insulin levels, which in turn makes it possible to lose weight. The protocol also helps to lower blood pressure.

First step: The prescription drug *metformin* (Glucophage), which lowers blood sugar and increases the cells' ability to use insulin. This drug is approved by the FDA as an oral treatment for diabetes.

Bonus: Published studies show that metformin reduces hunger and food cravings.

Typically, a doctor prescribes a low dose of metformin at the start, increasing the dosage gradually until the patient detects changes in appetite and a corresponding weight loss. Patients usually need long-term metformin therapy. Side effects are rare but can include stomach upset and/or diarrhea. Metformin can raise the risk for a rare but dangerous disorder called *lactic acidosis*, in which lactic acid builds up in the bloodstream—so the doctor should monitor the patient's kidney function.

Caution: People who have kidney problems should not take metformin. Alcohol should be avoided while taking metformin.

Second step: Adopt a unique eating regimen that reduces insulin levels, staves off hunger pangs and promotes weight loss. *It is based on six simple guidelines...*

1. Eliminate the "four Cs"—candy, cookies, cakes and cereal. Avoid foods that list sugar, sucrose, fructose, dextrose, sorbitol, corn syrup or honey among the first three ingredients. Artificially sweetened treats can be eaten in moderation.

2. Save complex carbohydrates—bread, rice, pasta—for after 4 pm *only*. Studies show that daily food intake increases far more when carbs are eaten early in the day.

3. Include vegetables with every meal and snack. Veggies at breakfast? Try an egg-white omelet or quiche made with chopped mushrooms, onions and spinach.

4. Eat lots of low-fat protein, such as skinless chicken or turkey, beans, lentils and egg whites. Fish is fine—but limit serving sizes to two or three ounces for fatty fishes, such as salmon, tuna and herring.

5. Choose low-fat dairy products, such as skim milk, fat-free yogurt and low-fat cottage cheese.

6. For dessert, select fruit—especially high-fiber ones, such as oranges, berries and the stone fruits (peaches, plums, apricots). Avoid fruit juices and dried fruits.

After the first two weeks on this regimen, allow yourself a once-a-week treat—a slice of pizza, a piece of pie—so you won't feel deprived. Otherwise, stick to the plan. It is very effective at triggering weight loss in women with Syndrome W, so you'll find that it is well worth the effort...and it is flexible enough to follow for the rest of your life.

Why Women Don't Sleep Well—and How to Get Your Rest!

Meir H. Kryger, MD, director of the St. Boniface Hospital Sleep Lab in Winnepeg, Canada. He is a past president of the American Academy of Sleep Medicine and the Canadian Sleep Society, and currently serves as vice president of the National Sleep Foundation. The author of *A Woman's Guide to Sleep Disorders*, McGraw-Hill, he has been researching and treating women's sleep problems for more than 25 years. *www.guidetosleep.com*.

Have you ever felt frustrated watching a man sleep like a baby while you tossed and turned? Do the strategies that used to help you sleep no longer seem to work? Blame biology, at least in part.

As early as adolescence, gender-based differences in physiology and lifestyle make sleep disorders far more common in women than in men. Because these factors change over time, a woman's sleep problems also change—usually for the worse. So it is no surprise that a 2007 poll from the National Sleep Foundation reported that 67% of American women frequently experience sleep problems.

Fortunately, once you understand what triggers your sleep problems, you can take steps toward a better night's rest.

THE CHILDBEARING YEARS

Some women don't need a calendar to remind them when their period is due. A group of symptoms called premenstrual syndrome (PMS)—including breast tenderness, headaches and joint and muscle pains—can cause sufficient discomfort to interfere with rest. A more severe form is *premenstrual dysphoric disorder* (PMDD), which can cause insomnia, depression, anxiety and fatigue.

Polycystic ovary syndrome is an endocrine disorder in which the ovaries are enlarged and have multiple cysts. Symptoms include infrequent or irregular periods, weight gain, excessive hair growth, prediabetes and infertility. About 30% to 40% of women with this syndrome experience sleep apnea, in which breathing repeatedly stops during sleep. Sleep apnea causes persistent fatigue and raises a person's risk for cardiovascular disease by increasing blood pressure and decreasing oxygen going to the heart.

During pregnancy, the quality of sleep worsens for more than 80% of women. The uterus presses on the bladder (so you wake up often to urinate) and on the stomach (causing heartburn that can keep you awake). Pregnant women also are prone to *restless legs syndrome* (RLS), an irresistible urge to move the legs, which interferes with sleep.

Sleep disturbances worsen after the baby is born, thanks to nighttime feedings. For the 10% to 15% of women who develop *postpartum depression* in response to the hormonal fluctuations that follow childbirth, sleeplessness often accompanies depression.

MIDLIFE AND BEYOND

The transition to menopause can be brief or may last as long as seven years. During this stage, called *perimenopause*, periods become irregular. Due to fluctuating hormone levels, many women experience hot flashes and night sweats caused by dilating blood vessels. These rapid changes in body temperature can awaken women repeatedly.

Menopause typically occurs between ages 48 and 55. One in two postmenopausal women experience a sleep disorder, such as sleep apnea or RLS, while nearly two in three have insomnia at least a few nights each week. One culprit is excess weight, particularly in the neck area. The average woman gains eight pounds after menopause, increasing her risk for sleep apnea.

Another sleep-disturbing factor may be the worry that accompanies caregiving, as women in midlife assume responsibility for aging parents and/or ailing husbands.

Advancing age increases a woman's risk for arthritis, diabetes, cardiovascular disease, depression and other chronic conditions. In addition to the discomforts of the diseases themselves, the medications used to treat such conditions can affect sleep.

FOR BETTER SLUMBER...

The good news: Several simple steps can yield big improvements.

•**Eat light—and early—in the evening.** A heavy, spicy or late-night dinner may trigger heartburn that keeps you awake.

Better: A light dinner two to four hours before bedtime. Limit beverages to avoid middle-of-the-night trips to the toilet.

•**Skip the buzz.** Limit caffeinated beverages to two per day, both before lunch. Have no more than one alcoholic drink per evening, at least three to four hours before bed. When you drink too much or too late, your blood alcohol level drops in the middle of the night, causing brain arousal that wakes you up.

•**Quit smoking.** Many smokers wake at night when blood levels of nicotine fall.

•**Argue in the morning (or not at all).** Evenings should be free from confrontations, strenuous workouts and heart-stopping action films that rev you up when you need to wind down.

Exception: Sex. It promotes sound sleep.

•**Avoid evening naps.** Nap only in the afternoon—and for no more than 45 minutes—to minimize disruptions of your nighttime sleep cycle.

•**Create a relaxing bedtime ritual.** Light reading or soothing music tells your brain it's almost bedtime. A warm bath is especially good because raising the body temperature promotes deep sleep.

•**Don't push back your bedtime.** Your body clock works best when bedtimes and awakening times are consistent.

•**Don't lie there staring at the clock if you can't sleep**—get up and do something boring. Tossing and turning is futile and frustrating. Fold some laundry or read something dull until you feel sleepy.

•**Get help for hormonal problems.** Consult your doctor if you suffer from PMS, PMDD or hot flashes.

•**Don't suffer in silence.** Talk to your doctor if your sleep disturbances are due to illness, stress or the side effects of drugs.

•**See a sleep specialist** if you have insomnia more than three times per week for more than a month…if you think you have RLS…or if reports of your snoring and gasping lead you to suspect sleep apnea. You may need an overnight evaluation in a sleep clinic. Your doctor can provide a referral.

For Women, a Happy Marriage Means Sweet Dreams

Wendy M. Troxel, PhD, psychologist, University of Pittsburgh.

Ana Krieger, MD, director, New York University School of Medicine, Sleep Disorders Center, New York City.

Associated Professional Sleep Societies annual meeting, Baltimore.

Trouble in your marriage can cause trouble in bed, but not necessarily the kind of trouble that first comes to mind.

New research has found that women in happy marriages tend to sleep more soundly than women in unhappy marriages. In fact, women with good marriages have about 10% greater odds of getting a decent night of shut-eye compared to women who aren't happy with their spouse.

"Marriage can be good for your sleep if it's a happy one. But, being in an unhappy marriage can be a risk factor for sleep disturbance," said the study's lead author, Wendy M. Troxel, PhD, a psychologist at the University of Pittsburgh.

THE STUDY

Dr. Troxel and her colleagues reviewed data on about 2,000 married women who participated in the Study of Women's Health Across the Nation (SWAN). The women were an average age of 46 years. Just over half were white, 20% were black, 9% were Hispanic, 9% were Chinese, and 11% were Japanese.

All of the women reported their sleep quality, the state of their marriage, how often they had difficulty falling asleep, if they stayed asleep, and how early they woke up.

THE RESULTS

Happily married women had less trouble getting to sleep, fewer sleep complaints, more restful sleep and were less likely to wake up early or awaken in the middle of the night than women whose marriages were less than ideal.

Even after the researchers adjusted the data to account for other factors known to disturb sleep, the researchers found that happily married women still slept more soundly. And, these findings appeared to hold up across racial lines. The only groups that the findings weren't statistically significant for were Chinese and Japanese women, but Dr. Troxel suspects this may be because there weren't as many Chinese or Japanese women in the study as white and black women.

IMPLICATIONS

"All marriages aren't created equal, and having a high quality marriage may be good for sleep, whereas an unhealthy marriage is a potent source of stress. You could be sleeping with the object of your hostility," Dr. Troxel said.

The million-dollar question, Dr. Troxel said, is which comes first—does the unhappy marriage lead to poor sleep, or does poor sleep contribute to a bad marriage?

"We have future studies planned, and we need to tease that out," she said. "If you're not sleeping, you're more irritable, have lower frustration and tolerance levels, so it's possible that could affect the marriage. But we suspect it's in the other direction," that the bad marriage is affecting the quality of sleep because you're trying to sleep next to someone you may be fighting with, and that's stressful.

"If you're stressed or anxious, it can have an effect on your sleep," agreed Ana Krieger, MD, director of the New York University Sleep Disorders Center in New York City.

RECOMMENDATIONS FOR SOUND SLEEP

If you have a lot of stress from your marriage or another source, such as your job, said Dr. Krieger, you need to try to fix the situation that is causing the anxiety. If you can't change the stressful situation, she recommended trying to change how you perceive the stress. Good ways to help you relax are meditation and yoga, she said.

Dr. Troxel said that if you're in an unhappy marriage, marriage therapy—or individual therapy if your spouse won't go to therapy—could be helpful.

She also recommended practicing good sleep habits, such as going to bed at the same time and waking up at the same time every day.

info For more advice on getting a good night's sleep, visit the Web site of the National Sleep Foundation, *www.sleepfoundation.org*, and search "healthy sleep tips."

The Very Latest Thinking On Hormone Therapy

JoAnn E. Manson, MD, DrPH, a professor of medicine and women's health at Harvard Medical School and chief of the division of preventive medicine at Brigham and Women's Hospital, both in Boston. A lead investigator for two important studies on women's health, Dr. Manson is coauthor of *Hot Flashes, Hormones & Your Health.* McGraw-Hill.

Hot flashes can strike at the most inconvenient moments, soaking your clothes and melting your makeup…night sweats can steal slumber…and vaginal dryness can make sex a pain. What's a menopausal woman to do?

Surprise: Hormone therapy (HT), nearly a pariah in the field of medicine, may be the best answer for some women. Five years after major studies indicated that HT increased the risk for cardiovascular problems and breast cancer—causing women across the US to toss their HT prescriptions in the trash—HT is on the rise again.

New findings: The latest studies along with reevaluations of earlier research have helped to clarify the benefits and the risks.

Fact: HT remains the most effective treatment for moderate to severe *vasomotor* symptoms (hot flashes and night sweats).

About 80% of women in the US experience hot flashes during menopause. The typical episode lasts one to five minutes—though hour-long "heat waves" can occur. During a major hot flash, a woman feels as if she's being consumed by an inner fire…her skin turns red and may drip with perspiration…her heart may pound… she may feel confused and/or light-headed… and she may experience a vague sense of dread. Menopausal women typically have several hot flashes per day, and some have 10 or more. Episodes usually persist for about four years—for a total of up to 15,000 hot flashes. I find that, for about one in four women, symptoms are severe enough to merit treatment.

BUT IS IT SAFE?

Among the general public, the term *hormone therapy* often is used to refer either to estrogen replacement alone or a combination of estrogen and a progesterone-like drug. Many people refer to all progesterone-replacement drugs as *progestins*, though the accurate all-inclusive term is *progestogens* (which covers natural and synthetic forms). Women who have had a hysterectomy can take estrogen alone. Otherwise, estrogen is given with a progestogen to protect against uterine cancer.

We have learned that timing is key to the safety of HT. The risks for heart attack, stroke and blood clots related to HT use are low in recently menopausal women who are in good cardiovascular health. Also, the risk for breast cancer does not increase appreciably until a woman has taken hormones for four to five years. Most women don't need HT for that long because menopausal symptoms often abate by then.

Essential factor: The amount of time since your final menstrual period. You reach menopause when you go 12 consecutive months without a period. The farther you are past menopause and the more risk factors you have for heart disease and breast cancer, the riskier HT is. (If you have hot flashes but still menstruate, low-dose oral contraceptives may be an option.)

Based on the large-scale Women's Health Initiative and other studies, here's a summary of the latest thinking…

•**Heart disease.** Women who began HT within five to 10 years after menopause and took it for five to seven years tended to have a lower risk for heart disease than women taking a placebo. Women who started HT more than 10 years after menopause tended to have an increased risk. The older a woman was, the greater the risk.

•**Stroke.** At all ages studied, HT increased the risk for stroke—but even so, for younger women, overall risk remained low. Among women who started HT in their 50s, stroke risk increased by about two cases per year for every 10,000 women.

•**Breast cancer.** Women of all ages who took estrogen with a progestin (a synthetic form of progestogen) had an increased risk for breast cancer after four years of use. The longer they used the combination therapy, the higher their risk. Women who took estrogen alone for seven years did not have an increased risk for breast cancer—however, estrogen-only therapy is appropriate only for women who have had a hysterectomy.

•**Bone fractures.** Estrogen (taken alone or in combination with a progestogen) clearly reduces bone fracture risk. However, this benefit would require long-term HT, so experts no longer recommend it as the first line of defense against osteoporosis.

TODAY'S OPTIONS

If you and your primary-care doctor or gynecologist decide HT is right for you, the next step is to consider the specific options.

•**Estrogen.** For relief from hot flashes, products include daily pills…or transdermal skin patches worn on the arm, abdomen or buttocks and replaced once or twice weekly…or a transdermal cream, gel or spray applied to the arms, legs or buttocks once or twice daily. Compared with oral estrogen, transdermal estrogen may be less likely to increase the risk for blood clots. Most doctors agree that with oral or transdermal estrogen, a progestogen also is needed unless a woman has had a hysterectomy.

•**Progestogen.** Oral progestogen may be used daily or taken 10 to 14 days of each month. Topical creams and vaginal gels also are available.

•**Vaginal options.** For women whose symptoms are limited to vaginal dryness and discomfort during sex, options include vaginal creams, rings or suppositories. In my view, it is prudent either to halt vaginal estrogen use for a few weeks every three to six months or to add a progestogen intermittently.

•**Bioidentical hormones.** These are laboratory-made hormones with the exact same molecular structure as hormones produced by the human body. Some people believe them to be safer than the more widely used conventional hormones—but no large-scale trials have yet been done to test this belief. If you prefer to use bioidentical hormones, feel free. However, for now it is prudent to assume that all hormone formulations confer a roughly similar balance of benefits and risks.

Usage guidelines: If you decide to use HT, start with the lowest recommended dose. If symptoms do not diminish within four weeks, talk to your doctor. You may need to increase your dose incrementally until you find the dosage that works for you. After one to three years, your doctor may recommend gradually reducing and then discontinuing HT. If symptoms return, resume HT for another six months. I recommend that the total time spent taking HT be less than five years.

■ **More from JoAnn E. Manson, MD…**

Is Hormone Therapy Right for You?

The answer depends on the intensity of the symptoms, how long ago you entered menopause and your overall health.

A good candidate…

•**Has hot flashes and/or night sweats severe enough to interfere with sleep** or reduce quality of life.

•**Had a final menstrual period less than five years ago** (some studies suggest up to 10 years).

•**Is at low risk** (based on personal and family history) for heart disease, stroke and blood clots.

•**Does not have a personal or strong family history of breast cancer.**

A poor candidate…

•**Is more than five years into menopause** (some studies say 10).

•**Has had, or is at high risk for,** heart disease, stroke or blood clots…or breast, uterine or ovarian cancer.

•**Currently has unexplained vaginal bleeding, diabetes, or liver or gallbladder disease.**

■ ■ ■ ■

What? Hormone Therapy May Damage Hearing

Women who are on hormone replacement therapy (HRT) with *progestin* are more prone to a type of hearing loss that makes understanding speech difficult than women on estrogen only or not on hormones.

Self-defense: Women on HRT who suffer hearing loss should discuss alternatives with their doctors.

Robert D. Frisina, PhD, professor and associate chair, department of otolaryngology, University of Rochester, NY, and leader of a study of 124 postmenopausal women, presented at the Proceedings of the National Academy of Sciences.

New Way to Cool Hot Flashes for Breast Cancer Patients

Eugene Lipov, MD, medical director, Advanced Pain Centers, Hoffman Estates, Illinois.

Joanne Mortimer, MD, vice chairwoman, medical oncology, City of Hope Comprehensive Cancer Center, Duarte, California.

Barbara A. Brenner, executive director, Breast Cancer Action, San Francisco

The Lancet Oncology.

An injection of a local anesthetic into the nerves of the neck that regulate temperature could give breast cancer patients long-term relief from hot flashes and sleep deprivation, a new study suggests.

Severe hot flashes often accompany treatment for breast cancer, especially among women taking anti-estrogen drugs. These hot flashes can become so severe that women stop taking their medication at the risk of the cancer returning. In fact, more than 50% of these women stop taking their medication after 180 days, researchers report.

"Breast cancer survivors can have very severe hot flashes, and this modality of treatment seems to resolve that without the usual problems of hormone treatments," said lead researcher Eugene Lipov, MD, medical director at Advanced Pain Centers in Hoffman Estates, Illinois.

THE STUDY

Dr. Lipov's team treated 13 breast cancer survivors who had severe hot flashes, using a stellate-ganglion block, a procedure in which an anesthetic is injected into the front of the neck to block sympathetic nerves. The researchers evaluated the number of hot flashes and the quality of the women's sleep one week before the injection and every week thereafter for 12 weeks.

The researchers found that the total number of hot flashes dropped from an average of 79.4 per week before the injection to an average of 49.9 per week in the first two weeks after treatment.

The number of hot flashes continued to decline over 12 weeks, reaching a mean of 8.1 hot flashes per week. The number of very severe hot flashes dropped to near zero by the end of the 12th week, according to the report.

In addition, the number of night awakenings dropped from an average of 19.5 per week before the injection to an average of 7.3 per week in the first two weeks after the procedure. Night awakenings continued to decrease, to an average of 1.4 per week by the end of the study.

IMPLICATIONS

"This is a big advance in treatment for women," Dr. Lipov said. "Women with severe hot flashes should really be treated with this."

In the future, this could be how all hot flashes are treated, Dr. Lipov said. "Fifty million older women in the United States have hot flashes, not just breast cancer survivors, but menopausal women," he said. "Five million of these women have severe hot flashes."

The current treatments for hot flashes, both alternative and conventional, including estrogen, all have dangers associated with them or are ineffective, Dr. Lipov said. "This is a safe approach that works very quickly and can last for up to three years," he said. "The average response time is three to eight months."

EXPERT REACTION

"Most women experience hot flashes," said Joanne Mortimer, MD, vice chairwoman of medical oncology at the City of Hope Comprehensive Cancer Center in Duarte, California. "And they interfere with normal activity in 15%."

"In addition, 65% of women treated for breast cancer experience hot flashes, and we cannot use estrogen replacement in these folks," Dr. Mortimer said. "Relief of symptoms is important for all women but is especially a need in the breast cancer survivor population."

Another expert was cautious about this treatment and said more data was needed before it could become widely accepted.

"While it looks like the study produced meaningful results for many of the women in the trial, this is a tiny study, which did not go on for very long," said Barbara A. Brenner, executive director of Breast Cancer Action.

Brenner added that because the procedure involves injections and the use of X-ray fluoroscopy to guide the needle, which involves a significant exposure to radiation, it might increase the risk of developing another cancer years later.

"While some women will be so seriously affected by hot flashes and disruptions of sleep to want to do this, it would be good to have considerably more data both in terms of numbers studied and long-term side effects before touting it," Brenner said.

info For more information on menopause and hot flashes, visit the Web site of the US National Library of Medicine at *www.nlm. nih.gov/medlineplus/menopause.html.*

■ ■ ■ ■

The Hot Flash–
High Blood Pressure Link

Women who experience hot flashes have much higher blood pressure, on average, than women who don't. Hot flashes—a symptom of menopause—are feelings of intense heat along with sweating and rapid heartbeat. Flashes usually last from two to 30 minutes each and may occur as often as a dozen times a day. Women who have hot flashes should have their blood pressure monitored closely.

Linda Gerber, PhD, professor of public health and medicine and director of the biostatistics and research methodology core, Weill Medical College of Cornell University, New York City, and author of a study on hot flashes and blood pressure, published in *Menopause.*

Ladies—You Need Testosterone, Too!

Elizabeth Lee Vliet, MD, author of *The Savvy Woman's Guide to Testosterone* (HER Place) and *Women, Weight and Hormones* (M. Evans). She also is founder and medical director of HER Place: Health Enhancement Renewal for Women, Inc., a medical practice specializing in women's health problems caused by hormone imbalances, with offices in Tucson, Arizona, and Dallas. *www.her place.com.*

From a woman's teens until menopause, her body makes a fair amount of the hormone testosterone in addition to estrogen. Testosterone affects sexual desire and response... muscle and bone formation...hair growth...immune function. In women, testosterone is produced by ovaries and adrenal glands.

•**Testosterone deficiency.** Many factors can inhibit testosterone production. These include aging...certain medications (antidepressants, birth control pills, hormone replacement therapy, corticosteroids)...removal of or damage to the ovaries...prolonged severe stress...obesity.

Potential symptoms of deficiency: Loss of interest in sex or loss of sexual sensation... unusual fatigue, low energy, feeling blue...diminished muscle tone and strength...reduced height due to bone loss...thinning hair...dry eyes...more frequent or intense headaches.

•**Diagnosis and treatment.** A blood test is the best way to assess testosterone levels. If you are diagnosed with low testosterone, work with a physician to create an individualized supplement program. FDA-approved testosterone therapies for women are limited. Currently, the medications must be compounded (specially prepared at a pharmacy). Testosterone can be taken by injection, as tablets, via a skin patch or as a cream. Non-oral methods minimize side effects. Excess testosterone can cause irritability, difficulty sleeping, violent dreams, acne, facial hair, loss of scalp hair, food cravings, increased blood pressure and suddenly high cholesterol. During treatment, have blood levels checked regularly.

Warning: The 1% to 2% concentrations often prescribed in testosterone creams are too strong—a safer dose starts at 0.1%. Testosterone therapy is not appropriate for women with breast or uterine cancer or cardiovascular or liver disease. It may interfere with certain drugs, such as the blood thinner *warfarin* (Coumadin) and the blood pressure medication *propanolol* (Inderal).

Why Your Joints Ache... And How to Make Them Feel Better

Beth E. Shubin Stein, MD, an orthopaedic surgeon and sports medicine specialist at the Women's Sports Medicine Center at the Hospital for Special Surgery and an assistant professor at Weill Medical College of Cornell University, both in New York City.

We women have lots of joint pain. A study by the Centers for Disease Control and Prevention found that nearly one-third of adult women in the US have some type of chronic joint pain or have been diagnosed with arthritis.

Despite what's commonly believed, it's not that women have more trouble with their joints than men, but that they tend to have different types of problems.

Example: In general, women's joints tend to be looser than men's joints, particularly the knees and shoulders, which can reduce stability and lead to damaging wear patterns and pain. *What you need to know to protect and soothe your joints...*

TROUBLE SPOT #1: KNEES

Women have inherent difficulties with their *patellofemoral joints*—or kneecaps—and how they join with the thigh bone. Because women have wider pelvises than men, the hip-to-knee line is less vertical than it is in men, which places more stress across women's knees.

Another reason for knee pain in women is that their soft tissue, including ligaments, is often more lax than that in men.

A third cause of female-specific problems is muscular imbalance. The muscles that support and control the knees include the quadriceps (front of the thighs), the hamstrings (back of the thighs) and the gluteal (buttocks) muscles. Many women do not have the proper balance of strength among these muscles to support and protect knees.

Weakness in the hamstrings may increase risk for traumatic injuries, such as tears to a ligament when jumping or turning quickly. Weakness in the quadriceps or gluteals may predispose a woman to "overuse" problems, such as pain in the kneecap area due to pressure from body weight on the knee.

What to do: Knee pain often can be treated with physical therapy to strengthen the proper muscles, including the quadriceps, hamstrings and gluteals—and followed up with a long-term exercise program incorporating the fundamentals learned in physical therapy. You can't change the alignment of your hips and knees or the laxity of your ligaments, but how strong you are is something you can control with exercise.

Also helpful: Don't wear high heels more often than necessary. They worsen knee pain by increasing pressure behind the kneecap.

For mild-to-moderate arthritis in the knee that doesn't respond well to physical therapy and home exercise, your doctor may want to try *viscosupplementation*, in which a lubricant based on *hyaluronic acid*, a substance that occurs naturally in the body, is injected into the knee. This procedure doesn't build back cartilage or bone damaged by injury or arthritis, but it can reduce pain for six months or more and appears to be safe.

TROUBLE SPOT #2: SHOULDERS

The rotator cuff, a group of muscles and tendons that attaches the arm to the shoulder joint and lets the arm rotate up and around, can become irritated and/or inflamed. Continued irritation can cause tendonitis and bursitis.

What to do: A physical therapist can help you strengthen muscles, often using a stretchy exercise band. It's a good idea to fortify the *scapulothoracic muscles* of the back (they come into play when you squeeze your shoulder blades together).

Smart: Don't carry a heavy purse or other bag on one side of your body—it will promote shoulder and back pain. Instead, distribute weight evenly across your body, such as with a backpack-style purse.

Best of all: A small bag on wheels with an extendable handle. That's what I use to take my patients' charts between my office and the hospital.

TROUBLE SPOT #3: THUMBS

The base of the thumb, called the *carpometacarpal* (CMC) joint, is prone to pain from arthritis in women. We aren't sure why, but it may have something to do with the fact that this joint is smaller in women than in men, yet women must bear the same loads while pinching and grasping as men do. With CMC arthritis, a simple activity like holding a fork can be painful.

What to do: A special splint fitted by an occupational therapist can support your thumb joint and limit the movement of your thumb and wrist. If a splint is going to help, you should start to feel relief within a few days to a week.

If a splint isn't effective, you may need an injection of a steroid into the thumb, which can reduce pain and inflammation. Although most such injections help for only six to eight weeks, that sometimes lets the joint heal enough to stay pain free.

If a steroid injection doesn't solve the problem and arthritis progresses to an advanced stage, you may require one of two kinds of surgery…

•**Replacement** of damaged portions of the joint with tendon from the wrist. This preserves mobility but usually results in a loss of strength.

•**Fusing** together of bones in your thumb. This maximizes strength but limits mobility.

AEROBICS–PLUS

It's important to get aerobic exercise at least several days a week. Besides improving your cardiovascular health, it will help keep your weight down—and excess weight is an enemy of your joints, especially your weight-bearing joints, such as the knees and ankles.

Best: Aerobic activities that also put joints through ranges of motion, such as using an elliptical training machine, swimming and cycling.

However, to slow age-related bone loss, it's vital to do weight-bearing exercise, which stimulates growth of new bone. The easiest way for most people is to walk a lot. To involve the upper body, use a variety of weight-lifting machines or handheld weights.

Why Women Get Shorter…and How to Stand Tall

Arthur H. White, MD, a retired orthopedic spine surgeon, Walnut Creek, California. He is author of *The Posture Prescription: The Doctor's Rx for Eliminating Back, Muscle, and Joint Pain.* Crown. He has published more than 200 medical journal articles related to spinal health.

The cliché of the hunched-over "little old lady" may seem outdated, yet it's true—we all get shorter as we age. However, a good portion of this shrinkage can be prevented or corrected. The following steps can help you stand tall…plus reduce your risk for back pain and slim your silhouette.

PROTECT YOUR SPINE

Each vertebra (an individual segment of the spine) is separated from adjoining ones by fluid-filled cushions called discs. In young adults, each disc is about one inch thick.

After about age 40, these discs begin to crack and tear, which causes them to dehydrate and flatten. Over two decades or so, each disc loses about one-twelfth of an inch of thickness—and with 23 discs, that equates to a loss of nearly two inches of height.

Any uncontrolled motion that stresses your back can increase the risk for disc damage. Activities that involve abrupt twisting—racket sports, skating, roughhousing with kids—can tear the discs, hastening dehydration. There is no need to quit these activities, but do try to avoid wrenching twists.

Three other reasons we shrink as we get older are poor posture, loss of abdominal muscle tone and reduced bone density. These problems tend to worsen over time. The humpback seen in many older women is an obvious manifestation of this, but there also are more subtle warning signs that you may be aging yourself prematurely.

PERFECT YOUR POSTURE

Poor attention to posture may be to blame for height loss if you…

•**Slump forward when sitting.**

•**Drive, watch TV or work at a computer with your head jutting forward** (instead of centered over your torso).

•**Experience back or neck pain.**

•**Have frequent headaches.**

•**Have pain in your arms or hands** (e.g. when pain radiates from the spine).

Chronic slouching weakens back and chest muscles, so you eventually may find it difficult to stand up straight even when you try. Large-breasted women are particularly prone to slouching. The extra weight they carry in front places more strain on the back muscles.

Simple steps to stand tall…

•**First, check your posture.** Have your picture taken while you stand sideways in a doorway. Stand as you normally do—if you don't, you'll only be fooling yourself. The door frame provides a visual guide to what is straight.

•**Second, stay aware.** Whenever you pass a mirror or window, notice the position of your neck, shoulders and back. If your head juts forward, pull it back in line with your shoulders…roll your shoulders back…squeeze your shoulder blades together more…draw in your tummy…and tuck your rear end under. Even if you can't remain that straight for long, these brief moments of awareness will help to realign your spine and strengthen your muscles.

•**Third, stretch and strengthen.** When muscles are not used, they get smaller and weaker. Prevent or reverse problem posture with a few simple exercises. The following are generally safe for everyone, though it is best to check with your doctor before beginning any new workout.

•**Secret stretch** can be done anywhere. It returns shoulders to their proper position and strengthens muscles between shoulder blades. Clasp your hands behind your back at the level of your buttocks. Push your hands down and roll your shoulders back (don't arch your back). Hold for 30 seconds. Repeat several times daily—while waiting for an elevator, watching TV or anytime you are idle.

Advanced: With hands clasped behind you and arms straight, raise arms up and away from your body.

•**Doorway stretch** strengthens your shoulders and back and stretches chest muscles. Stand in a doorway, a few inches behind the threshold, feet shoulder-width apart. Place one hand on each side of the doorjamb, level with your shoulders. Keeping your back straight, alean forward, feeling the stretch across your upper chest as your back muscles work to draw your shoulder blades together. Hold for 30 seconds…repeat five times. Do every other day or whenever you feel upper-back tension.

FABULOUS ABS HELP

When tightened, abdominal muscles act as a muscular corset, supporting the vertebrae and protecting them from injury.

Women generally have a more difficult time maintaining abdominal strength than men because pregnancy overstretches the belly muscles. A little effort goes a long way, however, in restoring the strength of the abs. *To try...*

•**Isometric crunch and hold.** Lie with knees bent and feet flat on the floor. Cross arms across your chest. Tuck chin slightly... then raise your head and shoulders (not your entire back) off the floor by tightening your belly muscles. Keep your head in line with your shoulders, and leave enough space to fit a tennis ball between your chin and chest. Hold the position as long as possible, then slowly lower down to the starting position. Repeat, continuing for three to  five minutes. At first, you may be able to hold the pose for only a second or two and may need to rest between each crunch. Keep at it daily—soon your abs will get stronger.

BUILD STRONGER BONES

Osteoporosis is a disease in which bones become so fragile that even the jolt of slipping off a curb can fracture the spine. In severe cases, the weight of the person's own body can slowly crush vertebrae, producing a rounded humpback that steals as much as four inches of height.

Because estrogen helps to build bone, women are susceptible to osteoporosis after menopause, when estrogen levels drop. Ask your doctor about taking supplements of calcium and vitamin D. *Also...*

•**Go for a walk.** Every time your foot hits the ground, the impact sends a vibration through your bones. This produces an electrical force (the *piezoelectric effect*) that strengthens bones by encouraging calcium deposits throughout the skeleton.

Alternatives: Running builds even more bone density but may injure joints—so stick to brisk walking unless you're already an accomplished runner. Weight lifting also builds bone through similar forces, but the effect is less because there is no impact or vibration.

Illustrations by Shawn Banner.

Tea Helps Strengthen Women's Hips

Amanda Devine, PhD, senior lecturer, nutrition program, School of Exercise, Biomedical and Health Science, Edith Cowan University, and adjunct senior lecturer, School of Medicine and Pharmacology, University of Western Australia, Perth.
American Journal of Clinical Nutrition.

New Australian research suggests that having a cuppa (tea, that is) may help strengthen older women's hips.

"This study suggests that drinking tea in moderation can actually benefit your bones," said lead researcher Amanda Devine, PhD, a senior lecturer in the nutrition program at the School of Exercise, Biomedical and Health Science, Edith Cowan University, and adjunct senior lecturer at the University of Western Australia's School of Medicine and Pharmacology in Perth.

"Those who drank tea in the study had a higher bone density over the four years that they were studied," she said.

"These women lost less bone than those who did not drink tea. More than three-quarters of the study participants were daily tea drinkers, and they consumed on average about three cups per day," she continued.

BACKGROUND

Prior research has suggested that drinking tea may improve bone mineral density in people at risk for osteoporosis, but the findings were not conclusive.

Fractures, especially hip fractures associated with osteoporosis, are a major source of disability in postmenopausal women.

Osteoporosis causes the bones to become fragile and more likely to break. Although it primarily affects older women, osteoporosis can affect others as well.

THE STUDY

The new study followed 1,500 elderly (70 to 85 years old) women in Australia. They participated in a five-year trial that measures the effects of the effect of calcium supplementation on osteoporotic hip fracture.

Preliminary information on tea consumption was collected at the beginning of the study for 275 participants, and all participants filled out a beverage consumption questionnaire at the end of the trial.

The researchers measured bone mineral density at the hip at year one and year five.

RESULTS OF THE
STUDY

By the end of the study, bone mineral density at the hip was 2.8% greater in tea drinkers than in non-tea drinkers, the researchers found.

Over four years, tea drinkers lost an average of only 1.6% of their total hipbone mineral density. Non-tea drinkers, on the other hand, lost 4%—consistent with previous studies.

There was, however, no relationship between the amounts of tea consumed and bone gains, which raises some questions about the mechanisms that might be responsible for the bone density effect.

"We didn't see a dose-response to tea drinking—that is, if you drank more tea, then your bones were even better," Dr. Devine said.

"The lack of such a relationship may be due to the small numbers of tea drinkers in each group, once we started examining these data. When we just look at the whole group, we have more power to see a difference," she noted.

THEORY

The authors speculated that certain components of tea, such as antioxidant flavonoids, might account for the benefit seen.

Flavonoids "have been shown to have a stimulatory effect on new cells that build bone in cell line studies," Dr. Devine explained.

Phytoestrogens, estrogen-like chemicals that are found in plants in tea may be beneficial. This may be especially true with regard to older women whose levels of estrogen are low, according to Dr. Devine.

info For more information on bone loss, visit the Web site of the National Osteoporosis Foundation, *www.nof.org.*

New Cures for Incontinence

Amy E. Rosenman, MD, assistant clinical professor of obstetrics and gynecology at the David Geffen School of Medicine of the University of California, Los Angeles. She also is the president of the American Urogynecologic Society Foundation (*www.augs.org*)...a founder of the Pacific Continence Center in Santa Monica...and coauthor of *The Incontinence Solution: Answers for Women of All Ages.* Fireside.

Did you ever laugh so hard that you accidentally leaked a little urine? That's normal. But if you involuntarily urinate a little or a lot nearly every time you laugh, cough, sneeze or exercise—or sometimes when you do nothing at all—that's incontinence. The condition affects an estimated 24 million American adults, the majority of whom are women. Its primary causes are either *anatomical* or *neurological*. If you have it, you know it—but you don't have to live with it.

While not a life-threatening health problem, incontinence certainly is a life-limiting social problem. An outing becomes a logistical hassle when it must be planned around finding the public restrooms—not to mention carrying absorbent pads wherever you go.

STRESS INCONTINENCE

With *stress incontinence*, damaged pelvic muscles allow urine to leak when a sudden action (a cough, an exercise) puts pressure on the bladder.

Injury while giving birth is a common anatomical cause. Vaginal delivery or a long labor before a C-section can weaken pelvic muscles and/or damage nerves around the vagina and the bladder. The longer the labor, the more children you had and the bigger your babies were, the more severe the damage may be.

It's common for mothers in their late 40s or 50s to develop stress incontinence, even if their children were born many years earlier. Muscle function normally decreases with age—and previously injured tissues are most susceptible.

Whether or not they ever give birth, as women age, incontinence risk increases. *Here's why...*

•**Declining estrogen** leads to thinning of the walls of the urethra (the tube that empties the bladder), creating a wider and weaker pipeline for urine to flow through.

- **Gravity's cumulative effects** over the decades pull the pelvic organs out of their proper position.

- **Some women inherit a genetic tendency toward weak *collagen,*** a fibrous protein that makes up the urinary tract support structure, exacerbating age-related weakening.

What helps…

- **Do Kegel exercises** to strengthen pelvic muscles.

- **Eat more fiber.** Constipation puts pressure on the bladder. High-fiber foods, such as beans and whole grains, improve regularity.

- **Avoid foods that irritate the bladder lining,** including caffeine, carbonated beverages, citrus fruits and juices, spicy foods and alcohol.

- **Take cranberry extract daily** to prevent bacteria from adhering to bladder walls and triggering infection. Avoid cranberry juice—its acidity irritates bladder tissues.

- **Don't force or limit fluids.** Let thirst be your guide for when and how much to drink. I recommend 32 ounces of fluids per day.

- **Lose excess pounds,** which can put extra pressure on pelvic organs.

- **Quit smoking.** Smoking impairs oxygen flow, weakening muscles.

- **See a physical therapist,** who may use low-current electrical stimulation and targeted exercises to improve pelvic muscle strength.

If these measures don't help, ask your doctor about surgical options, such as the *tension-free vaginal tape* (TVT) or the *transobturator sling* (TOT). In these procedures, a sling of nylon-like mesh tape is inserted via tiny incisions in the abdomen and/or vagina, creating a hammock that supports the urethra. When a cough or sneeze pushes the urethra down, the tube is forced against the hammock and closed off, so urine doesn't escape. Up to 90% of patients get immediate incontinence relief, and most remain leak-free after five years. Side effects may include slower or incomplete urination.

Another anatomical cause of stress incontinence is called *intrinsic sphincter deficiency* (ISD), in which urine leaks almost constantly—even when a woman's bladder is nearly empty. The urethral sphincter is made of muscles and soft tissue that surround the urethra where it meets the bladder, squeezing the tube closed. An injury to the muscles or soft tissue can weaken the watertight seal.

Breakthrough: A pastelike substance called *Coaptite* is injected into soft tissue near the sphincter, where it solidifies, partially blocking the urethra and helping the sphincter stay closed when you want it to. Unlike injections of collagen, Coaptite does not provoke allergic reactions and does not need to be repeated periodically because it isn't absorbed into the body.

URGE INCONTINENCE

This type of incontinence occurs when nerves in the brain, bladder or spine malfunction, causing the bladder to contract inappropriately—so you leak before reaching a bathroom. Often, simply thinking about a toilet triggers bladder contractions. Urge incontinence increases with age, perhaps due to changes in the nerves' chemical signals. *Dietary changes and surgery cannot help urge incontinence, but other treatments can…*

- **Bladder retraining.** You delay urination as long as you can—then, over several weeks, you try to gradually increase the interval between trips to the toilet.

- **Acupuncture.** This can reduce urinary urgency and frequency and increase bladder capacity—perhaps by altering the release of brain chemicals that affect the central nervous system's control over involuntary body functions.

- **Drugs that reduce bladder spasms,** such as *oxybutynin* (Ditropan) or *tolterodine* (Detrol). Side effects include dry mouth—which may make you drink more, increasing the need to urinate. Rarely, these drugs may cause confusion or other cognitive side effects.

- **Botox injections.** This halts contractions by temporarily paralyzing the bladder. Currently it is used only for very severe cases, such as in patients who have had a stroke or who have multiple sclerosis. Injections must be repeated every six months or so, and health insurance generally does not cover it.

If you're having problems with incontinence, see a urogynecologist.

Referrals: American Urogynecologic Society (202-367-1167, *www.augs.org*). Even if you've lived with it for decades, you're never too old to be helped. I recently performed surgery for incontinence on an 87-year-old—and she says it has changed her life.

■ ■ ■ ■

Acupuncture Halts Overactive Bladder

In a recent study, women with overactive bladder received either acupuncture using acupuncture points targeted for bladder control or general, nontargeted acupuncture once weekly for four weeks.

Result: Women who were given targeted, bladder-specific acupuncture had improved bladder capacity, while the other group reported no change in symptoms.

Theory: Targeted acupuncture decreases the excess nerve stimulation that causes the bladder to feel full even when it is not.

Sandra L. Emmons, MD, associate professor of obstetrics and gynecology, Oregon Health and Sciences University, Portland.

What to Do When It Hurts to Make Love

Barbara Bartlik, MD, assistant professor of psychiatry and sex therapist at New York-Presbyterian Hospital/Weill Cornell Medical College in New York City and a member of the *Bottom Line/Women's Health* advisory board. She is a medical adviser for the book *Extraordinary Togetherness: A Woman's Guide to Love, Sex and Intimacy.* Rodale.

When it comes to sex, sometimes the spirit is willing but the flesh says, "Ow!" If lovemaking has become painful, see your gynecologist. *Possible causes…*

•**Dryness.** Insufficient vaginal lubrication can be caused by dehydration…side effects of birth control pills or antidepressants…and decreased levels of the hormone estrogen after menopause.

What helps: Drink 64 ounces of water daily. Try an over-the-counter (OTC) lubricant, such as Replens, available at drugstores. Extend foreplay to give your body time to create lubrication.

Prescription topical estrogen also can help. It is less likely than oral estrogen to increase risk for cardiovascular problems and breast cancer. Options include a vaginal estrogen cream or an estrogen-containing ring inserted into the vagina.

New research: A low-dose vaginal estrogen suppository or ring (about 10 to 25 micrograms) is as effective as a higher-dose product for relieving dryness yet is less likely to cause side effects, such as headache and breast pain.

•**Endometriosis.** When tissue from the endometrium (uterine lining)—which should stay inside the uterus—instead attaches itself to organs outside the uterus, it causes pelvic pain and inflammation for women in their reproductive years.

Options: Take an OTC nonsteroidal anti-inflammatory drug (NSAID), such as *ibuprofen* (Advil, etc.) or *naproxen* (Aleve), starting the day before your period is due and continuing until bleeding stops. To halt disease progression, your doctor may prescribe oral contraceptives. In severe cases, surgery to remove endometrial tissue and adhesions can relieve pain while preserving fertility—though symptoms may recur. If pain is extreme and you are done having children, you may want to consider a hysterectomy.

•**Pelvic inflammatory disease (PID).** A bacterial infection of the reproductive organs, PID results from a sexually transmitted disease (such as chlamydia or gonorrhea) or other vaginal infection. Repeated douching and using an IUD (intrauterine device for birth control) can increase risk. PID symptoms include painful intercourse, vaginal discharge and abdominal or back pain.

Caution: PID can cause scarring that leads to infertility and chronic pain. Antibiotics cure the infection but cannot reverse damage.

Your partner: He must see his doctor, even if he has no symptoms of infection (such as pain or discharge from the penis)—without treatment, he could reinfect you.

•**Trichomoniasis.** This parasitic infection usually is transmitted sexually but in rare cases can occur if genitals come in contact with an object that harbors the parasite, such as a wet towel. It causes vaginal odor, yellow-green discharge, sores on vaginal walls, genital itching and pain during sex.

Cure: One large dose of an antibiotic, such as *metronidazole* (Flagyl), can work as well as a seven-day lower-dose course of treatment—but it increases risk for side effects, such as nausea and vomiting. Your partner also must be tested.

•**Uterine prolapse.** This occurs when weakened muscles and ligaments of the pelvic floor allow the uterus to drop into the vagina, creating pressure in the vagina or a lump at the vaginal opening. Contributing factors include pregnancy, childbirth, obesity, chronic constipation and decreased estrogen.

Self-help: Kegel exercises strengthen the pelvic floor. Contract vaginal muscles as if to stop the flow of urine…hold five seconds…relax… repeat. Aim for 30 repetitions daily.

Treatment: Your doctor may fit you with a pessary—a flexible plastic device worn in the vagina to reposition the uterus. Some can be worn during sex. If a bulge protrudes from the vagina, your doctor may recommend surgery to repair the pelvic floor…or a hysterectomy.

•**Vaginismus.** Involuntary spasms of the pubococcygeus (PC) muscles surrounding the vagina make intercourse extremely painful. Possible causes include pelvic or vaginal infection or injury…lingering pain (or fear of pain)…hormonal changes…or psychological issues.

Relief: Treat any underlying physical cause— with medication to cure an infection or with hormone therapy for low estrogen. Kegel exercises, physical therapy and biofeedback help relax the PC muscles.

In the privacy of your home: Your doctor may recommend vaginal dilators, phallic-shaped rods of various sizes. Starting with the smallest (tampon-size), you gently insert the dilator into your vagina, working up to larger dilators over time until you can comfortably accommodate penetration by your partner.

•**Vulvodynia.** This chronic condition is characterized by stinging or stabbing pain in the vagina or vulva. The cause may be related to genetics…infection…or injury to vulvar nerves,

such as during childbirth, especially if you had an incision or tear at the vaginal opening. There is no known cure.

What helps: Medication options include an anticonvulsant or tricyclic antidepressant to block pain signals…and injections of the anesthetic *lidocaine*. Physical therapy and biofeedback help relax pelvic muscles.

Avoid: Hot tubs, tight underwear, scented toilet paper and perfumed soaps.

For more comfortable sex: Apply a topical anesthetic, such as lidocaine cream, 30 minutes before intercourse. This will diminish sensations of pain (and also, unfortunately, of pleasure). Use a vaginal lubricant…and apply cold compresses after lovemaking.

▪ ▪ ▪ ▪

Get Tested Every Year

Annual HPV tests are recommended for women over age 40.

Recent finding: 25% of women who tested negative on a Pap test but positive for the *human papillomavirus* (HPV) developed abnormal cervical cells within five years. HPV is linked to cervical cancer.

Susanne Kjaer, MD, DMSc, head, department of viruses, hormones and cancer, Institute of Cancer Epidemiology, Danish Cancer Society, Copenhagen, Denmark, and leader of a study of 10,234 women, published in *Cancer Research*.

How to Get Rid of Spider Veins

Ronald Moy, MD, professor of dermatology at the David Geffen School of Medicine at the University of California, Los Angeles, and director of dermatology at the California Health & Longevity Institute, Westlake Village, California.

Spider veins are small blood vessels near the skin's surface that are visible as thin red, blue or purple lines. They most often appear on the legs but can occur on the face and elsewhere. In addition to being unsightly, spider veins may ache, swell or burn.

Age is one culprit—over time, veins get larger and skin gets thinner and more transparent. A propensity for spider veins runs in families, though they also can be caused by injury. Home remedies cannot get rid of spider veins, but two types of medical treatment can.

•**Sclerotherapy** involves injecting spider veins with a salt-based solution. This collapses the veins and cuts off blood flow, causing them to fade and easing discomfort. Pain is minor because the needle is very small. A single injection can destroy many interconnected veins, though you may need several injections and/or sessions. After sclerotherapy, healthy veins take over for destroyed ones.

Cost: Between $100 and $400 per treatment.

•**Laser treatment** uses pulses of energy to heat and shrink the veins. Each laser pulse feels like a rubber band snapping on your skin, and many pulses are needed along the length of the vein. Facial spider veins generally respond best to laser treatment.

Cost: $300 or more per treatment.

Both treatments may need to be repeated if new spider veins appear. Insurance seldom covers the cost. Side effects are rare but may include sores, red marks and scarring.

info To find a doctor, contact the American Academy of Dermatology, 866-503-7546, *www.aad.org*.

Are Big Breasts a Pain in Your Neck (and Back)?

Jason A. Spector, MD, assistant professor of plastic surgery at New York-Presbyterian Hospital/Weill Medical College of Cornell University, in New York City. He is a member of the Plastic Surgery Research Council and a diplomate of the American Board of Surgery.

Large-breasted women often have chronic back and neck pain from the sheer weight of their breasts…gouged shoulders from bra straps…discomfort during exercise…and feelings of self-consciousness. Breast reduction can provide relief.

•**Surgical techniques.** Surgery is done under general anesthesia and takes two to three hours. The "anchor" or "inverted T" technique typically is used when more than 1,000 grams (about two pounds) of tissue are removed from each breast. An incision is made around the nipple…down the front of the breast and under the breast. Excess tissue, fat and skin are then removed. Nipples may be repositioned to give a natural appearance. When removing less than two pounds from each breast, the vertical or short-scar technique is used. This incision runs around the areola and down the front of the breast only, resulting in less scarring.

•**Risks.** Breast reduction usually involves minimal blood loss. As with any surgery, there are risks from general anesthesia and possible infection. Complication rates generally are low, but can be higher for women who smoke, have diabetes or are overweight.

•**Postoperative care.** The patient usually goes home the same day, with a dressing over each incision. Sometimes small tubes are attached to the incision to drain fluid. These are removed after a few days. Discomfort can be managed with prescription or nonprescription pain medicine.

•**Recovery.** A supportive sports bra worn constantly for one month minimizes swelling. After one week or so, you can return to work and do moderate, low-impact exercise. You can resume all normal activities after a month.

•**Appearance.** Age, skin condition and breast size and shape affect results. At your initial consultation, ask the surgeon how your breasts will look after surgery. Generally, they will be smaller, rounder, higher and more symmetrical. Typically some scarring is visible.

•**Sensation.** Breasts usually feel numb overall for up to three months but return to normal within six months. It may take a year to regain normal nipple sensation, and in 10% to 15% of cases, it does not return completely. In some cases, if nipples were repositioned, they lose all sensation.

•**Cost.** Breast reduction ranges from $4,000 to $12,000. Usually health insurance covers reduction when medically necessary (for instance, to ease pain).

•**Breast-feeding.** Most women can nurse after reduction—but if breast-feeding is vital to you, delay surgery until you are done having children.

•**Cancer concerns.** Women age 40 or older and those at high risk for breast cancer due to family history should have a baseline mammogram first. Among high-risk women, reduction may diminish breast cancer risk. Breast reduction does not interfere with breast cancer detection or treatment.

info For more information about choosing a surgeon for a breast reduction, log on to the American Society of Plastic Surgeons Web site, *www.plasticsurgery.org.*

When Your Partner Has Erectile Dysfunction

Michael E. Metz, PhD, a clinical psychologist and marital and family therapist specializing in relationship and sex therapy. He is in private practice in Minneapolis-St. Paul. He is coauthor, with Barry W. McCarthy, PhD, of *Coping with Erectile Dysfunction: How to Regain Confidence and Enjoy Great Sex* (New Harbinger), which won the 2007 SSTAR (Society for Sex Therapy and Research) Consumer Book Award...*Coping with Premature Ejaculation* (New Harbinger)...and *Men's Sexual Health* (Routledge).

W e've all seen the commercials. An elegant couple strolls along the beach at dusk. He looks rugged and heroic... she looks content. Why? It's all because he takes a drug that lets him overcome erectile dysfunction (ED), the persistent inability to gain and maintain an erection sufficient for sexual intercourse.

If you are a woman whose partner suffers from ED, you may know all too well that the problem can extend beyond what any drug can remedy. ED doesn't affect only sex—it challenges a man's self-image (and sometimes a woman's), tests a woman's patience and can hurt a relationship much more broadly. *Women's most common questions...*

Does ED mean the end of our sex life?

It doesn't have to. Actually, your first question should be, "Does my man really have ED?" It is normal for a man to occasionally be unable to

start or complete intercourse, usually because of fatigue or excess alcohol consumption.

True ED means that a man can *seldom* or *never* get and sustain an erection or avoids sex because he's worried that he might not be able to. This condition affects more than 25 million American men, according to the American Urological Association.

Vicious cycle: Sometimes a man will have trouble once or twice—and then become terribly embarrassed or filled with dread that he has ED. His partner may become equally anxious, and they may feed each other's worries, compounding the stress. Such worry itself may cause "performance anxiety" that leads to ED.

What to do: If your partner has erection trouble, remain calm, supportive and, most of all, optimistic. Calmly focusing on relaxation and pleasure, as described below, can often bring erections back.

Is this happening because I'm not attractive to him anymore?

Even though a woman may know that ED can have medical and psychological causes that don't stem from the relationship, she still may find herself reacting emotionally to her partner's problem in one of two counterproductive ways...

•**She doubts herself.** She feels she isn't attractive or sexually skilled enough to excite him.

•**She gets angry.** She thinks, *I'm attractive. This is his fault.* If her partner starts to associate trying to have sex with being blamed for failing, he will avoid it—making her angrier. Anger destroys intimacy.

The truth: Your man's ED has little or nothing to do with your desirability. If he has had an erection with you before, you probably are attractive to him. If he simply weren't interested, he wouldn't go through the drama of trying and failing.

Next step: Don't lose your intimate connection. Not every sexual experience must end in intercourse. In fact, variations from the same kind of sex you've always had will probably increase your mutual desire, which may help reduce the psychological effects of ED.

Make it okay to snuggle, caress and kiss without expecting intercourse. Casually explore each other's bodies…ask him where and how he would like to be touched…tell him the same about you. If you both want it, he can bring you to orgasm manually or orally, but don't make this a pressured demand. Talk…laugh with each other…realize that you are simply facing one of life's many challenges together. At first, he may not believe that you are not angry or horribly disappointed. Be gently persistent in saying that this kind of contact is pleasurable to you.

When I try to talk about it, he shuts down. Why won't he get help?

You may think he is ignoring the problem or doesn't care.

The truth: He isn't ignoring his ED—it is plaguing him. But if you push him too hard, he may retreat even more. At this point, you have to walk a tightrope of gently addressing the issue without showing anger or blame. If he reacts negatively, don't take the bait. Instead, say, "I am sorry that it's painful for you, and I want to help you through this." Do your best to be accepting and nonjudgmental. When a man pushes back ("Leave me alone!"), he may in fact be testing you, wanting assurance that you still are on his side.

What can be done medically for ED?

He should go to his doctor, who may refer him to a urologist for a complete checkup to look into the following potential causes of ED…

•**A chronic condition that affects the nervous system or circulation,** such as diabetes.

•**A side effect of medication,** such as a blood pressure or chemotherapy drug, an antidepressant or an antihistamine.

•**A side effect of surgery,** such as for prostate cancer.

•**Injury to vascular, neurological, hormonal or muscular systems**—from an accident, for example.

•**Lifestyle issues,** such as alcohol misuse, long-term smoking or excessive fatigue.

There is a two-part method of dealing with ED caused by physical problems. It's called "fix and foster." First, with the physician, fix the underlying problem, if possible. (When ED is caused by damage to the neurological or circulatory systems, the ability to gain an erection sometimes can't be recovered.)

The couple also may need to work on the "foster" part—teaming up on healing the anxiety and other negative emotions caused by ED.

Beware: If a doctor prescribes an ED drug without first trying to diagnose the root cause, find another doctor. Treatment should progress in increments to find the least intrusive solution.

Direct treatment of ED can include an oral drug such as *sildenafil* (Viagra), *tadalafil* (Cialis) or *vardenafil* (Levitra)…or the drug *alprostadil* (Caverject) applied directly to the penis either by injection or penile suppository. All these drugs promote blood flow to the penis.

Some couples prefer nondrug equipment, such as a splint that supports the penis or a vacuum device that draws blood into the penis. A surgical penile implant is a treatment of last resort.

What if there's no physical cause?

Seek the help of a professional sex therapist, who can help both of you as a couple. This counselor can show you how to overcome ED and incorporate drugs or other treatment into your life—not just your sex life. He/she should work in cooperation with your partner's physicians.

info To find a marital/sex therapist, contact the American Association of Sexuality Educators, Counselors and Therapists (804-752-0026, *www.aasect.org*)…Association for Behavioral and Cognitive Therapies (212-647-1890, *www.abct.org*)…Society for Sex Therapy and Research (202-863-1644, *www.sstarnet.org*).

Index